David Alejandro Pelta, Natalio Krasnogor, Dan Dumitrescu, Camelia Chira, and Rodica Lung (Eds.)

Nature Inspired Cooperative Strategies for Optimization (NICSO 2011)

T0205451

Studies in Computational Intelligence, Volume 387

Editor-in-Chief

Prof. Janusz Kacprzyk
Systems Research Institute
Polish Academy of Sciences
ul. Newelska 6
01-447 Warsaw
Poland
E-mail: kacprzyk@ibspan.waw.pl

David Alejandro Pelta, Natalio Krasnogor,
Dan Dumitrescu, Camelia Chira,
and Rodica Lung (Eds.)

Nature Inspired Cooperative Strategies for Optimization (NICSO 2011)

 Springer

Editors

Prof. David Alejandro Pelta
University of Granada
Dept. of Computer Science and A.I.
E.T.S. Ingenieria Informatica y de
Telecomunicación, C/ Periodista Daniel
Saucedo Aranda s/n, 18071 Granada, Spain
E-mail: dpelta@decsai.ugr.es

Prof. Natalio Krasnogor
University of Nottingham
School of Computer Science
Jubilee Campus
Wollaton Road
Nottingham NG8 1BB, UK
E-mail: nxk@cs.nott.ac.uk

Prof. Dan Dumitrescu
Babes-Bolyai University of Cluj Napoca
Center for Cognitive and Neural Studies
(Coneural), Str. Ciresilor 29
400487 Cluj-Napoca, Romania
E-mail: ddumitr@cs.ubbcluj.ro

Dr. Camelia Chira
Babes-Bolyai University
Department of Computer Science
Kogalniceanu 1
400084 Cluj-Napoca, Romania
E-mail: cchira@cs.ubbcluj.ro

Prof. Rodica Lung
Babes-Bolyai University of Cluj Napoca
Faculty of Economics and Business
Administration, Str. Teodor Mihali,
Nr. 58-60, 400591 Cluj Napoca, Romania
E-mail: rodica.lung@econ.ubbcluj.ro

ISBN 978-3-642-26991-2

ISBN 978-3-642-24094-2 (eBook)

DOI 10.1007/978-3-642-24094-2

Studies in Computational Intelligence

ISSN 1860-949X

Typeset & *Cover Design:* Scientific Publishing Services Pvt. Ltd., Chennai, India.

Printed on acid-free paper

9 8 7 6 5 4 3 2 1

springer.com

We can only see a short distance ahead, but we can see plenty there that needs to be done.

Alan Turing

Preface

Biological and other natural processes have always been a source of inspiration for computer science and information technology. Many emerging problem solving techniques and algorithms integrate advanced evolution and cooperation strategies, encompassing a range of spatio-temporal scales for visionary conceptualization of information processing and computation.

We are very pleased to present the proceedings volume of the V International Workshop on Nature Inspired Cooperative Strategies for Optimization (NICSO 2011) which took place in Cluj-Napoca, Romania. The previous editions of NICSO were held in Granada, Spain (2006), Acireale, Italy (2007), Tenerife, Spain (2008), and again in Granada in 2010. NICSO evolved to be one of the most interesting and profiled workshops in nature inspired computing. In accordance with the NICSO tradition, each submission has been evaluated in an extensive peer review process. The top papers have finally been selected for the inclusion in this volume.

The aim of NICSO 2011 was twofold:

- To provide an inspiring environment for debating the state of the art ideas and techniques in nature inspired cooperative strategies, and
- To offer a comprehensive image on the most recent applications of these ideas and techniques.

The topics covered by the NICSO 2011 contributions include Swarm Intelligence (such as Ant and Bee Colony Optimization), Genetic Algorithms, MultiAgent Systems, Coevolution and Cooperation strategies, Adversarial Models, Synergic Building Blocks, Complex Networks, Social Impact Models, Evolutionary Design, Self Organized Criticality, Evolving Systems, Cellular Automata, Hybrid Algorithms, and Membrane Computing (P-Systems).

We are grateful to all authors who submitted their contributions and to all reviewers for their work in refereeing the papers.

We would like to thank the support given by several people and institutions. We acknowledge the support of the National Research Council Romania (CNCS) through project PN-II-TE320 (Emergence, auto-organization and evolution: New computational models in the study of complex systems) and the support received from the National Agency for Scientific Research. D. Pelta acknowledges the support of the Spanish Ministry of Science and Innovation (project TIN2008-01948) and the Andalusian Government (project P07-TIC-02970). We would also like to thank Babes-Bolyai University and the members of the Centre for the Study of Complexity for their help with the local organization tasks.

Spain   David Pelta
UK   Natalio Krasnogor
Romania   Rodica Ioana Lung
Romania   Dan Dumitrescu
Romania   Camelia Chira
                                                                    2011

Organization

Steering Committee

David A. Pelta   University of Granada, Spain
Natalio Krasnogor   University of Nottingham, UK

Programme Chair

Rodica Ioana Lung   Babeş-Bolyai University, Romania

Organizing Committee

Camelia Chira   Babeş-Bolyai University, Romania
Dan Dumitrescu   Babeş-Bolyai University, Romania
Anca Gog   Babeş-Bolyai University, Romania
Tudor-Dan Mihoc   Babeş-Bolyai University, Romania

Programme Committee

Davide Anguita   University of Genova, Italy
Cecilio Angulo   Technical University of Catalunya, Spain
Germán Terrazas   University of Nottingham, UK
Ignacio G. Del Amo   University of Granada, Spain

Contents

List of Contributors

Ignacio G. del Amo
DECSAI, ETSIIT, CITIC-UGR,
University of Granada,
C/Daniel Saucedo Aranda, s/n
E-18071, Granada, Spain
ngdelamo@decsai.ugr.es

K. V. Arya
ABV-Indian Institute of Information,
Technology and Management,
Gwalior
kvarya@gmail.com

Jagdish Chand Bansal
ABV-Indian Institute of Information,
Technology and Management,
Gwalior
jcbansal@gmail.com

John Mark Bishop
Goldsmiths, University of London,
New Cross,
London SE14 6NW, U. K.
m.majid@gold.ac.uk

Tim Blackwell
Goldsmiths, University of London,
New Cross,
London SE14 6NW, U.K.
m.majid@gold.ac.uk

Edmund K. Burke
Automated Scheduling, Optimisation
and Planning (ASAP)
Group School of CSIT,
University of Nottingham,
Nottingham, NG8 1BB, U.K.
ekb@cs.nott.ac.uk

Camelia Chira
Babes-Bolyai University,
Cluj-Napoca, Romania
cchira@cs.ubbcluj.ro

Joaquim de Ciurana
University of Girona,
Escola Politècnica Superior
Edifici PII,
Avd. Lluís Santaló, s/n 17071
Girona, Spain
quim.ciurana@udg.es

Emilio Corchado
University of Salamanca,
Salamanca, Spain
escorchado@usal.es

Marcel Cremene
Communications Department,
Technical University Cluj-Napoca,
Cluj, Romania
Cremene@com.utcluj.ro

Carlos Cruz
DECSAI, ETSIIT, CITIC-UGR,
University of Granada,
C/Daniel Saucedo Aranda, s/n,
E-18071, Granada, Spain
carloscruz@decsai.ugr.es

D. Dumitrescu
Centre for the Study of Complexity,
Babes-Bolyai University,
Cluj-Napoca, Romania
ddumitr@cs.ubbcluj.ro

Christopher Expósito Izquierdo
Dpto. de Estadística,
IO y Computación,
ETS de Ingeniería Informática,
Universidad de La Laguna, Spain
cexposit@ull.es

Noémi Gaskó
Babes-Bolyai University,
Cluj-Napoca, Romania
gaskonomi@cs.ubbcluj.ro

Mircea Giurgiu
Communications Department,
Technical University Cluj-Napoca,
Cluj, Romania
Mircea.Giurgiu@
com.utcluj.ro

Anca Gog
Babes-Bolyai University,
Cluj-Napoca, Romania
anca@cs.ubbcluj.ro

José Luis González Velarde
Centro de Manufactura y Calidad,
Tecnológico de Monterrey,
México
gonzalez.velarde@itesm.mx

Juan R. González
DECSAI, ETSIIT, CITIC-UGR,
University of Granada,
C/Daniel Saucedo Aranda, s/n,
E-18071, Granada, Spain
jrgonzalez@decsai.ugr.es

Manuel Graña Romay
Computational Intelligence Group
University of the Basque Country
ccpgrrom@gmail.com

Ştefan Holban
Department of Computer Science,
Politehnica University of Timişoara,
Bd. Vasile Pârvan 2,
300223 Timişoara, Romania
stefan@cs.upt.ro

Tom Holvoet
DistriNet Labs,
Department of Computer Science,
Katholieke Universiteit
Leuven

David Iclănzan
Department of Electrical Engineering,
Sapientia - Hungarian University
Transylvania,
540485,Tg-Mureş, Romania
david.iclanzan@gmail.com

Florentin Ipate
Department of Computer Science,
University of Pitesti,
Targu din Vale nr. 1,
110040, Pitesti, Romania
florentin.ipate@ifsoft.ro

Raluca Lefticaru
Department of Computer Science,
University of Pitesti,
Targu din Vale nr. 1,
110040, Pitesti, Romania
raluca.lefticaru@upit.ro

Coromoto León
Dpto. Estadística,
I. O. y Computación.
Universidad de La Laguna
La Laguna, 38271,
Santa Cruz de Tenerife, Spain
cleon@ull.es

Lenka Lhotská
CTU in Prague,
Technicka 2, Prague
lhotska@fel.cvut.cz

Andrei Lihu
Department of Computer Science,
Politehnica University of Timişoara,
Bd. Vasile Pârvan 2,
300223 Timişoara, Romania
andrei.lihu@gmail.com

Rinde R.S. van Lon
DistriNet Labs,
Department of Computer Science,
Katholieke Universiteit
Leuven
rinde.vanlon@cs.kuleuven.be

Rodica Ioana Lung
Faculty of Economics
and Business Administration
Babes-Bolyai University,
Cluj-Napoca, Romania
rodica.lung@econ.ubbcluj.ro

Evelyne Lutton
AVIZ Team,
INRIA Saclay - Ile-de-France,
Bat 490, Université Paris-Sud,
91405 ORSAY Cedex, France
Evelyne.Lutton@inria.fr

Muhammad Rezaul Karim
University of Limerick, Ireland
rezaul.karim@ul.ie

Natalio Krasnogor
ASAP Group,
School of Computer Science
University of Nottingham, UK
nxk@cs.nott.ac.uk

Martin Macaš
CTU in Prague,
Technicka 2, Prague
macas.martin@fel.cvut.cz

Belén Melián Batista
Dpto. de Estadística,
IO y Computación,
ETS de Ingeniería Informática,
Universidad de La Laguna, Spain
mbmelian@ull.es

Tudor-Dan Mihoc
Babes-Bolyai University,
Cluj-Napoca, Romania
mihoct@cs.ubbcluj.ro

J. Marcos Moreno-Vega
Dpto. de Estadística,
IO y Computación,
ETS de Ingeniería Informática,
Universidad de La Laguna, Spain
jmmoreno@ull.es

Denis Pallez
Laboratoire d'Informatique, Signaux,
et Systèmes de Sophia-Antipolis (I3S),
Université de Nice Sophia-Antipolis,
France
Denis.Pallez@unice.fr

David A. Pelta
Models of Decision and Optimization
Research Group, Department of Computer Science and Artificial Intelligence,
CITIC-UGR, ETSIIT,
University of Granada,
C/Periodista Saucedo Aranda,
18071 Granada, Spain
dpelta@decsai.ugr.es

Camelia-M. Pintea
G.Cosbuc N.College,
A.Iancu 70-72,
Cluj-Napoca, Romania
cmpintea@yahoo.com

Florin-Claudiu Pop
Communications Department,
Technical University of Cluj-Napoca,
Cluj, Romania
Florin.Pop@com.utcluj.ro

Laura Puigpinós
Fundación Privada Ascamm,
Avda. Universitat Autònoma,
23 08290 Cerdanyola del Vallés, Spain
lpuigpinos@ascamm.com

Mohammad Majid al-Rifaie
Goldsmiths, University of London,
New Cross, London SE14 6NW,
United Kingdom
m.majid@gold.ac.uk

Conor Ryan
University of Limerick, Ireland
conor.ryan@ul.ie

Léon J.M. Rothkrantz
Man-Machine Interaction Group,
Department of Mediamatics,
Delft University of Technology,
Netherlands

Israel Rebollo Ruiz
Informática 68 Investigación
y Desarrollo S.L.,
Computational Intelligence Group
University of the Basque Country,
beca98@gmail.com

Sorin V. Sabau
Tokai University,
Minamisawa 5-1-1-1,
Sapporo, Japan
sorin@tspirit.tokai-u.jp

Javier Sedano
Instituto Tecnológico de Castilla
y León, Poligono Industrial de
Villalonquejar, Burgos, Spain
javier.sedano@itcl.es

Eduardo Segredo
Dpto. Estadística,
I. O. y Computación.
Universidad de La Laguna,
38271, Santa Cruz de Tenerife, Spain
esegredo@ull.es

Carlos Segura
Dpto. Estadística,
I. O. y Computación.
Universidad de La Laguna,
38271, Santa Cruz de Tenerife, Spain
csegura@ull.es

Harish Sharma
ABV-Indian Institute of Information,
Technology and Management, Gwalior
harish.sharma0107@gmail.com

Andrei Sîrghi
Babeş-Bolyai University,
Cluj-Napoca, Romania,
andreisirghi@yahoo.com

Amr Soghier
Automated Scheduling, Optimisation
and Planning (ASAP)
Group School of CSIT,
University of Nottingham,
Nottingham, NG8 1BB, U.K.
azs@cs.nott.ac.uk

Adriana Stan
Communications Department,
Technical University of Cluj-Napoca,
Cluj, Romania
Adriana.Stan@com.utcluj.ro.

Giovanni Squillero
Politecnico di Torino
Dip. Automatica e Informatica,
C.so Duca degli Abruzzi 24
10129 Torino - Italy
Giovanni.Squillero@
polito.it

Germán Terrazas
ASAP Group,
School of Computer Science
University of Nottingham, UK
gzt@cs.nott.ac.uk

Alberto Tonda
ISC-PIF, CNRS CREA, UMR 7656,
57-59 rue Lhomond, Paris, France
Alberto.Tonda@gmail.com

Cristina Tudose
Department of Computer Science,
University of Pitesti, Targu din Vale nr.
1,110040, Pitesti, Romania
cristina.tudose@upit.ro

José R. Villar
University of Oviedo,
Campus de Viesques s/n 33204
Gijón, Spain
villarjose@uniovi.es

Pablo J. Villacorta
Models of Decision and Optimization
Research Group, Department of Com-
puter Science and Artificial Intelligence,
CITIC-UGR, ETSIIT,
University of Granada,

C/Periodista Saucedo Aranda, 18071
Granada, Spain
pjvi@decsai.ugr.es

Pascal Wiggers
Man-Machine Interaction Group,
Department of Mediamatics,
Delft University of Technology,
Netherlands

R. Qu
Automated Scheduling, Optimisation
and Planning (ASAP)
Group School of CSIT,
University of Nottingham,
Nottingham, NG8 1BB, U.K.
rxq@cs.nott.ac.uk

Chapter 1
Ant Colony Optimization for Automatic Design of Strategies in an Adversarial Model

Pablo J. Villacorta and David A. Pelta

Abstract. Adversarial decision making is aimed at determining optimal strategies against an adversarial enemy who observes our actions and learns from them. The field is also known as decision making in the presence of adversaries. Given two agents or entities S and T (the adversary), both engage in a repeated conflicting situation in which agent T tries to learn how to predict the behaviour of S. One defense for S is to make decisions that are intended to confuse T, although this will affect the ability of getting a higher reward. It is difficult to define good decision strategies for S since they should contain certain amount of randomness. Ant-based techniques can help in this direction because the automatic design of good strategies for our adversarial model can be expressed as a combinatorial optimization problem that is suitable for Ant-based optimizers. We have applied the Ant System (AS) and the Max-Min Ant System (MMAS) algorithms to such problem and we have compared the results with those found by a Generational Genetic Algorithm in a previous work. We have also studied the structure of the solutions found by both search techniques. The results are encouraging because they confirm that our approach is valid and MMAS is a competitive technique for automatic design of strategies.

1.1 Introduction

Adversarial decision making is aimed at determining optimal strategies against an adversarial enemy who observes our actions and learns from them. This situation arises in many areas of real life, with particular (but not the only one) interest in counter-terrorist combat and crime prevention [13, 20].

Pablo J. Villacorta · David A. Pelta
Models of Decision and Optimization Research Group, Department of Computer Science and Artificial Intelligence, CITIC-UGR, ETSIIT, University of Granada, C/Periodista Saucedo Aranda, 18071 Granada, Spain
e-mail: {pjvi,dpelta}@decsai.ugr.es

D.A. Pelta et al. (Eds.): NICSO 2011, SCI 387, pp. 1–19, 2011.
springerlink.com © Springer-Verlag Berlin Heidelberg 2011

The field is also known as decision making in the presence of adversaries and we may talk about problems within an adversarial domain when an "adversary" exists. Essentially, the focus is on technologies for opponent strategy prediction, plan recognition, deception discovery and planning, and strategy formulation that not only applies to security issues but also to game industry, business, transactions, etc. [12]. For example, patrolling strategies can be viewed as another application of adversarial decision making: the aim is to design routes for patrolling trying to minimize the chance that an enemy enters a security border. Several patrolling models ([1, 18]) have been proposed and solved following a game-theoretic approach. A more abstract adversarial framework which does not focus on patrolling has also been proposed in [19]. This model will be briefly reviewed later.

From an abstract point of view, decision making in presence of adversaries can be described as follows: given two agents or entities S and T (the adversary), both want to maximize their rewards that are inversely related. One defense for S is to make decisions that are intended to confuse T, although this will affect the ability of getting a more optimal reward. The question for S is how to define its decision strategies. Hand-made strategies could be fine, but the designer may omit interesting alternatives due to its inherent limited ability to search in the space of strategies.

In the last years, automatic design by means of evolutionary techniques is gaining increasing attention. For example, in the design of self-assembly systems [14], certain kind of neural networks [17], controllers for collective robotics [2], oriented-tree networks [21], just to cite a few. Ant-based techniques can also help in this direction in a similar way. As we will explain, the automatic design of good strategies for our adversarial model can be expressed as a combinatorial optimization problem that is suitable for Ant-based optimizers. Ant colony optimization techniques are widely used in a great variety of problems, although the Traveling Salesman Problem is probably the most well-known application because of its analogy with path exploration carried out by natural ants. However, it has proven useful in other optimization problems such as [3, 6, 11, 15] and, for that reason, we hypothesize it may work fine in this one too.

In this context, our aims are: (a) to adapt Ant-based optimization algorithms to the problem of finding good strategies, (b) to compare the performance of these techniques against a Generational Genetic Algorithm (GGA) previously applied to the same problem [23], and (c) to provide insights into the structure of the solutions found by Ant-based algorithms and compare these solutions with those found by the GGA.

This contribution is organized as follows: in Section 1.2, the model explained in [19] is briefly summarized. Section 1.3 motivates the use of automatic techniques for the design of strategies. In Section 1.4, the suitability of Ant-based optimizers for this problem is discussed. Section 1.5 is devoted to computational experiments and analysis of results, including insights of the solutions found. Finally, conclusions are discussed in Section 1.6.

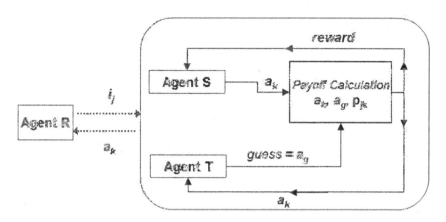

Fig. 1.1 Graphical representation of the model. Events i_j are issued by agent R while response or actions a_k are taken by agent W.

1.2   Adversarial Reasoning Model

The model we are dealing with has been first presented in [19]. It is based on two agents S and T (the adversary), a set of possible inputs or events $I = \{i_1, i_2, \ldots, i_n\}$ issued by a third agent R, and a set of potential responses or actions $A_i = \{a_1, a_2, \ldots, a_m\}$ associated with every event. We have a payoff or rewards matrix P :

$$P(n \times m) = \begin{pmatrix} p_{11} & p_{12} & \cdots & p_{1m} \\ p_{21} & p_{22} & \cdots & p_{2m} \\ p_{31} & p_{32} & \cdots & p_{3m} \\ & & & \\ p_{n1} & p_{n2} & \cdots & p_{nm} \end{pmatrix}$$

where p_{ij} is the reward or profit associated with action j to respond to the event i.

Agent S must decide which action to take given a particular input i_k and with a perfect knowledge of the payoff function P. His aim is to maximize the sum of the profits or rewards given a sequence of inputs. These are issued one at a time and they come from an external environment, represented by agent R. For the experiments, the inputs of the sequence are independent and generated randomly.

Agent T does not know the payoff function P but is watching agent S in order to learn from his actions. His aim is to reduce agent S payoff by guessing which action he will take as a response to each input of the sequence. Algorithm 1.1 describes the steps of the model, with E being the length of the sequence of inputs.

Given a new input i_j, S and T issue responses a_k and a_g respectively. Agent T keeps records of the actions taken by S using an observation matrix, O, with dimensions $N \times M$. O_{ij} stores the number of times that, in the past, agent S decided to take action j when the input was i.

Algorithm 1.1. Sequence of steps in the model.

for $j = 1$ to E do
    A new input i_j arises.
    Agent T "guesses" an action a_g
    Agent S determines an action a_k
    Calculate payoff for S
    Agent T records the pair i_j, a_k
end for

The reward calculation for S at stage c is defined as:

$$p' = p_{jk} \times F(a_g, a_k) \tag{1.1}$$

where F is:

$$F(a,b) = \begin{cases} 0 & \text{if a = b} \\ 1 & \text{otherwise} \end{cases} \tag{1.2}$$

This means that agent S gets no reward at all when agent T has predicted his response properly.

The behaviour pattern of both agents can vary from a totally deterministic way to a totally random one. Agent T can use his observation matrix to make predictions in several manners, such as always choosing the most frequently observed action in the past, or a random action with a probability that is proportional to the number of times that action was observed in the past, etc. Agent S can always choose the action with the highest reward, but this behaviour is very easy to learn for an observer when repeated along time. He can also choose randomly or with a probability that is proportional to the payoff, or randomly among some of the actions with higher payoff. There exist many other patterns that try to balance payoff and confusion.

In this contribution, agent T applies a simple frequency-based decision strategy: the probability of selecting an action a_i for responding to event e_j is proportional to O_{ij}. Despite its simplicity, this strategy is one of the hardest for agent S as it was shown in [19].

1.3   Automatic Design of Strategies

We need to provide a representation that will induce a search space that can be explored in an automatic way in order to find good decision strategies. For this, we define a *tactic* as a process with two stages. In the first stage, a number of different candidate actions are chosen into a "candidate set" according to certain criterion. The cardinal of the candidate set is a specific integer parameter of the tactic, $k \in \{1, 2, ..., K_{max}\}$, where K_{max} is the total number of possible actions. In the second stage, one single action is finally selected from the candidate set according to another different criterion. Therefore a tactic can be fully described in terms of three components: $< k >, < criterion >, < criterion >$. The values available for *criterion* are BEST (picks the k best actions according to their payoff), RANDOM (picks k actions randomly) and PROP-PAYOFF (picks k actions with a probability that is proportional to their payoff).

In the original model [19], the same tactic is used in every decision step. A new approach proposed in [23] consists in varying the tactic every time we have to make a decision. One way to do this is to have a predefined succession of tactics that can be used in a cyclic way. We call this succession of tactics a *strategy*. For instance, consider the strategy 3 | PROP-PAYOFF | RANDOM, 4 | BEST | PROP-PAYOFF, 2 | RANDOM | RANDOM, 2 | PROP-PAYOFF | PROP-PAYOFF. In more abstract way, a strategy can be viewed as a word generated by the following regular grammar:

$$S = AS \mid A$$
$$A = <K><CRIT><CRIT>$$
$$K = 1 \mid 2 \mid ... \mid K_{MAX}$$
$$CRIT = RANDOM|PROP-PAYOFF|BEST$$

We define the *length L* of a strategy as the number of tactics it contains. The previous example has length 4 as it is composed by 4 different tactics. When using the strategy in a simulation that consists in a succession of several inputs (most likely more than 4), the strategy is considered to be cyclic: agent S will response to the n-th input by using the $(n \, mod \, L)$ tactic.

Under these conditions, it is clear that automatic techniques are required in order to design a good strategy (succession of tactics) in the space of cyclic strategies of a given length L. This search space is large enough to make exhaustive enumeration computationally unfeasible when $L > 3$, even in a simple model instance with just five different stimuli and five possible actions. The number of different strategies is given by $8 \times (K_{max} - 2) + 3$, which yields 531.441 for $K_{max} = 5$ and $L = 4$. Recall that the evaluation of each of these strategies requires several independent executions to obtain a reliable result since the strategies are non-determinist. The exhaustive evaluation process is very time-consuming and in practice it turns computationally infeasible. For more details see [23].

1.4   Ant Colony Optimization for Combinatorial Optimization Problems

Ant Colony Optimization (ACO) algorithms are a powerful heuristic search technique, especially in combinatorial optimization problems [9]. They were first introduced in [10] and they are based on the natural metaphor of the intelligent behaviour of ant colonies. Ants are themselves blind agents but when they cooperate, they manage to solve complex tasks, mainly finding the shortest path between two locations. Due to this analogy, the first problem ACO algorithms attempted to solve was the classical Traveling Salesman Problem. However, ACO has been applied to many others optimization problems, provided that they can be expressed as a graph-path problem [3]. Special emphasis has been put in combinatorial optimization problems because they can be easily adapted as graph problems [7]. Some examples are the quadratic assignment problem [15], task sequencing [16], graph coloring [6], protein folding [11], and vehicle routing [5].

The Ant System explained in [10] was the first proposal, but some others variants have been introduced later, such as the Ant Colony System [8], the Best-Worst Ant System [4] and the Max-Min Ant System, which is currently one of the most competitive algorithms in the TSP problem [22].

1.4.1 ACO Algorithms for Automatic Design of Strategies

Designing good strategies in our adversarial model can be viewed as a combinatorial optimization problem in the space of strategies of length L. Although the optimal value for L is also unknown, a constant prefixed value has been used in all the experiments (i.e. paramenter L was not part of the optimization process). The problem consists in finding the combination (succession) of L strategies that maximizes the total payoff for agent S, or equivalently, that minimizes the *gap* between the total payoff that could have been attained if there were no adversary, and the actual payoff attained when there is an adversary who tries to guess our decisions. This gap is always measured as a percentage. Notice that the number of possible tactics at each step is the same for all the L steps of the strategy, and in fact this value is quite large in relation to the length of a strategy (4 in all our experiments). This particular feature of our problem also encouraged the use of Ant-based algorithms.

The problem of building a strategy of length L can be adapted as a graph-path problem in a quite straightforward manner. Each ant must build a path with L steps, starting from an extra (dummy) point called Z, which is not a tactic. The corresponding dummy step will be denoted e_0. At each step, the ant chooses a tactic to move to. The tactics available at each step are always all the possible combinations that can be created following the rules of the grammar indicated in Section 1.3, with K varying from 1 to K_{max}. The number of combinations is very high, as stated in that section. The ant can move from one tactic to the same one at the next step. Repetition is allowed since there can be strategies like for instance 3-PP-BEST | 3-PP-BEST | 2-RND-BEST. An example with $L = 3$ and 7 different tactics at each step (t_1 to t_7) is depicted in Fig. 1.2.

Probabilistic Transition Rule

Each ant follows a probabilistic transition rule that guides its movement from one tactic to another in the next step. Let r, s represent two tactics (different or not). The

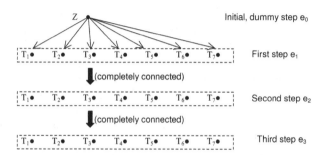

Fig. 1.2 Graph of a model with seven possible tactics $t_1, ..., t_7$ at each step and strategies of length 3

probability that an ant moves from tactic r at current step to tactic s at the next step is given by

$$p(r,s) = \frac{[\tau_{rs}]^{\alpha} \cdot [\eta_{rs}]^{\beta}}{\sum\limits_{u \in U} [\tau_{ru}]^{\alpha} \cdot [\eta_{ru}]^{\beta}} \tag{1.3}$$

where U is the set of all the existing tactics, i.e. all the combinations of $K, < CRIT > < CRIT >$; and α and β are the weights assigned to the pheromone and heuristic components, respectively.

Heuristic Information

Many alternatives are feasible to compute the heuristic component of the pheromone graph. Here we have used a simple approach that takes into account the performance (*gap*) exhibited by a single static tactic that does not change along time. This information has been collected by running previous experiments to exhaustively evaluate every tactic. We will provide details about this in the experiments section.

1.4.2   On the Comparison of Strategies

During the optimization process, in many situations it is necessary to compare the performance of two strategies. The evaluation of a strategy requires running a simulation, as described in Algorithm 1. Strategies contain non-deterministic elements, and also the external inputs are randomly generated, so the fitness of a strategy is inherently non deterministic. Instead of just repeating the simulation *many* times and taking the average of all samples, a statistical-based comparison technique described in [23] has been applied. The statistical-based comparison method allows for a better redistribution of the number of total simulations available and has shown better performance with the same total number of simulations than just comparing the average of the samples. Basically, it relies on the fact that when two individuals are very different, few simulations are needed to have a clear winner, so just the necessary repetitions are done. On the other hand, a larger number of repetitions is employed for comparing individuals that are very alike. The stopping criterion of all the algorithms is a fixed number of *simulations*, where a simulation can be defined as evaluating a strategy (i.e. running Algorithm 1 with a given input sequence) just once.

The criterion that determines if we can rely on the result of a comparison or not is a statistical Student-T test. To be safely applied, this parametrical test requires that two previous conditions hold. The first one is the normality of the data, and it can be tested with a Kolmogorov-Smirnov normality test. The second one is the homogeneity of variances (homocedasticidy) of the two populations being compared, and it can be tested with a Levene test. Both conditions were succesfully checked in [23] so a t-test can be safely applied to strategy comparison.

The comparison process carried out is the following. When two strategies are to be compared, each one is evaluated p times initially, which means running p independent simulations and annotate the gap obtained in every simulation (the lower the gap, the better the strategy). We call each of these gaps a *sample*. A confidence-interval is built for the difference of the means of both sets of samples. If this interval contains 0, then the winner of the comparison is not clear, so each individual is re-evaluated with another block of p independent simulations, and a new confidence-interval is built with $2p$ samples per strategy. This process is repeated until one of them is better with statistical signification, or until $3p$ simulations for each strategy are done. If the confidence-interval still contains 0 after $3p$ simulations, then the strategy with the lower average is considered the best, although in that case there is no statistical evidence. Note that all the samples associated with a strategy are attached with the strategy so that resampling is only applied to the strategy that has fewer samples.

1.5 Computational Experiments and Results

1.5.1 Preliminary Experiments to Compute the Heuristic Information

An exhaustive evaluation of all the existing simple strategies ($L = 1$) was made to compute heuristic information. In each experiment, one single tactic is used to choose the response to all the events of an input sequence. In other words, the tactic employed does not change during the experiment. Each experiment is composed of 4000 independent repetitions, where each repetition consists in evaluating the tactic using a different input sequence of 500-event length. At the end of the experiment, the gap of that tactic is annotated. Since 4000 independent input sequences are thus required, these sequences were generated just once, and used in all these experiments so the results are comparable. The payoff matrix used in the experiments is the first matrix M_1 among the 15 different matrices used in all the remaining

Table 1.1 An example of heuristic information in a model with 5 different events and 5 actions

Strategy	Gap (%)	Strategy	Gap (%)
1 — RANDOM — RANDOM	52,3701	4 — BEST — RANDOM	48,1297
1 — BEST — BEST	99,1985	2 — BEST — P-PAYOFF	55,5155
1 — P-PAYOFF — P-PAYOFF	46,0188	3 — BEST — P-PAYOFF	46,2783
2 — RANDOM — RANDOM	52,3731	4 — BEST — P-PAYOFF	45,2854
3 — RANDOM — RANDOM	52,3486	2 — P-PAYOFF — RANDOM	46,202
4 — RANDOM — RANDOM	52,4329	3 — P-PAYOFF — RANDOM	46,9754
2 — RANDOM — BEST	46,0911	4 — P-PAYOFF — RANDOM	49,1199
3 — RANDOM — BEST	53,3355	2 — P-PAYOFF — BEST	55,6893
4 — RANDOM — BEST	70,5899	3 — P-PAYOFF — BEST	75,3516
2 — RANDOM — P-PAYOFF	47,3443	4 — P-PAYOFF — BEST	91,5552
3 — RANDOM — P-PAYOFF	46,2999	2 — P-PAYOFF — P-PAYOFF	45,9018
4 — RANDOM — P-PAYOFF	46,0679	3 — P-PAYOFF — P-PAYOFF	45,8006
2 — BEST — RANDOM	54,6092	4 — P-PAYOFF — P-PAYOFF	45,616
3 — BEST — RANDOM	46,3449		

Table 1.2 The heuristic information of Table 1.1 sorted from the best to the worst tactic

Strategy	Gap (%)	Strategy	Gap (%)
4 — BEST — P-PAYOFF	45,2854	4 — P-PAYOFF — RANDOM	49,1199
4 — P-PAYOFF — P-PAYOFF	45,6160	3 — RANDOM — RANDOM	52,3486
3 — P-PAYOFF — P-PAYOFF	45,8006	1 — RANDOM — RANDOM	52,3701
2 — P-PAYOFF — P-PAYOFF	45,9018	2 — RANDOM — RANDOM	52,3731
1 — P-PAYOFF — P-PAYOFF	46,0188	4 — RANDOM — RANDOM	52,4329
4 — RANDOM — P-PAYOFF	46,0679	3 — RANDOM — BEST	53,3355
2 — RANDOM — BEST	46,0911	2 — BEST — RANDOM	54,6092
2 — P-PAYOFF — RANDOM	46,202	2 — BEST — P-PAYOFF	55,5155
3 — BEST — P-PAYOFF	46,2783	2 — P-PAYOFF — BEST	55,6893
3 — RANDOM — P-PAYOFF	46,2999	4 — RANDOM — BEST	70,5899
3 — BEST — RANDOM	46,3449	3 — P-PAYOFF — BEST	75,3516
3 — P-PAYOFF — RANDOM	46,9754	4 — P-PAYOFF — BEST	91,5552
2 — RANDOM — P-PAYOFF	47,3443	1 — BEST — BEST	99,1985
4 — BEST — RANDOM	48,1297		

experiments; more details are given in the next section. The number of strategies is small enough to be exhaustively evaluated; as an example, note that in a model with five different actions and five different events there are only 27 different tactics according to our 2-step encoding scheme explained in Section 1.3. The results of this evaluation are summarized in Tables 1.1 and 1.2. The lower the gap, the better the strategy.

1.5.2   Performance of the ACO Algorithms

The second set of experiments was devoted to testing the performance of two ACO algorithms, Ant-System (AS) and Max-Min Ant System (MMAS), and comparing it with the performance of a Generational Genetic Algorithm presented in [23]. They both share the probabilistic transition rule (transitionrule); the only difference between them is the rule employed to evaporate the pheromone, which is more aggressive in MMAS. See [22] for further details. A third algorithm called Ant Colony System (ACS) was tested in preliminary experiments but the results were similar to AS so they are not reproduced here. The implementation is an adaptation of Marco Dorigo's source code for ACOTSP to fit our adversarial problem[1]. Since the optimal parameters of the ACO algorithms were unknown, several different combinations of them were tested before running the whole set of experiments, as shown in Table 1.3 which also shows the best combination found. All these experiments were done with payoff matrix M_1. After this preliminary experiment, only the best parameter combination of each algorithm was used in the performance experiments. They were divided in an execution phase and a post-evaluation phase. In the execution phase, we made 100 independent runs of each algorithm and annotated the best solution found (not the gap of the solution but the solution itself). The stopping criterion for every run was reaching 150.000 simulations (recall that evaluating a strategy consumes at least 30 simulations). The parameters of the adversarial model were the same ones used in [23], which are summarized in Table 1.4. The optimal value for L should be properly studied, although preliminary experiments showed no improvement with high values so $L = 4$ was a good choice. Fifteen different payoff matrices were employed. All the rows of every payoff matrix are a permutation of a given set of payoffs for that matrix. The first row of every payoff matrix is shown in Table 1.5.

Table 1.3 Parameters tested in the ACO algorithms

Parameter	Description	Values tested	AS	MMAS
Ants	Number of ants	15	15	15
α	Weight of the pheromone component	$\{1.0\}$	1.0	1.0
β	Weight of the heuristic component	$\{0, 0.5, 1.0, 2.0\}$	2	0.5
ρ	Pheromone evaporation ratio	$\{0.1, 0.15, 0.2\}$	0.1	0.1

After the execution phase, we did a post-evaluation of the best solutions found, all in the same conditions (with 100 independent simulations each). This is the same testing methodology we used in [23]. The 100 results of the post-evaluation of the solutions found by every algorithm follow a normal distribution and have similar variances, so parametrical tests can be applied to compare the three algorithms over

---

[1] ACOTSP source code is available for download at http://www.aco-metaheuristic.org/aco-code/

the 15 payoff matrices. This kind of analysis is similar to those carried out in Machine Learning works, but in this case the payoff matrices resemble *datasets*. The absolute results are summarized in Table 1.6. The ranking columns indicate which algorithm performs better with each payoff matrix (1 is the best, 3 is the worst). T stands for a "tie" in the numerical results. In addition, an asterisk indicates whether this ranking has statistical significance (according to ANOVA tests with Tukey post-hoc analysis aided by SPSS software).

Table 1.4 Parameters of the adversarial model used in the fitness function

Parameter	Value
Number of different events and actions	5
Length of the strategies explored	4
Strategy of agent T	Prop. to observed frequency
Simulations used to evaluate a strategy	$p = 30$ in each block, up to 90
Confidence level for statistical assessment of strategies	95 %

Table 1.5 Set of payoffs associated to each 5x5 payoff matrix. Showing first row of every matrix.

Matrix	First row				
1	1	0,9	0,95	0,8	0,85
2	0,8	0,9	0,6	0,7	1
3	1	0,85	0,7	0,4	0,55
4	1	0,6	0,8	0,4	0,2
5	0,25	0,01	0,5	1	0,75
6	1,1	0,95	0,9	1,05	1
7	1,2	1	1,1	0,9	0,8
8	1,3	1	1,15	0,85	0,7
9	1,2	1,4	1	0,8	0,6
10	1,5	1	0,75	1,25	0,5
11	0,8	0,6	0,4	1,5	1
12	0,8	0,6	0,4	1,75	1
13	0,8	0,6	0,4	2	1
14	0,8	0,6	0,4	2,25	1
15	0,8	0,6	0,4	2,5	1

Several conclusions can be drawn from Table 1.6. Firstly, the Max-Min Ant System (MMAS) works better than the Ant-System (AS), as could be expected according to existing results on the classical TSP problem and other combinatorial problems. As it is well known, mechanisms such as a more aggressive evaporation rule enhance the optimization process and yields to better results in MMAS. Secondly, notice that MMAS is performing as well as the Generational Genetic Algorithm (GGA) evaluated in [23]. In fact, in some cases, the three algorithms are totally equivalent (matrices M_1, M_2, M_6 and M_7). These matrices are "easier" because the

Table 1.6 Gap (%) of AS, MMAS and GGA. Average results over 100 independent runs.

Poff matrix	AS Avg	AS Std dev	MMAS Avg	MMAS Std dev	GGA Avg	GGA Std dev	Ranking AS	Ranking MMAS	Ranking GGA
M_1	27.81	0.1461	27.76	0.1597	27.79	0.1743	3	1	2
M_2	34.89	0.1749	34.85	0.1572	34.85	0.1908	3	T	T
M_3	40.98	0.1794	40.88	0.1677	40.91	0.1780	3*	1	2
M_4	45.78	0.1998	45.54	0.1605	45.54	0.2067	3*	T	T
M_5	49.58	0.1927	49.47	0.2013	49.50	0.1622	3*	1	2
M_6	27.10	0.1603	27.08	0.1421	27.13	0.1594	3	1	2
M_7	32.64	0.1398	32.59	0.1489	32.61	0.1599	3	1	2
M_8	36.92	0.1572	36.86	0.1618	36.86	0.1812	3*	T	T
M_9	40.22	0.1902	40.13	0.1567	40.14	0.1677	3*	T	T
M_{10}	42.14	0.1797	42.00	0.1925	41.99	0.1837	3*	2	1
M_{11}	49.74	0.1933	49.55	0.1685	49.55	0.1720	3*	T	T
M_{12}	53.55	0.1834	53.42	0.1662	53.46	0.1635	3*	1	2
M_{13}	56.48	0.1799	56.31	0.1728	56.29	0.1522	3*	2	1
M_{14}	58.64	0.2178	58.51	0.1611	58.55	0.1744	3*	1	2
M_{15}	60.37	0.2112	60.26	0.1773	60.24	0.1759	3*	2	1
Min	27.10	0.1398	27.08	0.1421	27.13	0.1522			
Max	60.37	0.2178	60.26	0.2013	60.24	0.2067			
Avg	43.79	0.1804	43.68	0.1663	43.69	0.1735			
Std dev	10.80	0.0223	10.76	0.0153	10.75	0.0139			

payoffs are very similar to each other, so the election of an action becomes less important because the loss due to sub-optimal actions is small. As a result, the strategy employed is not so important. In the rest of the matrices, AS performs significantly worse than the others. Although in some matrices MMAS seems to work better if we look at the average results, actually there is no payoff matrix where GGA and MMAS are different from a statistical point of view.

Nevertheless, there are some advantages on the use of ACO algorithms instead of a GGA. Firstly, they are less time consuming (a GGA with 150000 simulations took about three minutes while ACO algorithms take about one minute, even using a "standard" source code). Secondly, they avoid the use of a crossover operator, which in this problem seems useless because the building blocks of a chromosome do not seem to exist. A tactic within a strategy is good or bad only in relation to what tactics precede and follow it.

1.5.3 Analysis of the Solutions Found by Each Algorithm

Although Genetic Algorithms and Ant-based Algorithms are quite different, it is interesting to provide insights in the solutions of both techniques, specially when a solution (a strategy) is easily interpretable and has a clear meaning, as in this case. For that reason, the solutions found by each algorithm were collected and carefully analysed in terms of their constituent tactics. The main objective is to look

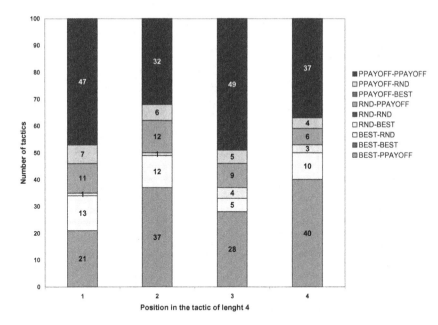

Fig. 1.3 Frequency of appearance of each criteria combination in the 100 solutions found by the GGA

for similarities between strategies found by different algorithms, to see if the area of the search space explored is the same or different.

A way to depict this similarity is by doing an "alignment" plot, similar to those used in bioinformatics to show sequence alignments, called sequence logos. This kind of graphic represent the common parts in two or more DNA (or protein) sequences. For our purposes, we have counted the number of times each criteria combination (e.g. BEST-RND, PPAYOFF-BEST, RND-BEST, etc) appears in every position of the strategy found by the algorithm. Since we have 100 solutions from each algorithm, they are enough to have a clear understanding of what kind of strategies are mostly found, in terms of frequency. If this frequency is graphically depicted for each possible criteria combination, it is easy to see which are the most frequent tactics and in what position of the cyclic strategy they use to arise.

Regarding Fig. 1.3 alone, one can see that two of the poosible combinations of criteria are much more frequent than the rest. The most frequent is PPAYOFF-PPAYOFF, in the top of the bars. This is an interesting fact, since the three existing combinations of PPAYOFF-PPAYOFF are also the ones that provide the best results when tested alone (as simple strategies of lenght 1) according to Table 1.2. Notice that the Genetic Algorithm does not know (nor it is required) the heuristic information shown in Table 1.1, but the solutions found are consistent with this information. In second place, two of the three existing combinations of BEST-PPAYOFF are also

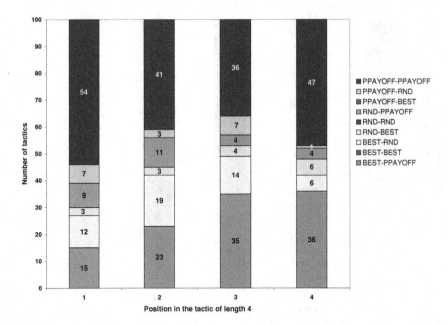

Fig. 1.4 Frequency of appearance of each criteria combination in the 100 solutions found by the best performing MMAS (weight of the heuristic component $\beta = 0.5$)

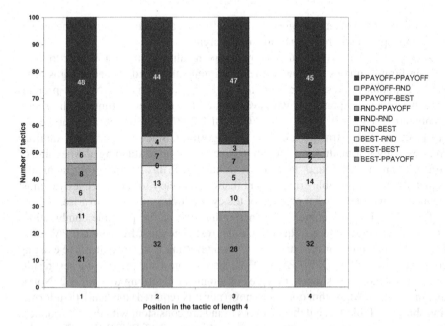

Fig. 1.5 Frequency of appearance of each criteria combination in the 100 solutions found by a MMAS algorithm with $\beta = 2$

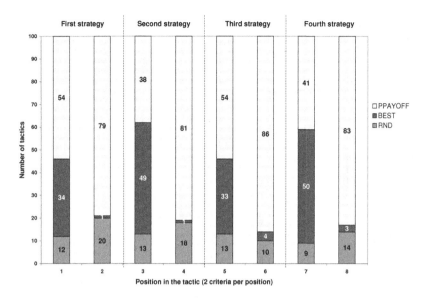

Fig. 1.6 Frequency of appearance of every criterium separately in the 100 solutions found by the GGA

very good, which is consistent with the second-most frequent combination BEST-PPAYOFF (at the bottom of the bars).

Notice that Fig. 1.3 is quite similar to Fig. 1.4 and also to Fig. 1.5. This means that both algorithms reach approximately the same conclusions because they converge to the same area (in general) of the search space. The proportion of the colors (combinations of criteria) is more or less the same in the three figures. An important detail can be seen in Fig. 1.5. In this variant, the heuristic component β has been enhanced with respect to Fig. 1.4 by using $\beta = 2$. As a result, the heuristic information of Table 1.1 is more trusted and the resulting solutions have always a higher number of the heuristically good combinations in all the positions. This is confirmed by the fact that the number of occurrences of every combination is much more balanced in the four positions of a strategy in Fig. 1.5 than in Fig. 1.3 or in Fig. 1.4.

Although the possible combinations have already been analysed, it is still possible to count how frequently each criterium appears on its own, disregarding the combination with others criteria. The results of this are shown in Fig. 1.6 and 1.7. There are 8 bars in each figure because there exist 2 criteria in each position and the strategy length is 4. The first conclusion that can be drawn from both figures is that they are both surprisingly alike, which confirms once more that the convergence of both a Genetic Algorithm and an Ant-based algorithm is to the same area. In addition to this, one can see a general tendence in both figures. In the majority of the cases, the first criterium of a tactic tends to be PPAYOFF (half of the times) or BEST to select the candidate actions, while the second criterium tends to be PPAYOFF in about 80 % of the tactics. There are very few tactics where the

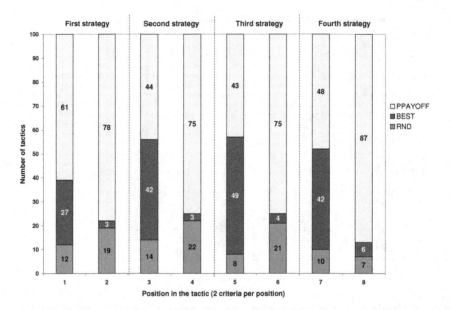

Fig. 1.7 Frequency of appearance of every criterium separately in the 100 solutions found by the best performing MMAS (weight of the heuristic component $\beta = 0.5$)

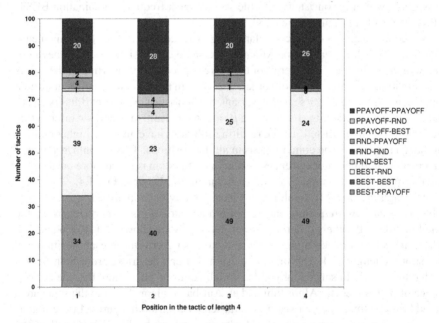

Fig. 1.8 Frequency of appearance of every criteria combination in the 100 solutions found by the MMAS (weight of the heuristic component $\beta = 0.5$) with an uniformly random matrix

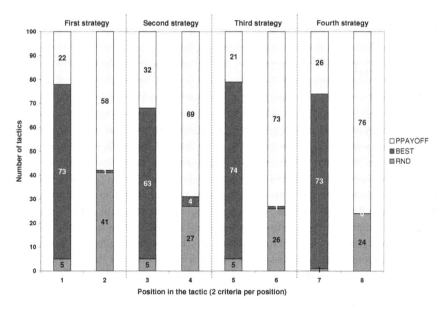

Fig. 1.9 Frequency of appearance of every criterium separately in the 100 solutions found by the MMAS (weight of the heuristic component $\beta = 0.5$) with an uniformly random matrix

second criterium is BEST, although it could be a good choice in cases where the first criterium was RND. Graphically, this phenomenon be observed as an alternating expanding-contracting yellow bar, since the same happens in the four positions of the strategy, and in both the Genetic and the Ant-based solutions.

An additional matter concerning the three decision criteria (BEST, RND, PPAY-OFF) is still pending. The above analysis has been done for a specific payoff matrix M_A that is not among the 15 matrices explained before. A good question would be if these results are also valid in other situations, with other payoff matrices. In order to answer to this question, the experiments of the ACO algorithms were repeated with a new random payoff matrix M_B with values randomly generated in [0, 1] following a uniform distribution.

The results are quite interesting. Firstly, Fig. 1.8 demonstrates that the payoff matrix has a heavy influence on the resulting tactics, because the proportions of this figure are very different from those in already commented figures. Combination PPAYOFF-PPAYOFF (in the top) is much less frequent, while combinations BEST-RND and BEST-PPAYOFF are more frequent than in the first matrix M_A. Secondly, Fig. 1.9 shows that the criterium BEST is more frequently used than in the graphics above: it is in fact the most frequently used criterium in the first step of a combination (in more than 70 % of the cases), while in the matrix M_A, the most used was PPAYOFF. The tendency of the second criterium is similar: PPAYOFF is still the most used, but the RND criterium is also more frequently employed. Finally, notice that the shape of the yellow bar (PPAYOFF) is again similar to the graphics above: an alternating expanding-contracting bar.

1.6 Conclusions

Two ant-based optimization algorithms have been applied to the problem of automatic design of strategies in an adversarial model, which can be tackled as a combinatorial optimization problem. The best solutions found by ant-based algorithms are similar to the best results obtained so far but the execution time is three times less. In addition, insights have been provided about the nature of the solutions found by the reference Genetic Algorithm and by our ant algorithms, showing that both techniques converge to approximately the same area of the search space and the solutions found are thus quite similar in terms of their constituent tactics. The influence of the payoff matrix on the solutions found by ant algorithms has also been proven because their constituent tactics are different in terms of the occurrence frequency over 100 independent runs. This confirms the heavy influence of the payoff matrix of each problem over the strategies that should be employed in each case.

Acknowledgements. This work was supported in part by the projects TIN2008 - 01948 and TIN2008 - 06872 - C04 - 04 from the Spanish Ministry of Science and Innovation and P07 - TIC - 02970 from the Andalusian Government.

References

[1] Amigoni, F., Gatti, N., Ippedico, A.: A game-theoretic approach to determining efficient patrolling strategies for mobile robots. In: Proceedings of the International Conference on Web Intelligence and Intelligent Agent Technology (IAT 2008), pp. 500–503 (2008)
[2] Baldassarre, G., Nolfi, S.: Strengths and synergies of evolved and designed controllers: A study within collective robotics. Artificial Intelligence 173(7-8), 857–875 (2009)
[3] Caro, G.D., Dorigo, M.: Antnet: Distributed stigmergetic control for communication networks. Journal of Artificial Intelligence Research 9, 317–365 (1998)
[4] Cordón, O., de Viana, I.F., Herrera, F., Moreno, L.: A new aco model integrating evolutionary computation concepts: The best-worst ant system. In: Proceedings of the Second International Workshop on Ant Algorithms, ANTS 2000, pp. 22–29 (2000)
[5] Corne, D., Dorigo, M., Glover, F., Dasgupta, D., Moscato, P., Poli, R., Price, K.V. (eds.): New ideas in optimization. McGraw-Hill Ltd, UK (1999)
[6] Costa, D., Hertz, A.: Ants can colour graphs. The Journal of the Operational Research Society 48(3), 295–305 (1997)
[7] Dorigo, M., Caro, G.D.: Ant algorithms for discrete optimization. Artificial Life 5(2), 137–172 (1999)
[8] Dorigo, M., Gambardella, L.M.: Ant colony system: A cooperative learning approach to the traveling salesman problem. IEEE Trans. on Evolutionary Computation 1(1), 53–66 (1997)
[9] Dorigo, M., Stützle, T.: Ant Colony Optimization. MIT Press, Cambridge (2004)

[10] Dorigo, M., Maniezzo, V., Colorni, A.: The ant system: Optimization by a colony of cooperating agents. IEEE Trans. on Systems, Man and Cybernetics - Part B 26(1), 1–13 (1996)

[11] Fidanova, S., Lirkov, I.: Ant colony system approach for protein solving. In: Proceedings of 2nd Int. Multiconf. on Computer Science and Information Technology, pp. 887–891 (2008)

[12] Kott, A., McEneany, W.M.: Adversarial Reasoning: Computational Approaches to Reading the Opponents Mind. Chapman and Hall/ CRC, Boca Raton (2007)

[13] Kott, A., Ownby, M.: Tools for real-time anticipation of enemy actions in tactical ground operations. In: Proceedings of the 10th International Command and Control Research and Technology Symposium (2005)

[14] Krasnogor, N., Terrazas, G., Pelta, D.A., Ochoa, G.: A critical view of the evolutionary design of self-assembling systems. In: Talbi, E.-G., Liardet, P., Collet, P., Lutton, E., Schoenauer, M. (eds.) EA 2005. LNCS, vol. 3871, pp. 179–188. Springer, Heidelberg (2006)

[15] Maniezzo, V., Colorni, A., Dorigo, M.: The ant system applied to the quadratic assignment problem. Tech. Rep. 94/28, IRIDIA, Université Libre de Bruxelles, Belgium (1994)

[16] McMullen, P.R.: An ant colony optimization approach to addressing a jit sequencing problem with multiple objectives. Artificial Intelligence in Engineering 15(3), 309–317 (2001)

[17] Park, H.S., Pedrycz, W., Oh, S.K.: Evolutionary design of hybrid self-organizing fuzzy polynomial neural networks with the aid of information granulation. Expert Systems with Applications 33(4), 830–846 (2007)

[18] Paruchuri, P., Pearce, J.P., Kraus, S.: Playing games for security: An efficient exact algorithm for solving bayesian stackelberg games. In: Proceedings of 7th Int. Conf. on Autonomous Agents and Multiagent Systems (AAMAS 2008), pp. 895–902 (2008)

[19] Pelta, D., Yager, R.: On the conflict between inducing confusion and attaining payoff in adversarial decision making. Information Sciences 179, 33–40 (2009)

[20] Popp, R., Yen, J.: Emergent Information Technologies and Enabling Policies for Counter-Terrorism. John Wiley and Sons, Hoboken (2006)

[21] Salcedo-Sanz, S., Naldi, M., Perez-Bellido, A.M., Portilla-Figueras, A., Ortiz-Garcia, E.G.: Evolutionary design of oriented-tree networks using cayley-type encodings. Information Sciences 179(20), 3461–3472 (2009)

[22] Stützle, T., Hoos, H.H.: Max-min ant system. Future Generation Computer Systems 16(8), 889–914 (2000)

[23] Villacorta, P., Pelta, D.: Evolutionary design and statistical assessment of strategies in an adversarial domain. In: Proceedings of the IEEE Conference on Evolutionary Computation (CEC 2010), pp. 2250–2256 (2010)

Chapter 2
Resource Allocation and Dispensation Impact of Stochastic Diffusion Search on Differential Evolution Algorithm

Mohammad Majid al-Rifaie, John Mark Bishop, and Tim Blackwell

Abstract. This work details early research aimed at applying the powerful resource allocation mechanism deployed in Stochastic Diffusion Search (SDS) to the Differential Evolution (DE), effectively merging a nature inspired swarm intelligence algorithm with a biologically inspired evolutionary algorithm. The results reported herein suggest that the hybrid algorithm, exploiting information sharing between the population, has the potential to improve the optimisation capability of classical DE.

2.1 Introduction

In the literature, nature inspired swarm intelligence algorithms and biologically inspired evolutionary algorithms are typically evaluated using benchmarks that are often small in terms of their objective function computational costs [9, 39]; this is often not the case in real-world applications. This paper is an attempt to pave the way for more effectively optimising computationally expensive objective functions, by deploying the SDS diffusion mechanism to more efficiently allocate DE resources via information-sharing between the members of the population.

Mohammad Majid al-Rifaie
Goldsmiths, University of London, New Cross, London SE14 6NW, United Kingdom
e-mail: m.majid@gold.ac.uk

John Mark Bishop
Goldsmiths, University of London, New Cross, London SE14 6NW, United Kingdom
e-mail: m.majid@gold.ac.uk

Tim Blackwell
Goldsmiths, University of London, New Cross, London SE14 6NW, United Kingdom
e-mail: m.majid@gold.ac.uk

D.A. Pelta et al. (Eds.): NICSO 2011, SCI 387, pp. 21–40, 2011.
springerlink.com   © Springer-Verlag Berlin Heidelberg 2011

The use of SDS as an efficient resource allocation algorithm was first explored in [21, 26, 28] and these results provided motivation to investigate the application of the information diffusion mechanism originally deployed in SDS[1] with DE.

Communication – social interaction or information exchange – observed in social insects is important in all swarm intelligence and evolutionary algorithms, including SDS and DE algorithms.

This work investigates the communication between the members of the population as the mean to maintain population diversity, which is faciliated by using the resource allocation and resource dispensation of SDS algorithm.

In a former work [3], SDS is merged with Particle Swarm Optimisation (PSO) algorithm and the promising results of this hybridisation alongside some statistical analysis of its performance are reported.

Although in real social interactions, not just the syntactical information is exchanged between the individuals but also semantic rules and beliefs about how to process this information [18], in typical swarm intelligence algorithms, only the syntactical exchange of information is considered.

In the study of the interaction of social insects, two important elements are the individuals and the environment, which will result in two integration schemes: the first one is the way in which individuals self-interact and the second one is the interaction of the individuals with the environment [6]. Self-interaction between individuals is carried out through recruitment strategies and it has been demonstrated that, typically, differing recruitment strategies are used by ants [12] and honey bees. These recruitment strategies are used to attract other members of the society to gather around one or more desired areas, either for foraging purposes or for moving to a new nest site.

In general, there are many different forms of recruitment strategies used by social insects; these may take the form of local or global strategies; one-to-one or one-to-many communication; and deploy stochastic or deterministic mechanisms. The nature of information exchange also varies in different environments and with different types of social insects. Sometimes the information exchange is quite complex where, for example it might carry data about the direction, suitability of the target and the distance; or sometimes the information sharing is simply a stimulation forcing a certain triggered action. What all these recruitment and information exchange strategies have in common is distributing useful information throughout their community [23].

In this paper, the swarm intelligence algorithm and the evolutionary algorithm are first introduced, followed by the hybridisation strategy. Afterwards, the results are reported and the performance of the hybrid algorithm is discussed.

---

[1] The 'information diffusion' and 'randomised partial objective function evaluation' processes enable SDS to more efficiently optimise problems with costly [discrete] objective functions; see Stochastic Diffusion Search Section for an introduction to the SDS metaheuristic.

2.2   Stochastic Diffusion Search

This section introduces SDS [5], a multi-agent global search and optimisation algorithm, which is based on simple interaction of agents (inspired by one species of ants, *Leptothorax acervorum*, where a 'tandem calling' mechanism (one-to-one communication) is used, where the forager ant which finds the food location, recruits a single ant upon its return to the nest, and therefore the location of the food is physically publicised [24]). A high-level description of SDS is presented in the form of a social metaphor demonstrating the procedures through which SDS allocates resources.

SDS introduced a new probabilistic approach for solving best-fit pattern recognition and matching problems. SDS, as a multi-agent population-based global search and optimisation algorithm, is a distributed mode of computation utilising interaction between simple agents [22].

Unlike many nature inspired search algorithms, SDS has a strong mathematical framework, which describes the behaviour of the algorithm by investigating its resource allocation [26], convergence to global optimum [27], robustness and minimal convergence criteria [25] and linear time complexity [29]. In order to introduce SDS, a social metaphor *the Mining Game* [1] is used.

2.2.1   The Mining Game

This metaphor provides a simple high-level description of the behaviour of agents in SDS, where mountain range is divided into hills and each hill is divided into regions:

> A group of miners learn that there is gold to be found on the hills of a mountain range but have no information regarding its distribution. To maximize their collective wealth, the maximum number of miners should dig at the hill which has the richest seams of gold (this information is not available a-priori). In order to solve this problem, the miners decide to employ a simple Stochastic Diffusion Search.

> - At the start of the mining process each miner is randomly allocated a hill to mine (his hill hypothesis, h).
> - Every day each miner is allocated a randomly selected region, on the hill to mine.

> At the end of each day, the probability that a miner is happy is proportional to the amount of gold he has found. Every evening, the miners congregate and each miner who is not happy selects another miner at random for communication. If the chosen miner is happy, he shares the location of his hill and thus both now maintain it as their hypothesis, h; if not, the unhappy miner selects a new hill hypothesis to mine at random.

As this process is isomorphic to SDS, miners will naturally self-organise to congregate over hill(s) of the mountain with high concentration of gold.

In the context of SDS, agents take the role of miners; active agents being 'happy miners', inactive agents being 'unhappy miners and the agent's hypothesis being the miner's 'hill-hypothesis'.

Algorithm 2.1. The Mining Game

```
Initialisation phase
Allocate each miner (agent) to a random
   hill (hypothesis) to pick a region randomly

Until (all miners congregate over the highest
   concentration of gold)

   Test phase
      Each miner evaluates the amount of gold
         they have mined (hypotheses evaluation)
      Miners are classified into happy (active)
         and unhappy (inactive) groups

   Diffusion phase
      Unhappy miners consider a new hill by
         either communicating with another miner
         or,if the selected miner is also
         unhappy, there will be no information
         flow between the miners; instead the
         selecting miner must consider another
         hill (new hypothesis) at random
End
```

2.2.2   SDS Architecture

The SDS algorithm commences a search or optimisation by initialising its population (e.g. miners, in the mining game metaphor). In any SDS search, each agent maintains a hypothesis, h, defining a possible problem solution. In the mining game analogy, agent hypothesis identifies a hill. After initialisation two phases are followed (see Algorithm 2.1 for these phases in the mining game; for high-level SDS description see Algorithm 2.2):

- Test Phase (e.g. testing gold availability)
- Diffusion Phase (e.g. congregation and exchanging of information)

Algorithm 2.2. SDS Algorithm

```
Initialising agents()
While (stopping condition is not met)
   Testing hypotheses()
   Diffusion hypotheses()
End
```

In the test phase, SDS checks whether the agent hypothesis is successful or not by performing a partial hypothesis evaluation which returns a boolean value. Later in the iteration, contingent on the precise recruitment strategy employed, successful hypotheses diffuse across the population and in this way information on potentially good solutions spreads throughout the entire population of agents.

In the Test phase, each agent performs *partial function evaluation, pFE*, which is some function of the agent's hypothesis; $pFE = f(h)$. In the mining game the partial function evaluation entails mining a random selected region on the hill, which is defined by the agent's hypothesis (instead of mining all regions on that hill).

In the Diffusion phase, each agent recruits another agent for interaction and potential communication of hypothesis. In the mining game metaphor, diffusion is performed by communicating a hill hypothesis.

2.2.3   Standard SDS and Passive Recruitment

In standard SDS (which is used in this paper), *passive recruitment mode* is employed. In this mode, if the agent is inactive, a second agent is randomly selected for diffusion; if the second agent is active, its hypothesis is communicated (*diffused*) to the inactive one. Otherwise there is no flow of information between agents; instead a completely new hypothesis is generated for the first inactive agent at random (see Algorithm 2.3).

Algorithm 2.3. Passive Recruitment Mode

```
For ag = 1 to No_of_agents
    If (ag is not active)
        r_ag = pick a random agent()
        If (r_ag is active)
            ag.setHypothesis(r_ag.getHypothesis())
        Else
            ag.setHypothesis(randomHypothsis())
End
```

2.2.4   Partial Function Evaluation

One of the concerns associated with many optimisation algorithms (e.g. Genetic Algorithm [11], Particle Swarm Optimisation [17] and etc.) is the repetitive evaluation of a computationally expensive fitness functions. In some applications, such as tracking a rapidly moving object, the repetitive function evaluation significantly increases the computational cost of the algorithm. Therefore, in addition to reducing the number of function evaluations, other measures can be used in an attempt to reduce the computations carried out during the evaluation of each possible solution, as part of the overall optimisation (or search) processes.

The commonly used benchmarks for evaluating the performance of swarm intelligence algorithms are typically small in terms of their objective functions computational costs [9, 39], which is often not the case in real-world applications. Examples of costly evaluation functions are seismic data interpretation [39], selection of sites for the transmission infrastructure of wireless communication networks and radio wave propagation calculations of one site [38] etc.

Costly objective function evaluations have been investigated under different conditions [14] and the following two broad approaches have been proposed to reduce the cost of function evaluations:

- The first is to estimate the fitness by taking into account the fitness of the neighbouring elements, the former generations or the fitness of the same element through statistical techniques introduced in [4, 7].
- In the second approach, the costly fitness function is substituted with a cheaper, approximate fitness function.

When agents are about to converge, the original fitness function can be used for evaluation to check the validity of the convergence [14].

Many fitness functions are decomposable to components that can be evaluated separately. In partial evaluation of the fitness function in SDS, the evaluation of one or more of the components may provide partial information to guide the subsequent optimisation process.

2.3 Differential Evolution

DE, one of the most successful Evolutionary Algorithms (EAs), is a simple global numberical optimiser over continuous search spaces which was first introduced by Storn and Price [32, 33].

DE is a population based stochastic algorithm, proposed to search for an optimum value in the feasible solution space. The parameter vectors of the population are defined as follows:

$$x_i^g = \left[x_{i,1}^g, x_{i,2}^g, ..., x_{i,D}^g \right], i = 1, 2, ..., NP \tag{2.1}$$

where g is the current generation, D is the dimension of the problem space and NP is the population size. In the first generation, (when $g = 0$), the ith vector's j^{th} component could be initialised as:

$$x_{i,j}^0 = x_{min,j} + r \left(x_{max,j} - x_{min,j} \right) \tag{2.2}$$

where r is a random number drawn from a uniform distribution on the unit interval $U(0,1)$, and x_{min}, x_{max} are the lower and upper bounds of the jth dimension, respectively. The evolutionary process (mutation, corssover and selection) starts after the initialisation of the population.

2.3.1   Mutation

At each generation g, the mutation operation is applied to each member of the population x_i^g (target vector) resulting in the corresponding vector v_i^g (mutant vector). Among the most frequently used mutation approaches are the following:

- DE/rand/1

$$v_i^g = x_{r_1}^g + F\left(x_{r_2}^g - x_{r_3}^g\right) \tag{2.3}$$

- DE/target-to-best/1

$$v_i^g = x_i^g + F\left(x_{best}^g - x_i^g\right) + F\left(x_{r_1}^g - x_{r_2}^g\right) \tag{2.4}$$

- DE/best/1

$$v_i^g = x_{best}^g + F\left(x_{r_1}^g - x_{r_2}^g\right) \tag{2.5}$$

- DE/best/2

$$v_i^g = x_{best}^g + F\left(x_{r_1}^g - x_{r_2}^g\right) + F\left(x_{r_2}^g - x_{r_3}^g\right) \tag{2.6}$$

- DE/rand/2

$$v_i^g = x_{r_1}^g + F\left(x_{r_2}^g - x_{r_3}^g\right) + F\left(x_{r_4}^g - x_{r_5}^g\right) \tag{2.7}$$

where r_1, r_2, r_3, r_4 are different from i and are distinct random integers drawn from the range $[1, NP]$; In generation g, the vector with the best fitness value is x_{best}^g and F is a positive control parameter for constricting the difference vectors.

2.3.2   Crossover

Crossover operation, improves population diversity through exchanging some components of v_i^g (mutant vector) with x_i^g (target vector) to generate u_i^g (trial vector). This process is led as follows:

$$u_{i,j}^g = \begin{cases} v_{i,j}^g, & \text{if } r \leq CR \text{ or } j = r_d \\ x_{i,j}^g, & \text{otherwise} \end{cases} \tag{2.8}$$

where r is a uniformly distributed random number drawn from the unit interval $U(0,1)$, r_d is randomly generated integer from the range $[1, D]$; this value guarantees that at least one component of the trial vector is different from the target vector. The value of CR, which is another control parameter, specifes the level of inheritance from v_i^g (mutant vector).

2.3.3 Selection

The selection operation decides whether x_i^g (target vector) or u_i^g (trial vector) would be able to pass to the next generation $(g + 1)$. In case of a minimisation problem, the vector with a smaller fitness value is admitted to the next generation:

$$x_i^{g+1} = \begin{cases} u_i^g, \text{ if } f\left(u_i^g\right) \le f\left(x_i^g\right) \\ \\ x_i^g, \text{ otherwise} \end{cases} \tag{2.9}$$

where $f(x)$ is the fitness function.

Algorithm 2.4 summarises the behaviour of DE algorithm

Algorithm 2.4. DE Pseudo Code

```
Initialise population

For ( generation = 1 to n )
   For ( agent = 1 to NP )
      Mutation : generate mutant vector
      Crossover: generate trial vector
      Selection: generate target vector for next generation
   End

   Find agent with best fitness value
End
```

DE, like other evolutionary algorithms, suffers from premature convergance where the population lose their diversity too early and get trapped in local optima, therefore performing poorly on problems with high dimension and many local optima.

DE is known to be relatively good in comparison with other EAs and PSOs at avoiding premature convergence. However, in order to reduce the risk of premature convergence in DE and to preserve population diversity, several methods have been proposed, among which are: multi-population approaches [8, 19, 20, 34, 35]; providing extra knowledge about the problem space [30, 37]; information storage about previously explored areas [13, 41]; utilising adapting and control parameters to ensure population diversity [40]; using CrowdingDE for tracking and maintaining multiple optima [31, 36].

This paper proposes information exchange and agent dispensation (SDS-led random restart) as methods to avoid premature convergence and preserve population diversity.

2.4   Merging SDS and DE Algorithms

The initial motivating thesis justifying the hybridisation of SDS and DE is the partial function evaluation deployed in SDS, which may mitigate the high computational overheads entailed when deploying a DE algorithm onto a problem with a costly fitness function. However, before commenting on and exploring this area – which remains an ongoing research – an initial set of experiments aimed to investigate if the information diffusion mechanism deployed in SDS may on its own improve DE behaviour. These are the results that are primarily reported in this paper.

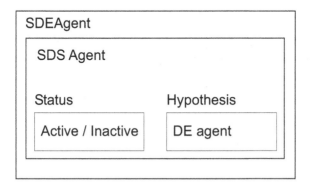

Fig. 2.1 Encapsulating SDS agent and DE agent as SDE-Agent

In this new architecture, a standard set of benchmarks are used to evaluate the performance of the hybrid algorithm. The resource allocation (or recruitment) and partial function evaluation sides of SDS (see Section 2.2.4) are used to assist allocating and dispensing resources (e.g. members of the DE population) after partially evaluating the search space.

Each DE agent has three vectors (target, mutant and trial vectors); and each SDS agent has one hypothesis and one status. In the experiment reported here (hybrid algorithm), every member of DE population is an SDS agent too – together termed *SDEAgents*. In SDEAgents, SDS hypotheses are defined by the DE target vector, and an additional boolean variable (status) determining whether the SDEAgent is active or inactive (see Figure 2.1). The behaviour of the hybrid algorithm in its simplest form is presented in Algorithm 2.5.

2.4.1   Test and Diffusion Phases in the Hybrid Algorithms

In the test-phase of a stochastic diffusion search, each agent has to partially evaluate its hypothesis. The guiding heuristic is that hypotheses that are promising are maintained and those that appear unpromising are discarded. In the context of the hybrid

Algorithm 2.5. Hybrid Algorithm

```
Initialise SDEAgents

For ( generation = 1 to generationsAllowed )

    For ( SDEAgent = 1 to NP )
        Mutation : generate mutant vector
        Crossover: generate trial vector
        Selection: generate target vector for next generation
    End For

    If ( generation counter MOD n == 0 )
    // START SDS
    // TEST PHASE
        For ag = 1 to NP
            r_ag = pick-random-SDEAgent()
            If ( ag.targetVecFitness() < r_ag.targetVecFitness() )
                ag.setActivity (true)
            Else
                ag.setActivity (false)
            End If
        End For

        // DIFFUSION PHASE
        For ag = 1 to No_of_SDEAgents

            If ( ag is not active )
                r_ag = pick-random-SDEAgent()
                If ( r_ag is active )
                    ag.setHypo( r_ag.getHypo() )*
                Else
                    ag.setHypo( randomHypo() )
                End If
            End If

        End for
    End If
    // END SDS

    Find SDEAgent with best fitness value

End For

* In setHypo() and getHypo(), Hypo refers to
the SDEAgent's hypothesis (target vector).
```

DE-SDS algorithm, it is clear that there are many different tests that could be performed in order to determine the activity of each SDEAgent. A very simple test is illustrated in Algorithm 2.5. Here, the test-phase is simply conducted by comparing the fitness of each SDEAgent's target vector against that of a random SDEAgent; if the selecting SDEAgent has a better fitness value, it will become active, otherwise it will be flagged inactive. On average, this mechanism will ensure 50% of SDEAgents remain active from one iteration to another.

In the Diffusion Phase, each inactive SDEAgent picks another SDEAgent randomly, if the selected SDEAgent is active, the selected SDEAgent communicates its hypothesis to the inactive one; if the selected SDEAgent is also inactive, the selecting SDEAgent generates a new hypothesis at random from the search space.

As outlined in the pseudo-code of the hybrid algorithm (see Algorithm 2.5), after each n generations, one full SDS cycle[2] is executed. The hybrid algorithm is called *SDSnDE*, where n refers to the number of generations before an SDS cycle should run.

In the next section, the experiment setup is reported and the results will follow.

2.5 Results

In this section, a number of experiments are carried out and the performance of one variation of DE algorithm (DE/best/1) is contrasted against the hybrid algorithm, *SDSnDE*.

2.5.1 Experiment Setup

The algorithms are tested over a number of benchmarking functions from Jones et al [15] and De Jong [16] test suite, preserving different dimensionality and modality (see Tables 2.1 and 2.2, where benchmark function equations, feasible bounds, the number of dimensions in which the benchmarks are used in the experiments, the optimum of each function which is known *a priori* and also the boundaries where particles are first initialised are presented).

The first two functions (Sphere/Parabola and Schwefel 1.2) have a single minimum and are unimodal functions; Generalised Rosenbrock for dimension D, where $D > 3$, is multimodal; Generalised Schwefel 2.6, Generalized Rastrigin, Ackley, Generalized Griewank, Penalised Function P8 and Penalised Function P16 are complex high-dimensional multi-modal problems with many local minima and a single global optimum; Six-hump Camel-back, Goldstein-Price, Shekel 5, 7 and 10 are lower-dimensional multi-modal problems with fewer local minima. Goldstein-Price, Shekel 5, 7 and 10 have one global optimum and Six-hump Camel-back has two global optima symmetric about the origin.

In order not to initialise the DE agents on or near a region in the search space known to have the global optimum, *region scaling* technique is used [10], which makes sure DE agents are initialised at a corner of the search space where there are no optimal solution.

The experiments are conducted with the population of 100 agents. The halting criterion for this experiment is when the number of generations reaches 2,000.

There are 30 independent runs for each benchmark function and the results are averaged over these independent trials.

Accuracy, which is used as performance measure, is defined by the quality of the best agent in terms of its closeness to the optimum position. If knowledge about

---

[2] Test Phase: decides about the status of each SDEAgent, one after another; Diffusion Phase: shares information according to the algorithm presented

Table 2.1 Benchmark Functions Equations

Function	Equation
Sphere/Parabola	$f_1 = \sum\limits_{i=1}^{D} x_i^2$
Schwefel 1.2	$f_2 = \sum\limits_{i=1}^{D} \left(\sum\limits_{j=1}^{i} x_j \right)^2$
Generalised Rosenbrock	$f_3 = \sum\limits_{i=1}^{D-1} \left\{ 100\left(x_{i+1} - x_i^2\right)^2 + (x_i - 1)^2 \right\}$
Generalised Schwefel 2.6	$f_4 = -\sum\limits_{i=1}^{D} x_i \sin\left(\sqrt{x_i}\right)$
Generalised Rastrigin	$f_5 = \sum\limits_{i=1}^{D} \left\{ x_i^2 - 10\cos\left(2\pi x_i\right) + 10 \right\}$
Ackley	$f_6 = -20\exp\left\{ -0.2\sqrt{\frac{1}{D}\sum\limits_{i=1}^{D} x_i^2} \right\} -$ $\exp\left\{ \frac{1}{D}\sum\limits_{i=1}^{D} \cos\left(2\pi x_i\right) \right\} + 20 + e$
Generalised Griewank	$f_7 = \frac{1}{4000}\sum\limits_{i=1}^{D} x_i^2 - \prod\limits_{i=1}^{D} \cos\left(\frac{x_i}{\sqrt{i}}\right) + 1$
Penalized Function P8	$f_8 = \frac{\pi}{D}\{10\sin^2(\pi y_1) + \sum_{i=1}^{D-1} (y_i - 1)^2 \{1 + 10\sin^2(\pi y_{i+1})\} + (y_D - 1)^2\} + \sum_{i=1}^{D} \mu(x_i, 10, 100, 4)$ $y_i = 1 + \frac{1}{4}(x_i + 1)$ $\mu(x_i, a, k, m) = \begin{cases} k(x_i - a)^m & x_i > a \\ 0 & -a \le x_i \le a \\ k(-x_i - a)^m & x_i < -a \end{cases}$
Penalized Function P16	$f_9 = 0.1\{\sin^2(3\pi x_1) + \sum_{i=1}^{D-1} (x_i - 1)^2 \{1 + \sin^2(3\pi x_{i+1})\} + (x_D - 1)^2 \times \{1 + \sin^2(2\pi x_D)\}\} + \sum_{i=1}^{D} \mu(x_i, 5, 100, 4)$
Six-hump Camel-back	$f_{10} = 4x_1^2 - 2.1x_1^4 + \frac{1}{3}x_1^6 + x_1 x_2 - 4x_2^2 + 4x_2^4$
Goldstein-Price	$f_{11} = \left\{ 1 + (x_1 + x_2 + 1)^2 \left(19 - 14x_1 + 3x_1^2 - 14x_2 + 6x_1 x_2 + 3x_2^2\right) \right\} \times$ $\left\{ 30 + (2x_1 - 3x_2)^2 \left(18 - 32x_1 + 12x_1^2 + 48x_2 - 36x_1 x_2 + 27x_2^2\right) \right\}$
Shekel 5	$f_{12} = -\sum_{i=1}^{5} \left\{ \sum_{j=1}^{4} \left(x_j - a_{ij}\right)^2 + c_i \right\}^{-1}$
Shekel 7	$f_{13} = -\sum_{i=1}^{7} \left\{ \sum_{j=1}^{4} \left(x_j - a_{ij}\right)^2 + c_i \right\}^{-1}$
Shekel 10	$f_{14} = -\sum_{i=1}^{10} \left\{ \sum_{j=1}^{4} \left(x_j - a_{ij}\right)^2 + c_i \right\}^{-1}$

the optimum position is known *a priori* (which is the case here), the following would define accuracy:

$$\text{Accuracy}(S,t) = \left| f\left(x_{best}^g\right) - f\left(x_{opt}\right) \right| \qquad (2.10)$$

where x_{best}^g is the best agent at generation g and x_{opt} is the position of the known optimum solution.

Table 2.2 Benchmark Functions Details

Function	D	Feasible Bounds	Optimum	Initialisation
f_1 Sphere/Parabola	30	$(-100,100)^D$	0.0^D	$(50,100)^D$
f_2 Schwefel 1.2	30	$(-100,100)^D$	0.0^D	$(50,100)^D$
f_3 Generalized Rosenbrock	30	$(-30,30)^D$	1.0^D	$(15,30)^D$
f_4 Generalized Schwefel 2.6	30	$(-500,500)^D$	420.9687^D	$(250,500)^D$
f_5 Generalized Rastrigin	30	$(-5.12,5.12)^D$	0.0^D	$(2.56,5.12)^D$
f_6 Ackley	30	$(-32,32)^D$	0.0^D	$(16,32)^D$
f_7 Generalized Griewank	30	$(-600,600)^D$	0.0^D	$(300,600)^D$
f_8 Penalized Function P8	30	$(-50,50)^D$	-1.0^D	$(25,50)^D$
f_9 Penalized Function P16	30	$(-50,50)^D$	1.0^D	$(25,50)^D$
f_{10} Six-hump Camel-back	2	$(-5,5)^D$	$(-0.0898,0.7126),$ $(0.0898,-0.7126)$	$(2.5,5)^D$
f_{11} Goldstein-Price	2	$(-2,2)^D$	$(0,-1)$	$(1,2)^D$
f_{12} Shekel 5	4	$(0,10)^D$	4.0^D	$(7.5,10)^D$
f_{13} Shekel 7	4	$(0,10)^D$	4.0^D	$(7.5,10)^D$
f_{14} Shekel 10	4	$(0,10)^D$	4.0^D	$(7.5,10)^D$

Another measure used, is *reliability*, which is the percentage of trials where swarms converge with a specified accuracy and it is defined by:

$$\text{Reliability} = \frac{n'}{n} \times 100 \qquad (2.11)$$

where n is the number of trials in the experiment and n' is the number of successful trials.

In this paper, *SDSnDE*, is presented with few variations of parameter, n (the number of generations before an SDS cycle is performed): $n = 5, 50$, and 200. These values were selected merely to provide a brief initial exploration of the behaviour of the new hybrid algorithm over three relatively widely separated parameter values; no claim is made for their optimality.

2.5.2   Results

Table 2.3 shows the performance of the various hybrid algorithms alongside DE algorithm. For each benchmark and algorithm, the table shows the accuracy measure. The overal reliability of each algorithm is also reported.

Algorithm 2.6. Hybrid Algorithm Control

```
// DIFFUSION PHASE
For ag = 1 to No_of_agents
  If ( ag is not active )
    ag.setHypo( randomHypo() )
  End If
End For
```

Table 2.3 Accuracy and Reliability Details

Accuracy ($\pm$standard error) is shown with two decimal places after 30 trials of 2,000 generations; and the reliability of each algorithm over all the benchmarks is given in the last row of the table. For each benchmark, algorithms which are **significantly** better (see Table 2.4) than the others are highlighted. Note that the highlighted algorithms do not significantly outperform each another.

	DE	H5: SDSnDE $n = 5$ generate Hypothesis	H50: SDSnDE $n = 50$ generate Hypothesis	H200: SDSnDE $n = 200$ generate Hypothesis	H50D: Dispenser $n = 50$ SDS Test + Dispense
f_1	1.06E-107$\pm$7.92E-108 (30)	5.29E-10$\pm$4.72E-10 (28)	5.52E-92$\pm$4.03E-92 (30)	4.70E-104$\pm$3.11E-104 (30)	2.03E-85$\pm$1.61E-85 (30)
f_2	1.20E-03$\pm$2.60E-04 (0)	1.21E+01$\pm$1.88E+00 (0)	2.55E-05$\pm$7.27E-06 (0)	1.48E-04$\pm$3.86E-05 (0)	8.58E-04$\pm$2.42E-04 (0)
f_3	3.66E+01$\pm$8.23E+00 (0)	4.40E+01$\pm$6.46E+00 (0)	**1.71E+00$\pm$5.36E-01** (0)	3.87E+00$\pm$2.29E+00 (0)	**1.26E+00$\pm$3.22E-01** (0)
f_4	5.00E+02$\pm$1.23E+02 (0)	**3.02E-02$\pm$8.28E-03** (0)	4.83E-01$\pm$4.37E-01 (0)	6.23E-01$\pm$2.39E-01 (0)	**2.59E-02$\pm$9.26E-03** (0)
f_5	1.61E+02$\pm$8.49E+00 (0)	**2.67E-01$\pm$8.15E-02** (2)	1.34E+01$\pm$7.49E+00 (0)	2.79E+01$\pm$1.74E+00 (0)	2.41E+01$\pm$1.00E+01 (9)
f_6	1.45E+01$\pm$1.34E+00 (0)	**2.36E-06$\pm$1.10E-06** (0)	1.02E-01$\pm$7.00E-02 (17)	3.23E-01$\pm$1.11E-01 (19)	**1.45E-01$\pm$1.34E-01** (21)
f_7	5.26E-02$\pm$1.05E-02 (1)	3.85E-02$\pm$1.43E-02 (6)	1.99E-02$\pm$4.40E-03 (5)	2.82E-02$\pm$6.76E-03 (4)	7.42E-02$\pm$5.50E-02 (2)
f_8	1.31E+01$\pm$3.07E+00 (3)	**5.66E-12$\pm$3.11E-12** (30)	1.96E-02$\pm$1.28E-02 (24)	1.05E-02$\pm$5.77E-03 (25)	7.00E-03$\pm$4.86E-03 (28)
f_9	3.24E+00$\pm$2.41E+00 (8)	1.51E-10$\pm$9.08E-11 (29)	5.27E-01$\pm$3.68E-01 (19)	1.03E-02$\pm$5.72E-03 (26)	3.50E+01$\pm$1.73E+01 (23)
f_{10}	1.90E-01$\pm$6.41E-02 (23)	**2.48E-04$\pm$2.34E-04** (28)	**4.44E-17$\pm$1.65E-17** (30)	5.92E-17$\pm$1.82E-17 (30)	**4.44E-17$\pm$1.65E-17** (30)
f_{11}	2.55E+02$\pm$5.97E+01 (1)	1.13E-08$\pm$1.13E-08 (29)	**0.00E+00$\pm$0.00E+00** (30)	2.96E-17$\pm$2.96E-17 (30)	**0.00E+00$\pm$0.00E+00** (30)
f_{12}	5.05E+00$\pm$6.73E-17 (0)	**1.25E+00$\pm$4.77E-01** (24)	3.02E+00$\pm$5.43E-01 (14)	3.37E+00$\pm$5.31E-01 (7)	4.80E+00$\pm$2.52E-01 (2)
f_{13}	5.27E+00$\pm$0.00E+00 (0)	**7.03E-01$\pm$3.33E-01** (23)	1.28E+00$\pm$4.33E-01 (11)	3.78E+00$\pm$5.56E-01 (0)	4.83E+00$\pm$3.09E-01 (1)
f_{14}	5.36E+00$\pm$6.02E-17 (0)	**3.57E-01$\pm$2.48E-01** (27)	**5.81E-01$\pm$3.26E-01** (13)	4.19E+00$\pm$4.86E-01 (0)	4.82E+00$\pm$2.99E-01 (0)
Σ	66	226	193	171	176
	15.71%	53.81%	45.95%	40.71%	41.90%

The focus of this paper is not finding the best n for *SDSnDE* (for this set of benchmarks), but rather investigate the effect of SDS algorithm on the performance of DE algorithm.

As Table 2.4 shows, over all benchmarks, other than f_2 in $(DE - H5)$, DE algorithm does not significantly outperform any of the hybrid algorithms SDSnDE ($n = 5, 50, 200$). On the other hand, in most cases (e.g. f_{3-6}, f_8 and f_{10-14}), the hybrid algorithms outperform the classical DE algorithm significantly.

As detailed in Table 2.3, in f_{1-3}, f_{11}, the performace of H5, which has the highest rate of information exchange, is weaker than the other hybrid algorithms with lower information sharing. This implies that the performance of some problems might be negatively affected by excessive information exchange (e.g. in f_1, $F_{H5} > F_{H50} > F_{H200}$, where F is the fitness value).

However in another set of problems, higher rate of information exchange (more communication between the agents) results in better outcome (e.g. f_{4-6}, f_{8-9}, f_{12-14}). More specifically, in f_{4-6} and f_{12-14} fewer communication between the agents, corresponds to worse performance of the hybrid algorithms ($F_{H5} < F_{H50} < F_{H200}$).

This demonstrates the importance of deploying the right frequency of communication and information exchange.

2.6   Discussion

The resource allocation process underlying SDS offers three closely coupled mechanisms to the algorithm's search component to speed its convergence to global optima. The first component is 'efficient, non-greedy information sharing' instantiated via positive feedback of potentially good hypotheses between agents; the second component is the dispensation mechanism – SDS-led random-restarts – deployed as part of the diffusion phase; the thrid component is random 'partial hypothesis evaluation', whereby a complex, computationally expensive objective function is broken down into 'k independent partial-functions', each one of which, when evaluated, offers partial information on the absolute quality of current algorithm search parameters. It is this mechanism of iterated selection of a *random* partial function that ensures SDS does not prematurely converge on local minimum.

The resource allocation and dispensation components of SDS in the hybrid algorithm are executed in the 'Diffusion Phase', where information is shared (diffused) among SDEAgents (see Algorithm 2.3). Analysis of the performance of the hybrid algorithm (see results above) demonstrates that adding the SDS resource allocation and dispensation mechanisms to the classical DE architecture improves the overall performance of the algorithm (i.e. it enhances algorithm accuracy and reliability, as defined herein).

To further analyse the role of SDS in the hybrid algorithms, the Diffusion Phase of SDS algorithm is modified (see Algorithm 2.6) to investigate the dispensation effect caused by randomising a selection of agent hypotheses after a number of DE function evaluations (effectively instantiating a DE with SDS-led random-restarts).

In other words, after the SDS test-phase, the hypothesis of each inactive SDEAgent is randomised.

As detailed in Table 2.3, although, information sharing plays an important role in the performance of hybrid DE algorithm, the significance of dispensation mechanism (in randomly restarting some of the agents) in improving the performance of DE algorithm cannot be discarded.

In few cases ($f_{3,4,8}$), solely the dispensation mechanism (H50D), which is facilitated by the test-phase of the SDS algorithm, demonstrates a slightly better performance compared to the hybrid algorithm (see Table 2.3). However, in the majority of the cases, the hybrid algorithms outperform the modified algorithm: $f_{1,2}, f_{5-7}, f_9, f_{12-14}$, out of which f_9 and f_{12-14} are performing significantly better (see Table 2.4). Also it is shown that the algorithm with modified diffusion phase is less reliable than its corresponding hybrid algorithm.

The results show the importance of coupling the SDS-led restart mechanism (dispensation mechanism) and the communication of agents which are both deployed in SDS algorithm.

Table 2.4 TukeyHSD Test Results for Accuracy

Based on TukeyHSD Test, if the difference between each pair of algorithms is significant, the pairs are marked. X–o shows that the left algorithm is significantly better than the right one; and o–X shows that the right algorithm is significantly better than the one, on the left.

	DE-H5	DE-H50	DE-H200	DE-H50D	H5-H50	H5-H200	H5-H50D	H50-H200	H50-H50D	H200-H50D
f_1	–	–	–	–	–	–	–	–	–	–
f_2	X–o	–	–	–	o–X	o–X	o–X	–	–	–
f_3	–	o–X	o–X	o–X	o–X	o–X	o–X	–	–	–
f_4	o–X	o–X	o–X	o–X	–	–	–	–	–	–
f_5	o–X	o–X	o–X	o–X	–	–	–	–	–	–
f_6	o–X	o–X	o–X	o–X	–	–	–	–	–	–
f_7	–	–	–	–	–	–	–	–	–	–
f_8	o–X	o–X	o–X	o–X	–	–	–	–	–	–
f_9	–	–	–	X–o	–	–	X–o	–	X–o	X–o
f_{10}	o–X	o–X	o–X	o–X	–	–	–	–	–	–
f_{11}	o–X	o–X	o–X	o–X	–	–	–	–	–	–
f_{12}	o–X	o–X	o–X	–	X–o	X–o	X–o	–	X–o	–
f_{13}	o–X	o–X	o–X	–	–	X–o	X–o	X–o	X–o	–
f_{14}	o–X	o–X	–	–	–	X–o	X–o	X–o	X–o	–

The third SDS component feature, which is currently only implicitly exploited by the hybrid algorithm, is 'randomised partial hypothesis evaluation'. In the Mining Game (see Section 2.2.1), "At the start of the mining process each miner maintains a [randomly allocated] hypothesis - their current belief of 'best hill' to mine"; and each miner mines one small randomly selected area of this hill rather than the entirety of it (i.e. revealing a partial estimate of the the gold content of the entire hill); following this approach, each miner forms a partial view of the gold content of their hill hypothesis (which is merely part of the overall mountain range: the entire search space).

In typical optimisation algorithms, the search process iterates the evaluation of one point in the n-dimensional search space (iterating an objective function evaluation). In DE population, in addition to this information, each agent has implicit partial knowledge from other agents (derived from the mutation, crossover and selection mechanisms) comprising the historical evidence implicit in the prior [m] objective-function evaluations the population has performed. Thus, since each agent finds its current position by using this implicit knowledge, it has partial knowledge of the full search space.

In the hybrid algorithm each SDEAgent maintains a fitness value which is the best objective function value it has currently found, based on its exploration of the search space so far. Thus constituted, each SDEAgent's target vector defines a 'partial view' of the entire search space (via the partial interaction it has with the rest of the population through mutation, crossover and selection). Hence, when the fitness values of two SDEAgents' target vector are compared in the test-phase of the hybrid algorithm, two partial views of the entire search space are contrasted. This is analogous to the 'test' process of the Mining Game as in both processes, agents become active or inactive contingent upon the agent's evaluation of a randomised partial view of the entire search space.

In both the Mining Game and the new hybrid SDSnDE algorithm, the notion of partial-function evaluation differs importantly from that traditionally deployed in a simple discrete partial function SDS, where, for a given set of parameter values (the agent hypothesis) a complex objective function is broken into m components, only one randomly selected of which will be evaluated and the subsequent agent-activity is based on this. Clearly, as this process merely evaluates $1/m$ of the total number of computations required for the full hypothesis evaluation, it concomitantly offers a potentially significant performance increase. Whereas in the new hybrid SDSnDE algorithm, the objective function is evaluated in-toto, using a given set of parameter values (the agent's hypothesis) and the subsequent agent-activity is based on this. In the former case, the agent exploits knowledge of the partial objective function and in the process gains a potential partial-function performance dividend; in the latter the agent merely exploits partial knowledge of the search space without the concomitant explicit partial-function performance increase. Ongoing work, on computationally more complex benchmark problems, seeks to exploit this 'partial-function dividend' with the hybrid SDSnDE algorithm; if successful, this offers further, potentially significant, performance improvements for the new hybrid algorithm.

2.7 Conclusions

This paper presents a brief overview about the potential of integration of DE with SDS. Here, SDS is primarily used as an efficient resource allocation and dispensation mechanism responsible for facilitating communication between DE agents. Additionally, an initial discussion of the similarity between the hypothesis test employed in the hybrid algorithm and the test-phase in SDS algorithm is presented.

Results reported in this paper have demonstrated that initial explorations with the hybrid SDSnDE algorithm outperform the performance of (one variation of) classical DE architecture, even when applied to problems with low-cost fitness function evaluations (the benchmarks presented).

This work, further investigated an earlier work [3] attempting to integrate PSO with SDS[3]. In ongoing research, in addition to investigating the performance of the hybrid algorithm in other sets of problems (e.g. CEC2005 or some real-world problems), further theoretical work seeks to develop the core ideas presented in this paper on problems with significantly more computationally expensive objective functions, where the performance improvement (relative to classical DE) is anticipated to be much greater.

References

[1] al-Rifaie, M.M., Bishop, M.: The mining game: a brief introduction to the stochastic diffusion search metaheuristic. AISB Quarterly (2010)

[2] al-Rifaie, M.M., Bishop, M., Aber, A.: Creative or not? birds and ants draw with muscles. In: AISB 2011: Computing and Philosophy, University of York, York, U.K, pp. 23–30 (2011a), ISBN: 978-1-908187-03-1

[3] al-Rifaie, M.M., Bishop, M., Blackwell, T.: An investigation into the merger of stochastic diffusion search and particle swarm optimisation. In: GECCO 2011: Proceedings of the, GECCO conference companion on Genetic and evolutionary computation. ACM, New York (2011)

[4] el Beltagy, M.A., Keane, A.J.: Evolutionary optimization for computationally expensive problems using gaussian processes. In: Proc. Int. Conf. on Artificial Intelligence 2001, pp. 708–714. CSREA Press (2001)

[5] Bishop, J.: Stochastic searching networks. In: Proc. 1st IEE Conf. on Artificial Neural Networks, London, UK, pp. 329–331 (1989)

[6] Bonabeau, E., Dorigo, M., Theraulaz, G.: Inspiration for optimization from social insect behaviour. Nature 406, 3942 (2000)

[7] Branke, J., Schmidt, C., Schmeck, H.: Efficient fitness estimation in noisy environments. In: Spector, L. (ed.) Genetic and Evolutionary Computation Conference. Morgan Kaufmann, San Francisco (2001)

[8] Brest, J., Zamuda, A., Boskovic, B., Maucec, M., Zumer, V.: Dynamic optimization using self-adaptive differential evolution. In: IEEE Congress on Evolutionary Computation, CEC 2009, pp. 415–422. IEEE, Los Alamitos (2009)

---

[3] The artistic applications of merging SDS with PSO (falling into the category of generative art) are under further investigation and the early results are reported in [2].

[9] Digalakis, J., Margaritis, K.: An experimental study of benchmarking functions for evolutionary algorithms. International Journal 79, 403–416 (2002)

[10] Gehlhaar, D., Fogel, D.: Tuning evolutionary programming for conformationally flexible molecular docking. In: Evolutionary Programming V: Proc. of the Fifth Annual Conference on Evolutionary Programming, pp. 419–429 (1996)

[11] Goldberg, D.E.: Genetic Algorithms in Search, Optimization and Machine Learning. Addison-Wesley Longman Publishing Co., Inc., Boston (1989)

[12] Holldobler, B., Wilson, E.O.: The Ants. Springer, Heidelberg (1990)

[13] Huang, V., Suganthan, P., Qin, A., Baskar, S.: Multiobjective differential evolution with external archive and harmonic distance-based diversity measure. School of Electrical and Electronic Engineering Nanyang, Technological University Technical Report (2005)

[14] Jin, Y.: A comprehensive survey of fitness approximation in evolutionary computation. Soft Computing 9, 3–12 (2005)

[15] Jones, D.R., Perttunen, C.D., Stuckman, B.E.: Lipschitzian optimization without the lipschitz constant. J. Optim. Theory Appl. 79(1), 157–181 (1993)

[16] Jong, K.A.D.: An analysis of the behavior of a class of genetic adaptive systems. PhD thesis, University of Michigan, Ann Arbor, MI, USA (1975)

[17] Kennedy, J., Eberhart, R.C.: Particle swarm optimization. In: Proceedings of the IEEE International Conference on Neural Networks, vol. IV, pp. 1942–1948. IEEE Service Center, Piscataway (1995)

[18] Kennedy, J.F., Eberhart, R.C., Shi, Y.: Swarm intelligence. Morgan Kaufmann Publishers, San Francisco (2001)

[19] Kozlov, K., Samsonov, A.: New migration scheme for parallel differential evolution. In: Proceedings of the International Conference on Bioinformatics of Genome Regulation and Structure, pp. 141–144 (2006)

[20] Mendes, R., Mohais, A.: DynDE: a differential evolution for dynamic optimization problems. In: The 2005 IEEE Congress on Evolutionary Computation CEC 2005, vol. 3, pp. 2808–2815 (2005)

[21] de Meyer, K.: Explorations in stochastic diffusion search: Soft- and hardware implementations of biologically inspired spiking neuron stochastic diffusion networks. Tech. Rep. KDM/JMB/2000/1, University of Reading (2000)

[22] de Meyer, K., Bishop, J.M., Nasuto, S.J.: Stochastic diffusion: Using recruitment for search. In: McOwan, P., Dautenhahn, K., Nehaniv, C.L. (eds.) Evolvability and interaction: evolutionary substrates of communication, signalling, and perception in the dynamics of social complexity, Technical Report 393, pp. 60–65 (2003)

[23] de Meyer, K., Nasuto, S., Bishop, J.: Stochastic diffusion optimisation: the application of partial function evaluation and stochastic recruitment in swarm intelligence optimisation. In: Abraham, A., Grosam, C., Ramos, V. (eds.) Swarm Intelligence and Data Mining, ch. 12. Springer, Heidelberg (2006)

[24] Moglich, M., Maschwitz, U., Holldobler, B.: Tandem calling: A new kind of signal in ant communication. Science 186(4168), 1046–1047 (1974)

[25] Myatt, D.R., Bishop, J.M., Nasuto, S.J.: Minimum stable convergence criteria for stochastic diffusion search. Electronics Letters 40(2), 112–113 (2004)

[26] Nasuto, S.J.: Resource allocation analysis of the stochastic diffusion search. PhD thesis, University of Reading, Reading, UK (1999)

[27] Nasuto, S.J., Bishop, J.M.: Convergence analysis of stochastic diffusion search. Parallel Algorithms and Applications 14(2) (1999)

[28] Nasuto, S.J., Bishop, M.J.: Steady state resource allocation analysis of the stochastic diffusion search. csAI/0202007 (2002)

[29] Nasuto, S.J., Bishop, J.M., Lauria, S.: Time complexity of stochastic diffusion search. Neural Computation NC98 (1998)

[30] Smuc, T.: Improving convergence properties of the differential evolution algorithm. In: Proceedings of the MENDEL 2002 - 8th International Conference on Soft Computing, pp. 80–86 (2002)

[31] Stoean, C., Preuss, M., Stoean, R., Dumitrescu, D.: Multimodal optimization by means of a topological species conservation algorithm. IEEE Transactions on Evolutionary Computation 14(6), 842–864 (2010)

[32] Storn, R., Price, K.: Differential evolution - a simple and efficient adaptive scheme for global optimization over continuous spaces TR-95-012 (1995), http://www.icsi.berkeley.edu/storn/litera.html

[33] Storn, R., Price, K.: Differential evolution - a simple and efficient heuristic for global optimization over continuous spaces. J. Global Optim. 11, 341–359 (1997)

[34] Tasgetiren, M., Suganthan, P.: A multi-populated differential evolution algorithm for solving constrained optimization problem. In: IEEE Congress on Evolutionary Computation CEC 2006, pp. 33–40. IEEE, Los Alamitos (2006)

[35] Tasoulis, D., Pavlidis, N., Plagianakos, V., Vrahatis, M.: Parallel differential evolution. In: Congress on Evolutionary Computation CEC 2004, vol. 2, pp. 2023–2029. IEEE, Los Alamitos (2004)

[36] Thomsen, R.: Multimodal optimization using crowding-based differential evolution. In: Congress on Evolutionary Computation, CEC 2004, vol. 2, pp. 1382–1389. IEEE, Los Alamitos (2004)

[37] Weber, M., Neri, F., Tirronen, V.: Parallel Random Injection Differential Evolution. In: Applications of Evolutionary Computation, pp. 471–480 (2010)

[38] Whitaker, R., Hurley, S.: An agent based approach to site selection for wireless networks. In: 1st IEE Conf. on Artificial Neural Networks. Proc. ACM Symposium on Applied Computing, Madrid Spain. ACM Press, New York (2002)

[39] Whitley, D., Rana, S., Dzubera, J., Mathias, K.E.: Evaluating evolutionary algorithms. Artificial Intelligence 85(1-2), 245–276 (1996)

[40] Zaharie, D.: Control of population diversity and adaptation in differential evolution algorithms. In: Proc. of 9th International Conference on Soft Computing, MENDEL, pp. 41–46 (2003)

[41] Zhang, J., Sanderson, A.: JADE: adaptive differential evolution with optional external archive. IEEE Transactions on Evolutionary Computation 13(5), 945–958 (2009)

Chapter 3
An Adaptive Multiagent Strategy for Solving Combinatorial Dynamic Optimization Problems

Juan R. González, Carlos Cruz, Ignacio G. del Amo, and David A. Pelta

Abstract. This work presents the results obtained when using a decentralised multi-agent strategy (Agents) to solve dynamic optimization problems of a combinatorial nature. To improve the results of the strategy, we also include a simple adaptive scheme for several configuration variants of a mutation operator in order to obtain a more robust behaviour. The adaptive scheme is also tested on an evolutionary algorithm (EA). Finally, both Agents and EA are compared against the recent state of the art adaptive hill-climbing memetic algorithm (AHMA).

3.1 Introduction

There is a very active research on the field of Dynamic Optimization Problems (DOPs) [2, 3, 5, 14]. DOPs are problems where the fitness landscape changes with time and its main interest relies on its closeness to real world where the problems are rarely static (trade-market prediction, weather forecast, robot motion control, ...) . Since the problem changes generally occur on a gradual manner, the algorithms for DOPs try to reuse the information obtained in previous stages to better solve the problem at its current state and to try and track the movement of the optima. This approach is generally better and faster than re-optimizing.

Most of the research for DOPs has been focused on evolutionary algorithms (EAs), but we were able to improve the results of EAs on several usual test problems, such as the moving peaks [3], by using an algorithm with multiple agents cooperating on a decentralised manner to improve a set of solutions on a matrix [7, 9, 10]. This algorithm, called simply *Agents*, has only been tested on continuous DOPs.

Juan R. González · Carlos Cruz · Ignacio G. del Amo · David A. Pelta
DECSAI, ETSIIT, CITIC-UGR, University of Granada,
C/Daniel Saucedo Aranda, s/n, E-18071, Granada, Spain
e-mail: {jrgonzalez,carloscruz,ngdelamo,dpelta}@decsai.ugr.es

D.A. Pelta et al. (Eds.): NICSO 2011, SCI 387, pp. 41–55, 2011.
springerlink.com © Springer-Verlag Berlin Heidelberg 2011

Therefore, the **first goal** of this work is to present a new version of *Agents* for combinatorial DOPs to see if it can obtain results as good as in the continuous case.

Additionally, the performance of an heuristic depends on the parameters used and no specific parameters can work well across different problems and instances [11]. Moreover, the optimal parameters may vary as the search process is conducted. The difficulties increase with DOPs, since the problem is going to change and these changes may affect the validity of the learning done. Therefore, a **second goal** of this paper is to test a simple adaptive scheme for learning among several mutation operator variants and to see if it can improve the results even on this difficult scenario of combinatorial DOPs. To further test this adaptive scheme it will also be applied to a standard evolutionary algorithm (*EA*). An additional **third goal** of this paper will be to compare both *Agents* and *EA* algorithms with the adaptive scheme against the state of the art hill-climbing memetic algorithm (*AHMA*) [12].

To achieve the previous goals, this paper is structured as follows. Firstly, Section 3.2 presents the algorithms used through the paper: the new version of the *Agents* algorithm for combinatorial DOPs, the *EA* implemented to further test the adaptive scheme, and the state of the art *AHMA* used for comparison. Secondly, in Section 3.3 we describe the adaptive scheme and how it is incorporated to both *Agents* and the *EA*. Then, Section 3.4 describes the combinatorial DOPs that will be used to test the algorithms. After that, Section 3.5 describes the experiments done and their results. Finally, the paper concludes at Section 3.6 with the conclusions and future work.

3.2 Algorithms

3.2.1 Agents

The multiagents algorithm presented here is a decentralised cooperative strategy that has been previously applied to continuous dynamic optimization problems [7, 9, 10]. The strategy makes use of a matrix of solutions and a group of agents that move through the matrix trying to improve the solutions stored on the matrix cells they visit. The cooperation is based on the fact that the solutions that an agent improves may be later improved by other agents that arrive at the same cells on the matrix.

Algorithm 3.1 presents the pseudocode of *Agents*. Basically, the algorithm is run until the stop condition is met, which will normally be when the resources are exhausted (such as the time or the number of evaluations/changes of the objective function). For each iteration, the function *detectChanges()* is called to reevaluates all the solutions if a change on the problem is detected. The change detection is done by recomputing the fitness value of the best solution and comparing it with the previous one to see if it has changed. Then, for each iteration of the inner loop, an agent is selected, in a circular fashion, and it is moved to the best neighbor solution (in terms of horizontal and vertical adjacency on the matrix) and it tries to improve this solution. The algorithm performs as many iterations of the inner loop as solutions on the matrix. In this way, the number of iterations of the *for* loop is similar to a generation of the other population-based algorithms considered on the paper.

Algorithm 3.1. Agents algorithm.

---

procedure *Agents*()
 1: Initialise the *matrix* of *solution* and the vector of *agents*.
 2: while (stop condition not met) do
 3: *detectChanges*().
 4: *agentNum* ← 0.
 5: for ($i = 0; i < size(matrix); i++$) do
 6:   *agent* ← *agents*[*agentNum*].
 7:   Move *agent* to the best cell on the neighborhood of its current position on the *matrix*. If there is no best cell on the neighborhood, the *agent* is moved randomly.
 8:   *prevSol* ← Solution on the agent position.
 9:   *newSol* ← *mutation*(*prevSol*).
 10:   Evaluate *newSol*.
 11:   if (*newSol* is better or equal than *prevSol*) then
 12:    Store *newSol* on the agent position at the *matrix*.
 13:    Update *bestSol* with *newSol* if needed.
 14:   end if
 15:   *agentNum* ← (*agentNum* + 1) % *agentNum*.
 16: end for
 17: end while

To improve a solution, the agent uses the *mutation()* function on the *prevSol* solution to generate a new solution called *newSol*. *newSol* will replace *prevSol* on the matrix if an improvement is produced. The first difference with respect to the continuous DOPs pseudocode [7, 9, 10] is that solutions with equal fitness are also accepted as replacement solutions for the matrix, and not only better ones. The reason for this change is that on many combinatorial problems the fitness values vary on a discrete manner, so it is more common to reach solutions with the same objective value. If the solutions of the matrix were not replaced with the new ones on this case, the modification effort will be lost, and the search process could stagnate. When we allow the replacement of solutions with the same quality, we allow the search process to evolve so it is able to escape from flat zones on the search landscape.

Moreover, on the continuous DOPs for which the *Agents* algorithm was originally developed, the *mutation()* function was implemented as the generation of a solution inside of a hypersphere of radius r centered at the solution being modified. In the case of combinatorial DOPs, this approach is no longer useful and it becomes necessary to implement a different way of modifying solutions. In this work, the implementation done for *mutation()* is focused on the binary-encoded problems we are going to tackle (Section 3.4) and it allows to perform changes to a given number of consecutive bits (*numBits*) by flipping each of these bits with a given probability (*flipProb*). The starting bit for the *numBits* will be chosen randomly and if the last bit of the solution is reached, the remaining bits are taken from the initial solution bits (in a circular fashion). Each of the *numBits* will then be flipped independently with *flipProb* probability. In this way, if *numBits* = 1 and *flipProb* = 1, the *mutation()* function will change just one random bit of the solution. If *numBits* = 3 and *flipProb* = 0.8, the *mutation()* function will choose three consecutive random bits and flip each one with an independent 80% probability.

3.2.2 Evolutionary Algorithm (EA)

For comparison purposes and to further test the adaptive scheme that will be presented in Section 3.3 we introduce here a typical implementation of an EA for solving combinatorial DOPs. Since the development of this *EA* is not the main goal of this paper, we will just give a short description of its main characteristics.

Algorithm 3.2. Evolutionary Algorithm (EA).

---

procedure *EA*()
  Create an initial *solutionPopulation*.
  while (stop condition not met) do
    detectChanges().
    Create empty *newSolutionPopulation*.
    for (*i* = 0; *i* < *size*(*solutionPopulation*)/2; *i* + +) **do**
      parentSols = **selection**(*solutionPopulation*).
      offspringSols = **crossover**(*parentSols*).
      mutation(*offspringSols*).
      evaluate(*offspringSols*).
      Update *bestSol* **from** *offspringSols* **if needed.**
      newSols = **elitism**(*parentSols*, *offspringSols*).
      newSolutionPopulation.**add**(*newSols*).
    end for
  end while

The pseudocode for the *EA* is included in algorithm 3.2. It follows a standard EA structure plus a similar adaptation to DOPs than in the Agents case, that is, the *EA* is run until a stop condition is met, and at the beginning of each generation it calls the *detectChanges()* function, that reevaluates all the *solutionPopulation* if a change on the problem is detected. The change detection used is the same defined for *Agents* that is based on checking changes on the fitness value of the best solution. Then, for each generation, the *solutionPopulation* is replaced by a new one through a loop of selection, crossover, mutation and survival operators. The selection uses the negative/positive assortative mating [8] to choose two parents and it also ensures that the best solution is selected as parent at least once per generation. Then, two offspring are produced from the two parents using a standard single point crossover operator. The offspring solutions are then mutated using the same mutation operator that was described for the *Agents* algorithm. Finally, elitism is applied to select for survival the two best solutions among parents and offsprings.

3.2.3   Adaptive Hill Climbing Memetic Algorithm

The Adaptive Hill Climbing Memetic Algorithm (*AHMA*) from Yang [12] was selected as the third algorithm for the purpose of this paper. It is one of the state of the art algorithms for combinatorial DOPs so it is useful as a performance base to test how well the adaptive versions of the *EA* and *Agents* algorithms perform.

AHMA is a memetic algorithm that incorporates an adaptive hill climbing strategy for the local search. The algorithm tries to learn which of two different hill climbing procedures to use: a greedy crossover-based hill climbing (GCHC) and a steepest mutation hill climbing operator (SMHC). The GCHC performs local search using a special crossover operator between the current solution and the best solution of the population. On the other hand, the SMHC tries to improve the current solution by mutation of some random bits. The probability to select either GCHC or SMHC is adapted based on the improvements obtained by each local search operator and a measure of the population convergence.

There are several alternate versions of *AHMA*. We have build our implementation based on Yang's paper and on the reference code they kindly provided to us, but adapted to use the exact same problem-related classes that we used on the other algorithms. The version implemented here corresponds to the more complete *AHMA1* version of Yang's paper. This version includes two additional population diversity mechanism: adaptive dual mapping (ADM) and triggered random immigrants (TRI). Further details can be found on their published paper [12].

3.3   Adaptive Scheme

As we anticipated in the introduction, the use of fixed parameters for the configuration of an heuristic does not lead to robust results on different problems and instances. A detailed description of different approaches for adaptive evolutionary algorithms on combinatorial problems can be seen on [11]. While most of the ideas presented in that paper are feasible to be applied also to *Agents*, additional challenges appear when we are dealing with DOPs. Since DOPs change with time, it makes it more difficult to properly find out the best parameter settings for the whole search process. Moreover, it is generally not feasible to test values for every possible parameter when the available time between two consecutive problem changes is short. Despite that, the use of some adaptive scheme will probably improve the robustness of the algorithms and one of the goals of this paper is to verify this claim.

We will use a credit-based adaptive scheme implemented as a generic *Learning* class to discriminate among a set of configurations (values) of a given criterion. Each configuration of the criterion will have an associated index or value in the natural numbers. To implement this task, the class will have three main methods. **learn(value, credit)** assigns an additional credit to the configuration of the criterion represented by the value. **rouletteWheelSelection()** applies roulette wheel selection to return one of the learned values. That is, the probability of choosing each value corresponds with the quotient between its credit and the sum of credits for all the values of the criterion. Finally, **clearLearning**() deletes learned values and credits.

Since it is not feasible to learn every possible parameter of the algorithms, our approach will be to focus the learning on the configuration of the mutation operator (the configuration for the *mutation()* function on the *Agents* and *EA* algorithms' pseudocode). To achieve this, what we do is to consider the selection of the configuration to use for the mutation operator as the criteria and to assign a numerical value to every configuration variant that is going to be considered.

Initially, we will set the same credit for each configuration variant in order for all of them to have an equal non-zero chance of being selected. Then, each time the mutation operator is applied with a given operator configuration, the fitness change between the received solution and the mutated solution is computed. If the fitness was increased, the increment is added as an additional credit for that operator configuration (using the *learn* function). Besides that, to try to adjust the learning to the dynamic problem changes, when a change on the environment is detected, all the credits are restored to their initial values using a call to *clearLearning()* and repeating the initialisation.

Finally, since the offspring solutions on the *EA* are generated through a combination of both the crossover and the mutation, this introduces a difficulty for knowing the fitness change produced by the configuration of the mutation operator alone. There are basically two solutions for this: to recompute the fitness between the application of crossover and mutation (leading to more precise credits) or to consider the crossover and the mutation together (so the credits refer to the combination of the two steps). Since performing two evaluations per offspring solution will be a big disadvantage when compared to the single evaluation done by both *Agents* and *AHMA*, we have chosen to use the second option. Additional analysis for this decision will be presented on the experiments section.

3.4 Problems

The problems used on this paper are all binary-encoded combinatorial dynamic optimization problems that are generated from base static functions by means of the XOR generator technique [13]. The base functions have been chosen from some of the most used on the recent literature [12]: OneMax, Plateau, Royal Road and Deceptive. All of them consists of finding solutions that match all the bits of a target optimal solution. This target solution is initially considered to be the solution where all its bits are set to 1. To evaluate a solution, it is considered to be made up of blocks of 4 bits where each block contributes a given amount to the full objective value of the solution. The contribution of each block of 4 bits for each of the considered functions is computed as follows:

- **OneMax**: Each matched bit adds 1 to the fitness.
- **Plateau**: Three matched bits add 2 to the fitness while four matched bits add 4 and any other amount of bits matched leads to a 0 contribution.

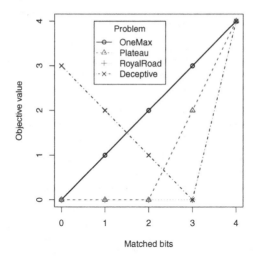

Fig. 3.1 Contribution to the fitness function (objective value) for each block of 4 bits on a solution with respect to the number of correctly matched bits

- **RoyalRoad**: Each perfectly matched block adds 4 to the fitness. Partially matched blocks have 0 fitness.
- **Deceptive**: Fitness is 4 if all the 4 bits are matched. If not, the fitness for the block is 3 minus the number of matched bits.

For further reference and clarity, Fig. 3.1 shows the contribution of each block of 4 bits for each of the base functions with respect to the number of correctly matched bits. Since all these problems are defined for bit strings with a length l (where l is divisible by 4), and given the aforementioned rules to compute the fitness, the optimum value for any of the problems is also l. The XOR generator [13] takes the base functions as a starting point and generates a problem change after a given period P defined as a number of evaluations. The changes consist on applying random masks with a given number of ones to the target solution. In this way, when a change on the problem is generated, it consists on doing a XOR operation between the current target solution and the new mask. This leads to a new target solution that differs from the previous one on as many bits as the number of ones that are present in the mask. Therefore, the number of ones in the mask controls the severity of the changes.

3.5   Experiments

To better assess the quality of the adaptive scheme and the proposed *Agents* and *EA* algorithm variants, the experiments done on this paper are focused on the exact same problem instances and dynamism setup available on the published paper for the state of the art *AHMA* algorithm. All the problems of the previous section are used with a solution size of 100 bits (25 blocks of 4 bits). For each problem, the experiments performed consider different periods of change ($P \in \{1200, 6000, 12000\}$) and different severities ($\rho \in \{0.1, 0.2, 0.5, 0.9\}$). The period ($P$) is given as the number of evaluations of the objective function between two consecutive changes. The severity of change (ρ) is used when a change is produced (every P evaluations) to control how many bits of the optimal solution are flipped. If l is the length of the solutions for the problem, the new optimal solution after a change will be generated flipping $\rho * l$ bits on the previous target optimal solution.

The extension of this paper does not allow to show the results for all the details of the experimentation, so just a short summary of the fixed parameters used for each algorithm is included now. First of all, we chose several values of the number of agents and the dimension of the matrix for the *Agents* algorithm on the basis of experience and the statistical results published for the continuous DOPs case [7, 9, 10]. These values were retested for the combinatorial DOPs and the statistical analysis showed that the best configuration overall was to use 4 agents and a matrix of size 2×2. These results are similar to the values obtained for the continuous DOPs but with a smaller matrix size, that can be explained by the nature of the problems tested, where it is probably better to do a good intensification than to have a big diverse population. Secondly, the use of just 10 individuals was proven to be statistically better in the case of the *EA*. The crossover and elitism chosen required no parameters, and since the population size was small all the individuals were used for performing the negative/positive assortative mating selection. Finally, since the problems used here corresponds to the paper where *AHMA* was published, we simply set the exact same original parameter values published on [12].

To assess the performance of the algorithms, we have used the off-line performance [4], that is defined as follows:

$$\text{off-line performance}(T) = \frac{1}{T} \sum_{t=1}^{T} e_t' \tag{3.1}$$

$$\text{with } e_t' = max\{e_\tau, e_{\tau+1}, \ldots, e_t\} \tag{3.2}$$

where e_t is the fitness of the solution evaluated by the algorithm at time t, and T is the total number of time instants considered. τ represents the first time instant right after the last change on the environment (on the problem) occurred. This measure represents the average of the best values obtained on each time instant up to the time T. If we further average the off-line performance for all the evaluations and all the runs of an algorithm, we obtain the overall off-line performance, that gives a good idea of the average performance of the algorithm through all the optimisation process. We have performed 30 independent runs with 100 fitness function changes, for each and every problem, period, severity and algorithm variant.

Given the above configurations of problems and algorithms, the goal will be to analyze the contribution of the adaptive scheme on both the *Agents* and the *EA* algorithms, and then to compare them with the state of the art *AHMA*.

We have considered all the mutation operator variants that are shown on Table 3.1, each of which uses a different configuration of the solution mutation operator (see the description of the *mutation* function on Section 3.2.1). The values for the operators have been selected with an increasing *numBits* and a decreasing *flipProb*. In this way, as the number of consecutive bits affected gets bigger, the configurations allow for a higher probability of leaving some of the affected bits unchanged. The last variant (*Adaptive*) is a special variant that, instead of using just a single configuration, uses all of them coupled with the adaptive scheme (Section 3.3). Additionally, when using the adaptive scheme, the initial credit assigned to each operator was set to the 10% of the sum of the fitness of solutions in the current population plus the 10% of the number of bits on a solution. This last addition is only included to guarantee that a non-zero initial credit is assigned to each operator configuration even when all the solutions of the population have 0 fitness. Since the time for learning is limited by the change period for the problems, we have reduced the mutation operator variants to just 4 configurations, so it becomes easier to learn the proper operator to use in a faster way. In this way, there are less values to learn, but we still have several different mutation operator configurations and it is expected that the learning will find the best performing ones for each algorithm / problem.

Table 3.1 Mutation operator variants

Variant	Configuration
MutOp1	$numBits = 1$ and $flipProb = 1$
MutOp2	$numBits = 2$ and $flipProb = 0.9$
MutOp3	$numBits = 3$ and $flipProb = 0.8$
MutOp4	$numBits = 4$ and $flipProb = 0.7$
Adaptive	Uses the adaptive scheme on all the previous variants

Finally, since the results for every possible algorithm, problem configuration and mutation operator variant would be difficult to interpret if displayed as a numerical table, we propose a ranking scheme to see the relative performance of the algorithms on a graphical way. The idea is to compare algorithms using statistical tests over the offline performances for every run of each algorithm and configuration. As recommended in [1, 6], non-parametric tests are used, first checking for differences among the results of *all* algorithms using a Kruskal-Wallis test. If there is enough evidence for overall statistical differences, then a test is performed to assess individual differences among each pair of algorithms. In this case, a Mann-Whitney-Wilcoxon test is used, with Holm's correction to compensate for multiple comparisons. In both tests, the significance level used is 0.05. If the test concludes that there is enough evidence for statistical differences, the algorithm with the highest overall offline performance adds 1 to its rank, and the other adds -1. In case of a tie, both receive a 0. The range of the rank values for n algorithms for any specific problem, period and severity will therefore be in the $[-n+1, n-1]$ interval. The higher the rank obtained, the better an algorithm can be considered in relation to the other ones. These ranks will be displayed as colors with the highest rank value $(n-1)$ being displayed as white and the lowest rank value $(-n+1)$ having the darkest color. The remaining rank values will be assigned a progressively darker color as the rank increases. If we then group together the ranks of an algorithm for a given problem with every possible different period and severity we can obtain a colored matrix, where it is easy to get a good idea of how the algorithm performs for that specific problem. A white color in a given cell indicates that the algorithm is statistically better than all the other algorithms for that specific problem configuration. As the color gets darker it means that the algorithm starts to be statistically equal or worse than some other algorithms. The worse case for a given algorithms occurs when its cell has the darkest possible color, what means that the algorithm is statistically worse than all others for that problem configuration.

The main advantages of this graphical representation are:

- It is able to compress a lot of the numerical information of the results and present it in a visual way, which is more meaningful to humans.
- Rankings are also meaningful *per se*, since positive ranks indicate "above the average" algorithms for a given configuration, while negative ones indicate "below the average".
- Moreover, the final numerical rank for an algorithm indicates "how many other algorithms it is better than" (there is statistical evidence for a significantly better performance).
- Finally, this compressed, visual way of presenting the results not only allows to clearly identify the best and worst performing algorithms, but also to perform an overall analysis of the behavior of these methods over several environmental conditions.

The results for the experiments regarding the adaptive scheme for both the *Agents* and *EA* algorithms on all the problems, periods and severities are displayed using the above explained representation on Figure 3.2.

Fig. 3.5 shows the results for the *Agents* algorithm with every single mutation operator and the variant with the adaptive scheme applied over all the other operators. It can be seen that the best individual operator is not the same for the different problems. *MutOp1* is clearly the best for the OneMax problem and *MutOp4* obtains the best ranks (excluding the *Adaptive* variant) for the Royal Road and Deceptive problems, while some mixed results are obtained for the different Plateau configurations. Focusing on the *Adaptive* variant it can be seen that it is the best variant for all the Royal Road problem configurations and most of the Plateau ones while it obtains the second place on all the other configurations. In this way, it can be easily concluded that the adaptive scheme provides a much more robust performance for the *Agents* algorithm with a ranking that is always at the top ones among all the mutation operator variants. We have also checked the relative credit ratios achieved when using the *Adaptive* variant. First of all it has been observed that the average ratios for any operator configuration at the end of each problem change are between 10% and 40%. In this way, it seems that all operators contribute to improve the solutions from time to time, so none of them gets very low ratios and all of them are preserved. But the more interesting part is that the ratios achieved are closely related to the performance of the best individual operator variants. For instance, when solving OneMax problem, *MutOp1* is always the operator getting higher ratios (30% − 40%) while *MutOp4* gets the higher ratios for the Royal Road and Deceptive problems.

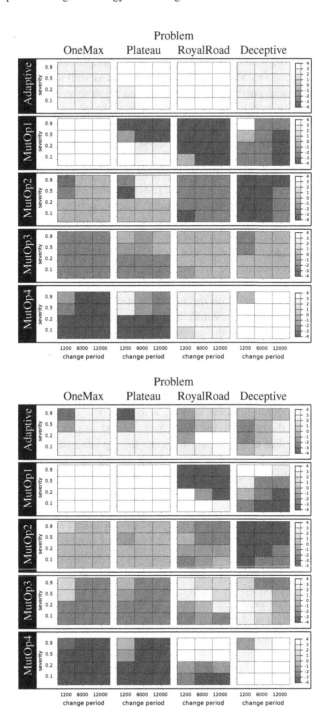

Fig. 3.2 Rank results for all the variants of the *Agents* and *EA* algorithms

Fig. 3.5 shows the results for the *EA* algorithm with the same mutation operator and adaptive variants that in the results for the *Agents* algorithm of Fig. 3.5. In the *EA* case, *MutOp1* variant is the best on most configurations including all the configurations for the OneMax and Plateau problems as well as some of the Royal Road experiments with small period of change and small severity; and also on some of the configurations with small period of change and high severity for the Deceptive problem. *MutOp4* is the other remarkable variant for the *EA* algorithm with a clearly better performance on most of the Deceptive problem configurations as well as the configurations of Royal Road with higher severities. If we now focus on the *Adaptive* variant we see that it still obtains first or second place on the rankings for most configurations, but it is ranked first a fewer number of times than in the adaptive *Agents* and also has some third and fourth places, including even a case where it obtains the worse result when considering the Plateau problem with the smallest period of change and the highest severity. The higher variability of the *Adaptive* variant on the *EA* algorithm experiments when compared with the *Agents* algorithm could be explained because the modification of solutions on *Agents* basically relays just on the mutation operator while the crossover applied on the *EA* algorithm is probably interacting both with the mutation operator performance and with the adaptive scheme, making it more difficult for the learning to work well on dynamic environments. In other words: the credits assigned to a mutation operator can actually be an effect of the crossover and not the mutation itself, leading to a partially incorrect learning. Despite that, if we compare the *Adaptive* variant with the other operator variants for *EA*, we see that it outperforms *MutOp2* and *MutOp2* on most cases. Moreover, while *MutOp1* and *MutOp4* are the best alternatives for several specific problems / configurations, they are also the worse variants for several other configurations. On the contrary, the *Adaptive* variant shows more robust results.

If we look together at the results for both the *Agents* and the *EA* algorithms, we can conclude that the adaptive variants are more likely to obtain good results independently of the problem configuration and the type of dynamism. And it is also much less likely that the adaptive variants obtain the worse result on any case. Since on real problems we probably will not know the exact "type of the problem" and the exact type of dynamism, it will not be easy to select a single operator and to hope that it will be a good one for that problem / instance. Therefore, it will be generally a better idea to always use the adaptive scheme so it becomes more probable to obtain good solutions without the unpredictable performance of blindly choosing a single operator.

We have also compared *Agents* and *EA* algorithms against the state of the art *AHMA* algorithm presented on Section 3.2.3 on all the problems and configurations. In order to simplify things we just show a single version of each algorithm where we have focused on the *Adaptive* variants for both *Agents* and *EA*. The results are shown on Fig. 3.3 where it it can be observed that the *EA* algorithm obtains the worse rank (red color) for most of the problem configurations so it is clearly worse than the more recent algorithms: *Agents* and *AHMA*. The *AHMA* algorithm is proven again to be a very good algorithm that outperforms the *EA* on most cases, even when using the adaptive scheme. But more importantly, the adaptive *Agents* algorithm that

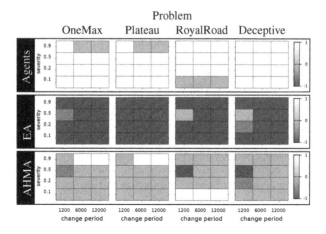

Fig. 3.3 Rank results of the Adaptive Agents and EA variants against AHMA

was the main goal of this paper is able to clearly outperform the state of the art *AHMA* on almost all scenarios tested. *AHMA* only beats *Agents* on the higher periods and higher severities for both OneMax and Plateu and also on the lowest severity configurations for the Royal Road problem. Every other configuration shows an advantage for the *Agents* algorithm. Nevertheless, we should also note that despite obtaining better rankings, the overall offline performance values obtained by *Agents* and *AHMA* are usually similar and quite close to the maximum possible value of 100 so we can state that both have quite good performances. Additionally, we should say that we have not obtained the exact same *AHMA* performance values that were reported on the published paper [12]. Our *AHMA* results are generally a bit better than the published ones and we obtain particularly higher performances on the worse configurations published by Yang, but on a few of the best configurations reported on Yang's paper, our results were slightly worse. Therefore, although the rankings we obtained are quite significant, we do not discard that the use of another implementation or finding improved parameter settings for *AHMA* could lead to improved rankings for this algorithm.

Finally, it should be noted that the *Agents* algorithm is the faster one with execution times between two and three times smaller than the *EA* and *AHMA* algorithms, being *EA* slightly faster than *AHMA*.

3.6 Conclusions and Future Work

This paper has presented an adaptive scheme for the *Agents* algorithm modified to tackle combinatorial DOPs. The adaptive behaviour is achieved with an adaptive scheme to select among several different configurations of a solution mutation operator. The same adaptive scheme was also tested on an evolutionary algorithm (*EA*).

The results showed that while the individual operator variants can obtain the best results for some problems and configurations, these high performances usually come at the cost of being one of the worse variants for other problem configurations. On the contrary, the *Adaptive* variant always obtained one of the top two rank positions with the *Agents* algorithm for any problem and configuration. The results of the *Adaptive* variant for the *EA* were a bit worse than the ones obtained for *Agents* but it still probed to be much more robust than any single mutation operator configuration alone. We can conclude that the adaptive scheme helps to ensure reasonably good solutions without the need to select a specific individual operator configuration. The relative operator ratios found for the *Adaptive* variant were also seen to be in good relation to the best performing individual variants. Therefore, the adaptive scheme reduces the risk of making a bad operator selection and allows to find good operator configurations at the same time than avoiding or reducing the occurrence of the configurations that are not well suited for a specific problem / instance.

Despite being the first version of *Agents* adapted to combinatorial DOPs, the results obtained are very good, with values of overall off-line performance that are close to the maximum possible value of 100 for all the test problems considered. Moreover, the results obtained are very good when compared with the state of the the art *AHMA* algorithm.

As future work we want to implement improvements to the adaptive scheme since a faster good credit assignment for the configurations of the mutation operator will probably lead to higher rankings of the adaptive algorithm variants, making their results closer to the ones of the best individual operators. It will also be possible to outperform single operators on most cases, by means of a richer exploration and exploitation of the search space that could emerge from a synergy among the several operator configurations. We will also study the dynamic evolution of the learning in terms of the distribution (credits) of operators as the number of evaluations of the objective function increases. This will allow to know how the probabilities to choose a specific individual mutation operator evolve as the search progresses and how the dynamism affects the quality and convergence of the adaptive scheme.

Acknowledgements. This work has been partially funded by the project TIN2008-01948 from the Spanish Ministry of Science and Innovation (70% are FEDER funds), and P07-TIC-02970 from the Andalusian Government. We also acknowledge support from CITIC-UGR.

References

[1] Bartz-Beielstein, T.: Experimental Research in Evolutionary Computation: The New Experimentalism. Natural Computing Series. Springer, Heidelberg (2006)

[2] Blackwell, T., Branke, J.: Multiswarms, exclusion, and anti-convergence in dynamic environments. IEEE Transactions on Evolutionary Computation 10(4), 459–472 (2006)

[3] Branke, J.: Memory Enhanced Evolutionary Algorithms for Changing Optimization Problems. In: Angeline, P.J., Michalewicz, Z., Schoenauer, M., Yao, X., Zalzala, A. (eds.) Proceedings of the Congress on Evolutionary Computation, vol. 3, pp. 1875–1882. IEEE Press, Los Alamitos (1999)

[4] Branke, J., Schmeck, H.: Designing evolutionary algorithms for dynamic optimization problems. In: Advances in Evolutionary Computing: Theory and Applications, pp. 239–262 (2003)

[5] Cruz, C., González, J., Pelta, D.: Optimization in dynamic environments: a survey on problems, methods and measures. Soft Computing 15, 1427–1448 (2011)

[6] García, S., Molina, D., Lozano, M., Herrera, F.: A study on the use of non-parametric tests for analyzing the evolutionary algorithms' behaviour: a case study on the cec'2005 special session on real parameter optimization. Journal of Heuristics 15(6), 617–644 (2009)

[7] González, J.R., Masegosa, A.D., del Amo, I.G.: A cooperative strategy for solving dynamic optimization problems. Memetic Computing 3, 3–14 (2011)

[8] Ochoa, G., Mädler-Kron, C., Rodriguez, R., Jaffe, K.: Assortative mating in genetic algorithms for dynamic problems. In: Applications on Evolutionary Computing, pp. 617–622 (2005)

[9] Pelta, D., Cruz, C., González, J.R.: A study on diversity and cooperation in a multiagent strategy for dynamic optimization problems. Int. J. of Intelligent Systems 24, 844–861 (2009)

[10] Pelta, D., Cruz, C., Verdegay, J.L.: Simple control rules in a cooperative system for dynamic optimisation problems. Int. J. of General Systems 38(7), 701–717 (2009)

[11] Smith, J.: Self-adaptation in evolutionary algorithms for combinatorial optimisation. In: Cotta, C., Sevaux, M., Sörensen, K. (eds.) Adaptive and Multilevel Metaheuristics. SCI, vol. 136, pp. 31–57. Springer, Heidelberg (2008)

[12] Wang, H., Wang, D., Yang, S.: A memetic algorithm with adaptive hill climbing strategy for dynamic optimization problems. Soft Computing 13(8-9), 763–780 (2009)

[13] Yang, S., Yao, X.: Experimental study on population-based incremental learning algorithms for dynamic optimization problems. Soft Computing 9(11), 815–834 (2005)

[14] Yang, S., Ong, Y.S., Jin, Y. (eds.): Evolutionary Computation in Dynamic and Uncertain Environments. SCI, vol. 51. Springer, Heidelberg (2007)

Chapter 4
Interactive Intonation Optimisation Using CMA-ES and DCT Parameterisation of the F0 Contour for Speech Synthesis

Adriana Stan, Florin-Claudiu Pop, Marcel Cremene, Mircea Giurgiu, and Denis Pallez

Abstract. Expressive speech is one of the latest concerns of text-to-speech systems. Due to the subjectivity of expression and emotion realisation in speech, humans cannot objectively determine if one system is more expressive than the other. Most of the text-to-speech systems have a rather flat intonation and do not provide the option of changing the output speech. We therefore present an interactive intonation optimisation method based on the pitch contour parameterisation and evolution strategies. The Discrete Cosine Transform (DCT) is applied to the phrase level pitch contour. Then, the genome is encoded as a vector that contains 7 most significant DCT coefficients. Based on this initial individual, new speech samples are obtained using an interactive Covariance Matrix Adaptation Evolution Strategy (CMA-ES) algorithm. We evaluate a series of parameters involved in the process, such as the

Adriana Stan
Communications Department, Technical University of Cluj-Napoca, Cluj, Romania
e-mail: Adriana.Stan@com.utcluj.ro

Florin-Claudiu Pop
Communications Department, Technical University of Cluj-Napoca, Cluj, Romania
e-mail: Florin.Pop@com.utcluj.ro

Marcel Cremene
Communications Department, Technical University of Cluj-Napoca, Cluj, Romania
e-mail: Cremene@com.utcluj.ro

Mircea Giurgiu
Communications Department, Technical University of Cluj-Napoca, Cluj, Romania
e-mail: Mircea.Giurgiu@com.utcluj.ro

Denis Pallez
Laboratoire d'Informatique, Signaux, et Systèmes de Sophia-Antipolis (I3S), Université de Nice Sophia-Antipolis, France
e-mail: Denis.Pallez@unice.fr

D.A. Pelta et al. (Eds.): NICSO 2011, SCI 387, pp. 57–71, 2011.
springerlink.com   © Springer-Verlag Berlin Heidelberg 2011

initial standard deviation, population size, the dynamic expansion of the pitch over the generations and the naturalness and expressivity of the resulted individuals. The results have been evaluated on a Romanian parametric-based speech synthesiser and provide the guidelines for the setup of an interactive optimisation system, in which the users can subjectively select the individual which best suits their expectations with minimum amount of fatigue.

4.1 Introduction

Over the last decade text-to-speech (TTS) systems have evolved to a point where in certain scenarios, non-expert listeners cannot distinguish between human and synthetic voices with 100% accuracy. One problem still arises when trying to obtain a natural, more expressive sounding voice. Several methods have been applied ([17], [20]), some of which have had more success than others and all of which include intonation modelling as one of the key aspects. Intonation modelling refers to the manipulation of the pitch or fundamental frequency (F0). The expressivity of speech is usually attributed to a dynamic range of pitch values. But in the design of any speech synthesis system (both concatenative and parameteric), one important requirement is the flat intonation of the speech corpus, leaving limited options for the synthesised pitch contours.

In this paper we propose an interactive intonation optimisation method based on evolution strategies. Given the output of a synthesiser, the user can opt for a further enhancement of its intonation. The system then evaluates the initial pitch contour and outputs a small number of different versions of the same utterance. Provided the user subjectively selects the best individual in each set, the next generation is built starting from this selection. The *dialogue* stops when the user considers one of a generation's individual satisfactory. The solution for the pitch parameterisation is the Discrete Cosine Transform (DCT) and for the interactive step, the Covariance Matrix Adaptation-Evolution Strategy (CMA-ES).

This method is useful in the situation where non-expert users would like to change the output of a speech synthesiser to their preference. Also, under resourced languages or limited availability of speech corpora could benefit from such a method. The prosodic enhancements selected by the user could provide long-term feedback for the developer or could lead to a *user-adaptive* speech synthesis system.

4.1.1 Problem Statement

In this subsection we emphasise some aspects of the current state-of-the-art speech synthesisers which limit the expressiveness of the result:

Issue #1: Some of the best TTS systems benefit from the prior acquisition of a large speech corpus and in some cases extensive hand labelling and rule-based intonation.

But this implies a large amount of effort and resources, which are not available for the majority of languages.

Issue #2: Most of the current TTS systems provide the user with a single unchangeable result which can sometimes lack the emphasis or expressivity the user might have hoped for.

Issue #3: If the results of a system can be improved, it usually implies either additional annotation of the text or a trained specialist required to rebuild most or all of the synthesis system.

Issue #4: Lately, there have been studies concerning more objective evaluations of the speech synthesis, but in the end the human is the one to evaluate the result and this is done in a purely subjective manner.

4.1.2   Related Work

To the best of our knowledge, evolution strategies have not been previously applied to speech synthesis. However, the related genetic algorithms have been used in articulatory [1] or neural networks based [11] speech synthesisers. A study of interactive genetic algorithms applied to emotional speech synthesis is presented in [8]. The authors use the XML annotation of prosody in Microsoft Speech SDK and try to convert neutral speech to one of the six basic emotions: *happiness, anger, fear, disgust, surprise* and *sadness*. The XML tags of the synthesised speech comprise the genome. Listeners are asked to select among 10 speech samples at each generation and to stop when they consider the emotion in one of the speech samples consistent with the desired one. The results are then compared with an expert emotional speech synthesis system. Interactive evolutionary computation has, on the other hand, been applied to music synthesis [10], and music composition [3], [9].

4.2   DCT Parameterisation of the F0 Contour

In text-to-speech one of the greatest challenges remains the intonation modelling. There are many methods proposed in order to solve this problem, some taking into account a phonological model [2], [15] and others simply parameterising the pitch as a curve [18]. Curve parameterisation is a more efficient method in the sense that no manual annotation of the text to be synthesised is needed and thus not prone to subjectivity errors.

Because in this study we are not using prior text annotations or additional information, we chose a parameterisation based on the DCT, that partially adresses Issue #1 of the Problem Statement section.

DCT is a discrete transform which expresses a sequence of discrete points as a sum of cosine functions oscillating at different frequencies with zero phase. The are several forms, but the most common one is DCT-II (Eq. 4.1). The coefficients are computed according to Eq. 4.2.

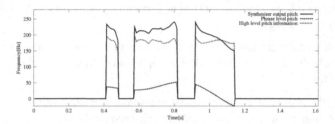

Fig. 4.1 An example of a pitch contour for the utterance "Ce mai faci" ("How are you"), the phrase level contour based on the inverse DCT of DCT1-DCT7 coefficients and the high level pitch information.

$$X_k = \frac{1}{2}x_0 + \sum_{n=0}^{N-1} x_n cos\left[\frac{\pi}{N}(n+\frac{1}{2})k\right], k = 0, 1...N-1 \tag{4.1}$$

$$c_n = \sum_{x=0}^{M-1} s(x)cos\left[\frac{x}{M}n(x+\frac{1}{2})\right] \tag{4.2}$$

DCT applied to pitch parameterisation has been extensively studied in [7], [13] and [19]. These works prove that DCT is an efficient way to parameterise the pitch with minimum error. Also, the principle behind DCT adheres to the superpositional aspect [14] of the fundamental frequency. The principle states that the pitch can be broken down into separate layers of realisation, heuristically named phrase, word, syllable and phoneme, in the sense that the cognitive process of speech derives a phrase contour unto which the rest of the layers are overlapped. Another important aspect of the DCT is its direct inverse transform. This is needed in the re-synthesis of the pitch contour from the DCT coefficients (Eq. 4.1).

The method we propose addresses the issue of modelling the phrase level intonation, or trend. Starting from a flat intonation, we would like to derive more dynamic and expressive contours. Therefore, we consider the phrase layer to be represented by the inverse DCT transform of the DCT1 to DCT7 coefficients of the pitch DCT. This assumption is also supported by the results presented in [19]. DCT0 represents the mean of the curve and in our case it is speaker dependent. Using DCT0 in the genome encoding would undesirably change the pitch of the speaker, our focus being on the overall trend of the phrase intonation. The phrase level is then subtracted from the overall contour, and the result is retained and will be referred to as *high level pitch information*. Fig. 4.1 presents an example of a pitch contour, the phrase level contour based on the inverse DCT of the DCT1-DCT7 coefficients and the high level pitch information. It can be observed that the phrase level contour represents the relative trend of the voiced segments intonation, while the high level information has a relatively flat contour with variations given by the word, syllable and phoneme levels.

Because DCT cannot parameterise fast variations with a small number of coefficients, the unvoiced segments of the F0 contour were interpolated using a cubic function (Eq. 4.3). During the interactive step we apply the inverse DCT transform over the winner's genome, add the high level pitch information and synthesise the speech using the resulted F0 contour.

$$f(x) = ax^3 + bx^2 + cx + d \tag{4.3}$$

4.3 Optimisation Using CMA-ES

CMA-ES (Covariance Matrix Adaptation - Evolution Strategy) was proposed by Hansen and Ostermeier [5] as an evolutionary algorithm to solve unconstrained or bounded constraint, non-linear optimisation problems defined in a continuous domain. In an evolutionary algorithm, a *population* of genetic representations of the solution space, called *individuals*, is updated over a series of iterations, called *generations*. At each generation, the best individuals are selected as *parents* for the next generation. The function used to evaluate individuals is called the *fitness* function.

The search space is explored according to the genetic operations used to update the individuals in the parent population and generate new offspring. In the case of evolution strategy (ES), the selection and mutation operators are primarily used, in contrast to the genetic algorithm (GA) proposed by Holland [6], which considers a third operator – crossover. Also, in GA the number of mutated genes per individual is determined by the *mutation probability*, while in ES mutation is applied to all genes, slightly and at random.

If mutation is according to a multivariate normal distribution of mean m and covariance matrix C, then CMA-ES is a method to estimate C in order to minimise the search cost (number of evaluations). First, for the mean vector $m \in \mathbb{R}^n$, which is assimilated to the preferred solution, new individuals are sampled according to the normal distribution described by $C \in \mathbb{R}^{n \times n}$:

$$x_i = m + \sigma y_i \tag{4.4}$$

$$y_i \sim N_i(0, C), i = 1..\lambda$$

where λ is the size of the offspring population and $\sigma \in \mathbb{R}_+$ is the step size.

Second, sampled individuals are evaluated using the defined fitness function and the new population is selected. There are two widely used strategies for selection: $(\mu + \lambda)$-ES and (μ, λ)-ES, where μ represents the size of the parent population. In $(\mu + \lambda)$-ES, to keep the population constant, the λ worst individuals are discarded after the sampling process. In (μ, λ)-ES all the parent individuals are discarded from the new population in favour of the λ new offspring.

Third, m, C and σ are updated. In the case of (μ, λ)-ES, which is the strategy we chose to implement our solution, the new mean is calculated as follows:

$$m = \sum_{i=1}^{\mu} w_i x_i \qquad (4.5)$$

$$w_1 \geq .. \geq w_\mu, \sum_{i=1}^{\mu} w_i = 1$$

where x_i is the i-th ranked solution vector $(f(x_1) \leq .. \leq f(x_\lambda))$ and w_i is the weight for sample x_i.

The covariance matrix C determines the shape of the distribution ellipsoid and it is updated to increase the likelihood of previously successful steps. Details about updating C and σ can be found in [4] .

CMA-ES is the proposed solution for Issues #2, #3 and #4 through the generation of several individuals (i.e. speech samples) the user can chose from, the extension of the coefficients' space and the subjective fitness function for the interactive step.

4.4 Proposed Solution

Combining the potential of the DCT parameterisation and evolution strategies, we introduce an interactive solution for the intonation optimisation problem, which requires no previous specific knowledge of speech technology. To achieve this, three problems need to be solved: *1)* generate relevant synthetic speech samples for a user to chose from, *2)* minimise user fatigue and *3)* apply the user feedback to improve the intonation of the utterance.

We solve the first problem by using CMA-ES to generate different speech samples, normally distributed around the baseline output of a Romanian speech synthesis system [16] based on HTS (Hidden Markov Models Speech Synthesis System) [21]. We consider a *genome* encoded using a vector of 7 genes, where each gene stores the value of a DCT coefficient, from DCT1 to DCT7. We start with an initial mean vector m that stores the DCT coefficients of the F0 phrase level generated by the HTS system and an initial covariance matrix $C = I \in \mathbb{R}^{7 \times 7}$. In each generation, new individuals are sampled according to Eq. (4.4).

In the next step, the user needs to evaluate generated individuals. If the population size is too large, the user may get tired before a suitable individual is found or might not spot significant differences between the individuals. On the other hand, if the population size is too small and the search space is not properly explored, a suitable individual may not be found. CMA-ES is known to converge faster even with smaller population than other evolutionary algorithms, but it was not previously applied to solve interactive problems. On the other hand, interactive genetic algorithms (IGA) have been extensively studied, but do not converge as fast as CMA-ES for non-linear non-convex problems. Faster convergence means fewer evaluations, therefore reducing user fatigue.

For the interactive version of CMA-ES, we used a *single elimination tournament* fitness [12]. In this case, the individuals are paired at random and play one game per pair. Losers of the game are eliminated from the tournament. The process repeats

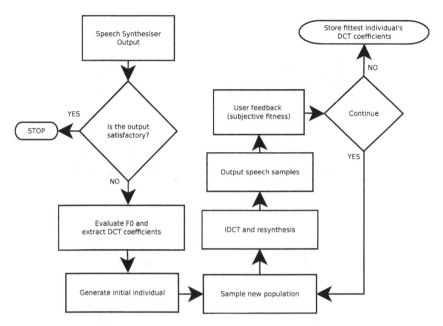

Fig. 4.2 Proposed method flow chart

until a single champion is left. The fitness value of each individual is equal to the number of played games. Each pair of individuals is presented to the user in the form of two speech samples. Being a subjective evaluation, the choice would best suit the user's requirements, thus giving the winner of a population.

The fitness value is used by CMA-ES to update mean vector m, the covariance matrix C and the standard deviation σ. A new population of individuals is sampled based on the updated values and the process repeats. The flow chart of the proposed method is presented in Fig. 4.2.

4.5   Results

The results presented below focus on establishing the correct scenario for the interactive application and on the ease of use on behalf of the listeners/users. This implies the evaluation of several parameters involved, such as: *initial standard deviation of the population* – gives the amount of dynamic expansion of pitch –, *the population size* – determines the number of samples the user has to evaluate in each generation, *the expressivity and naturalness of the generated individuals* – assures correct values for the pitch contour.

As a preliminary step in defining the standard deviation of the population, we employed an analysis of all the DCT coefficients within the *rnd1* subset of the Romanian Speech Synthesis corpus [16]. *rnd1* comprises 500 newspaper sentences read by a native Romanian female speaker. The number of phrases within this

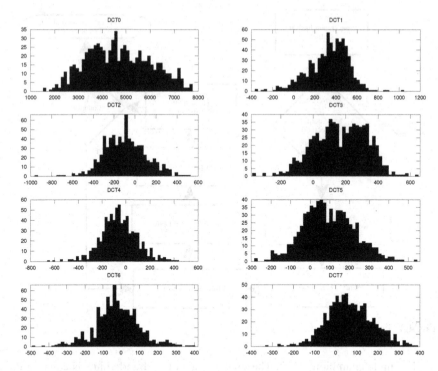

Fig. 4.3 The histograms of the first 8 DCT coefficients of the rnd1 subset of the RSS speech corpus. The 0x axis represents the values of the DCT coefficients separated in 50 equally spaced bins. The 0y axis is the number of coefficients equal to the values within the domain bin.

subset is 730 with an average length of 1.7 seconds. The intonation of the speech is flat, declarative. The histograms of the first 8 DCT coefficients of the phrases in *rnd1* are presented in Fig. 4.3. We included DCT0 as well for an overall view as it represents the mean of the pitch contour and it is speaker dependent. This coefficient was not used in the estimation of the phrase level contour. The means and standard deviations of the coefficients are presented in Table 4.1. The average pitch contour resulted from the mean values of the DCT coefficients and the average duration of the *rnd1* subset is shown in Fig. 4.4.

DCT1 has the most important influence in the F0 contour after DCT0. The mean value of the DCT1 coefficient is 331.75 with a standard deviation of 185.85 and the maximum F0 variation is given by the +*1 std. dev.* (i.e. 331.75+185.85 = 517.6) of around 40 Hz. One of the issues addressed in this paper is the expansion of the pitch range. This means that having a standard deviation of the flat intonation speech corpus, we should impose a higher value for it while generating new speech samples, but it should not go up to the point where the generated pitch contours contain F0 values which are not natural. In Fig. 4.5 we compare the third generation for an initial standard deviation of 150 and 350 respectively. We can observe in the

Table 4.1 Means and standard deviation of the DCT coefficients in *rnd1* subset with corresponding variations in Hz for an average length of 1.7 seconds.

'Coefficient	Mean	Mean F0 [Hz]	Standard deviation	Maximum F0 deviation [Hz]	
				- 1 std dev	+1 std dev
DCT0	4690.300	251-257	1318.300	179-186	322-329
DCT1	331.750	± 4	185.850	±12	±40
DCT2	-95.087	±7	197.470	±22	±7
DCT3	168.270	±12	161.030	±0.55	±25
DCT4	-57.100	±4	151.600	±16	±7
DCT5	94.427	±7	130.150	±2	±17
DCT6	-22.312	±1	123.020	±11	±7
DCT7	67.095	±5	110.370	±3	±13

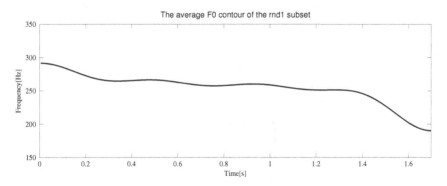

Fig. 4.4 The average pitch contour resulted from the mean values of the DCT0-DCT7 coefficients for the average length of 1.7 seconds in the *rnd1* subset.

350 case that individual 3 has F0 values going as low as 50 Hz – unnatural, while for a standard deviation of 150, the F0 contours do not vary too much from the original one and lead to a less dynamic output. Given these results, we selected a standard deviation of 250. An important aspect to be noticed from Table 4.1 is that all the 7 coefficients have approximately the same standard deviation. This means that imposing a variation based on DCT1 does not exceed natural values for the rest of the coefficients.

The single elimination tournament fitness we used to evaluate the individuals requires the user to provide feedback for $n - 1$ games, where n is the population size. So that the population size has a great importance in setting up the interactive application. Several values have been selected for it and the results are shown in Fig. 4.6. Although the highest the number of individuals the more samples the user

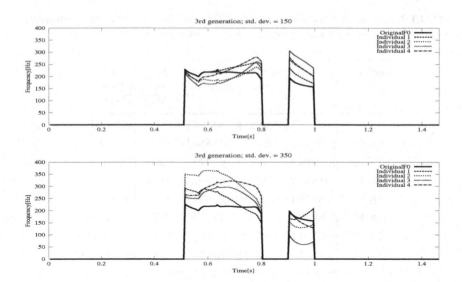

Fig. 4.5 The 3rd generation population of the F0 contour for the phrase "Ce mai faci?" ("How are you?"), with an initial standard deviation of 150 and 350 respectively. Original F0 represents the pitch contour produced by the synthesiser.

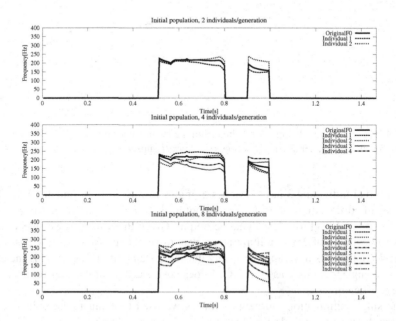

Fig. 4.6 Variation in the population size. Phrase "Ce mai faci?" ("How are you?"). Original F0 represents the pitch contour produced by the synthesiser.

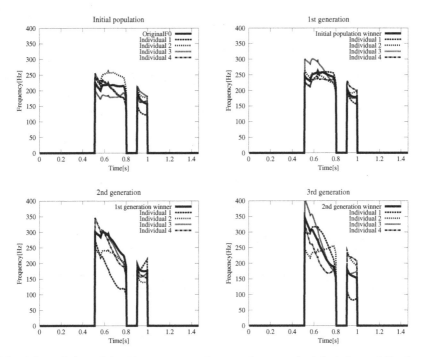

Fig. 4.7 Evolution of the F0 contour over 3 generations, standard deviation = 250, phrase "Ce mai faci?" ("How are you?"). Original F0 represents the pitch contour produced by the synthesiser.

can choose from, this is not necessarily a good thing in the context of user fatigue. But having only 2 individuals does not offer enough options for the user to choose from. We therefore suggest the use of 4 individuals per generation as a compromise between sample variability and user fatigue.

Another evaluation is the observation of the modification of the pitch contour from one generation to the other. Fig. 4.7 presents the variation of F0 from the initial population to the third. It can be observed that starting with a rather flat contour, by the third generation the dynamics of the pitch are much more expanded, resulting a higher intonation variability within and between generations. It is also interesting to observe the phrase level contours (Fig. 4.8). This is a more relevant evaluation as it shows the different trends generated by CMA-ES and the trend selected by the user in each generation. The selected trend can be used in the adaptation of the overall synthesis. In our example, the user selected an intonation with a high starting point and a descending slope afterwards, while another user could have chosen individual 1 which contains an initial ascending slope.

In order to establish the naturalness of the generated individuals and the enhanced expressivity of the winners of each generation, a small listening test was conducted.

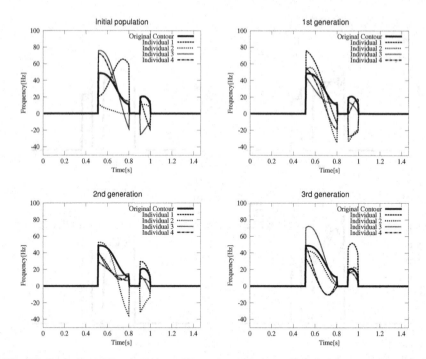

Fig. 4.8 Evolution of the phrase contour trend over 3 generations for the utterance "Ce mai faci" ("How are you"). Original contour represents the pitch contour produced by the synthesiser.

At first, a user was asked to select the winners over 4 generations for 10 phrases. Initial standard deviation was 250 and with a population size of 4. Then 10 listeners had to attribute Mean Opinion Scores (MOS) for the samples in two categories: *Naturalness* – the generated samples were compared to original recordings on a scale of [1 - Unnatural] to [5 - Natural]. All the individuals of the four generations were presented. *Expressivity* – the winners of each generation were compared to the correspondent synthesised versions of them. The listeners had to mark on a scale of [1-Less expressive] to [5-More expressive] the generated samples in comparison to the synthesiser's output. The results of the test are presented in Fig. 4.9. In the naturalness test, all the generations achieved a relatively high MOS score, with some minor differences for the 4th generation. The expressivity test reveals the fact that all the winning samples are more expressive than the originally synthesised one. The test preliminary conclude the advantages of this method. While maintaining the naturalness of the speech, its expressivity is enhanced.

Examples of speech samples generated by our method can be found at http://www.romaniantts.com/nicso2011.

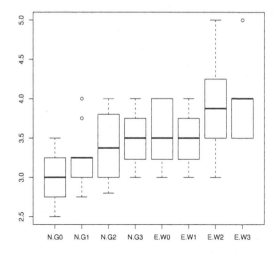

Fig. 4.9 Box plots results of the listening test. N-Gx represent the results for the natural-ness test of each generation and E-Gx represent the results for the expressivity test of each generation. The median is represented by a solid bar across a box showing the quartiles and whiskers extend to 1.5 times the inter-quartile range.

4.6   Conclusions

We introduced a new method for intonation optimisation of a speech synthesis sys-tem based on CMA-ES and DCT parameterisation of the pitch contour. The interac-tive manner of the optimisation allows the users to select an output which best suits their expectations. The novelty of the solution consists in using no prosodic anno-tations of the text, no deterministic rules and no predefined speaking styles. Also, to the best of our knowledge, this is one of the first applications of CMA-ES for an interactive problem.

The evaluation of the system's parameters provide the guidelines of the setup for an interactive application. The proposed solutions ensure an optimal value for stan-dard deviation and population size in order to concurrently maintain the naturalness of the speech samples, while expanding the dynamics of the pitch. The latter indica-tors have been evaluated in the listening test. The listening test also determined the enhancement of the expressivity of the samples.

One drawback to our solution is the lack of individual manipulation of each of the 7 DCT coefficients in the genome, unattainable in the context of the evolution-ary algorithm chosen. However the coefficients' statistics showed that the average standard deviation is similar and thus the choice for the initial standard deviation does not alter the higher order coefficients.

As the results obtained in this preliminary research have achieved a high-level of intonational variation and user satisfaction, a web-based application of the interac-tive optimisation is under-way. The application would allow the user to select the entire utterance or just parts of it – i.e., phrases, words or even syllables – for the optimisation process to enhance. For a full prosodic optimisation, we would like to include the duration of the utterance in the interactive application as well.

One interesting development would be a user-adaptive speech synthesiser. Based on previous optimisation choices, the system could adapt in time to a certain prosodic realisation. Having set up the entire workflow, testing different types of fitness functions is also of great interest.

Acknowledgements. This work has been funded by the European Social Fund, project POSDRU 6/1.5/S/5 and the national project TE 252 financed by CNCSIS-UEFISCSU.

References

[1] D'Este, F., Bakker, E.: Articulatory Speech Synthesis with Parallel Multi-Objective Genetic Algorithms. In: Proc. ASCI (2010)

[2] Fujisaki, H., Ohno, S.: The use of a generative model of F0 contours for multilingual speech synthesis. In: ICSLP- 1998, pp. 714–717 (1998)

[3] Fukumoto, M.: Interactive Evolutionary Computation Utilizing Subjective Evaluation and Physiological Information as Evaluation Value. In: Systems Man and Cybernetics, pp. 2874–2879 (2010)

[4] Hansen, N.: The CMA evolution strategy: A tutorial. Tech. rep., TU Berlin, ETH Zurich (2005)

[5] Hansen, N., Ostermeier, A.: Adapting arbitrary normal mutation distributions in evolution strategies: the covariance matrix adaptation. In: Proceedings of IEEE International Conference on Evolutionary Computation, pp. 312–317 (1996)

[6] Holland, H.: Adaptation in Natural and Artificial Systems. University of Michigan Press (1975)

[7] Latorre, J., Akamine, M.: Multilevel Parametric-Base F0 Model for Speech Synthesis. In: Proc. Interspeech (2008)

[8] Lv, S., Wang, S., Wang, X.: Emotional speech synthesis by XML file using interactive genetic algorithms. In: GEC Summit, pp. 907–910 (2009)

[9] Marques, V.M., Reis, C., Machado, J.A.T.: Interactive Evolutionary Computation in Music. In: Systems Man and Cybernetics, pp. 3501–3507 (2010)

[10] McDermott, J., O'Neill, M., Griffith, N.J.L.: Interactive EC control of synthesized timbre. Evolutionary Computation 18, 277–303 (2010)

[11] Moisa, T., Ontanu, D., Dediu, A.-H.: Speech synthesis using neural networks trained by an evolutionary algorithm. In: Alexandrov, V.N., Dongarra, J., Juliano, B.A., Renner, R.S., Tan, C.J.K. (eds.) ICCS-ComputSci 2001. LNCS, vol. 2074, pp. 419–428. Springer, Heidelberg (2001)

[12] Panait, L., Luke, S.: A comparison of two competitive fitness functions. In: Proceedings of the Genetic and Evolutionary Computation Conference, GECCO 2002, pp. 503–511 (2002)

[13] Qian, Y., Wu, Z., Soong, F.: Improved Prosody Generation by Maximizing Joint Likelihood of State and Longer Units. In: Proc. ICASSP (2009)

[14] Sakai, S.: Additive modelling of English F0 contour for Speech Synthesis. In: Proc. ICASSP (2005)

[15] Silverman, K., Beckman, M., Pitrelli, J., Ostendorf, M., Wightman, C., Price, P., Pierrehumbert, J., Hirschberg, J.: ToBI: A standard for labeling English prosody. In: ICSLP-1992, vol. 2, pp. 867–870 (1992)

[16] Stan, A., Yamagishi, J., King, S., Aylett, M.: The Romanian speech synthesis (RSS) corpus: Building a high quality HMM-based speech synthesis system using a high sampling rate. Speech Communication 53(3), 442–450 (2011), doi:10.1016/j.specom.2010.12.002

[17] Tao, J., Kang, Y., Li, A.: Prosody conversion from neutral speech to emotional speech. IEEE Trans. on Audio Speech and Language Processing 14(4), 1145–1154 (2006), doi: 10.1109/TASL,876113

[18] Taylor, P.: The tilt intonation model. In: ICSLP 1998, pp. 1383–1386 (1998)

[19] Teutenberg, J., Wilson, C., Riddle, P.: Modelling and Synthesising F0 Contours with the Discrete Cosine Transform. In: Proc. ICASSP (2008)

[20] Yamagishi, J., Onishi, K., Masuko, T., Kobayashi, T.: Acoustic modeling of speaking styles and emotional expressions in hmm-based speech synthesis. IEICE - Trans. Inf. Syst. E88-D, 502–509 (2005)

[21] Zen, H., Nose, T., Yamagishi, J., Sako, S., Tokuda, K.: The HMM-based speech synthesis system (HTS) version 2.0. In: Proc. of Sixth ISCA Workshop on Speech Synthesis, pp. 294–299 (2007)

Chapter 5
Genotype-Fitness Correlation Analysis for Evolutionary Design of Self-assembly Wang Tiles

Germán Terrazas and Natalio Krasnogor

Abstract. In a previous work we have reported on the evolutionary design optimisation of self-assembling Wang tiles. Apart from the achieved findings [11], nothing has been yet said about the effectiveness by which individuals were evaluated. In particular when the mapping from genotype to phenotype and from this to fitness is an intricate relationship. In this paper we aim to report whether our genetic algorithm, using morphological image analyses as fitness function, is an effective methodology. Thus, we present here fitness distance correlation to measure how effectively the fitness of an individual correlates to its genotypic distance to a known optimum when the genotype-phenotype-fitness mapping is a complex, stochastic and non-linear relationship.

5.1 Introduction

Self-assembly systems are characterized by inorganic or living entities that achieve global order as the result of local interactions within a particular closed environment. Self-assembly is a key cooperative mechanism in nature. Surprisingly, it has received (relatively) very little attention in computer science. In [11], we defined the *self-assembly Wang tiles system T_{sys}* as a computational model of self-assembly. This system comprises a set of square tiles with labelled edges that randomly move across the Euclidean plane forming aggregates or bouncing off as result of their

Germán Terrazas
ASAP Group, School of Computer Science
University of Nottingham, UK
e-mail: gzt@cs.nott.ac.uk

Natalio Krasnogor
ASAP Group, School of Computer Science
University of Nottingham, UK
e-mail: nxk@cs.nott.ac.uk

D.A. Pelta et al. (Eds.): NICSO 2011, SCI 387, pp. 73–84, 2011.
springerlink.com   © Springer-Verlag Berlin Heidelberg 2011

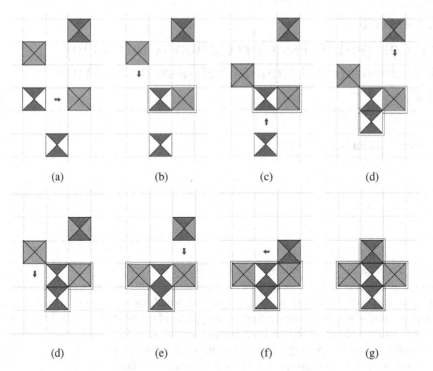

Fig. 5.1 A step-by-step (a to g) aggregate formed as the result of the interaction between five self-assembly Wang tiles performing random walk across a lattice.

interaction (see Fig. 5.1). Cooperativity is an emergent feature of this system where the combination of a certain number of tiles is required to initiate self-assembly [5][9].

Formally speaking, T_{sys} is defined as follows:

$$T_{sys} = (\mathscr{T}, \Sigma, g, \mathscr{L}, \tau)$$
$$\mathscr{T} = \{t | t = (c_0, c_1, c_2, c_3)\} \text{ where } c_0, c_1, c_2, c_3 \in \Sigma$$
$$g :: \Sigma^2 \rightarrow \mathscr{Z}^+ \cup \{0\}$$
$$\tau \in \mathbb{Z}^+ \cup \{0\} \qquad (5.1)$$

In this system, $\mathscr{T}$ is a finite set of non-empty Wang tile types t defined as a 4-tuple (c_0, c_1, c_2, c_3) which indicates the associated labels at the north, east, south and west edges of a tile, Σ is a set of symbols representing glue type labels, g is called the *glue function*, $\mathscr{L}$ is a lattice and τ is a threshold modelling the kinetic energy of the system. The glue function g is defined to compute the strength associated to a given pair of glue types. The lattice $\mathscr{L}$ is a two-dimensional surface with size $W \times H$ composed by a finite set of interconnected unit squared cells where tiles belonging to $\mathscr{T}$ are randomly located and perform random walks. Thus, when two or more tiles collide, the strength between the glue types at their colliding edges is calculated and compared to τ. If the resulting strength is bigger than τ, tiles self-assemble forming aggregates, otherwise they bounce off and keep moving.

Finding the appropriate combination of autonomous entities capable of arranging themselves together into a target configuration is a challenging open problem for the design and development of distributed cooperative systems. In [11] we addressed the self-assembly Wang tiles designability problem by means of artificial evolution. Our interest in combining self-assembly Wang tiles with evolutionary algorithms lays on the use of a method for the automated construction of supra-structures that emerge as the result of tiles interaction. In particular, we pursued to answer the following:

Given a collective target configuration, is it possible to automatically design, e.g. with an evolutionary algorithm, the local interactions so they self-assemble into the desired target?

In order to address this question, our research was centred in the use of a genetic algorithm (GA) to evolve a population of self-assembly Wang tile families. Broadly speaking, a self-assembly Wang tile family is a descriptor comprising a set of glue types each of which is associated to one of the four sides of a self-assembly Wang tile. Thus, each tile family is instantiated with equal number of tiles which are randomly located into a simulation environment where they drift and interact one another self-assembling in aggregates until the simulation runs it course. Once the simulation finishes, aggregates are compared for similarity to a user defined (target) structure employing the Minkowski functionals [6] [7]. This assembly assessment returns a numerical representation that is considered as the fitness value (Fitness_i) of each individual. Thus, individuals capable of creating aggregates similar to the specified target are better ranked and become the most likely to survive across generations. This process, together with one-point crossover and bitwise mutation operators, is applied to the entire population and repeated for a certain number of generations. In particular, our experiments comprised four increasingly rich/complex simulation environments: deterministic, probabilistic, deterministic with rotation and probabilistic with rotation. An illustration summarising our approach and its components is shown in Fig. 5.2.

The achieved results supported our evolutionary design approach as a successful engineering mechanism for the computer-aided design of self-assembly Wang tiles. Early evidence of our research in this topic is available in [10] where we employed a very simple evaluation mechanism composed by a lattice scanner fitness function and later, in [11], where morphological image analyses brought a more accurate and efficient way to collectively assess the assembled aggregates towards the target. Since the mapping from genotype to phenotype and from this to fitness is clearly a complex, stochastic and non-linear relationship, would be possible to analyse the effectiveness of Minkowski functionals as fitness function ? In order to address this question, we first introduce fitness distance correlation. Next we present how this statistical-based protocol is applied to analyse our method together with experiments and results. Finally, conclusions and discussions follow.

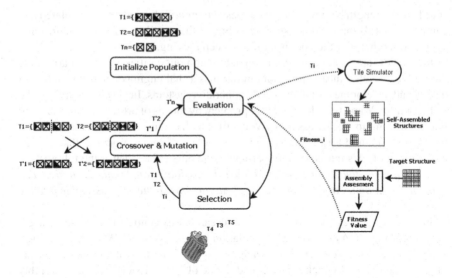

Fig. 5.2 Evolutionary approach for the evolutionary design optimisation of self-assembly Wang tiles. A population of self-assembling Wang tiles family (genotype) is randomly initialised. After that, each individual is set up into a tiles simulator from where the emerging self-assembled aggregations (phenotypes) are compared against a target structure for similarity. This comparison returns in the fitness of the individual. Later on, the application of genetic operators follows where the best ranked individuals are likely to pass throughout selection, crossover and mutation stages.

5.2 The Genotype-Fitness Assessment Protocol

The evolutionary design of self-assembly Wang tiles is characterised as a problem in which the mapping from genotype to phenotype and then from phenotype to fitness is a highly complex, non-linear and in some cases stochastic relationship. It is non-linear because different genotypes (tile sets) with small differences may lead to widely diverging phenotypes. While the same genotype, due to random effects, might produce a variety of end-products. This intricate relationship (see Fig. 5.3) makes the assessment of the genotype very difficult since the same (different) fitness value could be assigned to different (the same) genotypes. Hence, in order to analyse the efficiency by which individuals were evolved, we employed Fitness Distance Correlation (FDC) [3] [2] to measure how effectively the fitness of an individual correlates to its genotypic distance to a known optimum.

   FDC is a summary statistic that performs a correlation analysis in terms of a known optimum and samples taken from the search space, predicting whether a GA will be effective for solving an optimisation problem. Thus, when facing a minimisation (maximisation) problem, a large positive (negative) correlation indicates that a GA may successfully treat the problem or that the problem is straightforward, whereas a large negative (positive) value suggests that employing a GA may not be effective or that the problem is misleading. However, a correlation around zero,

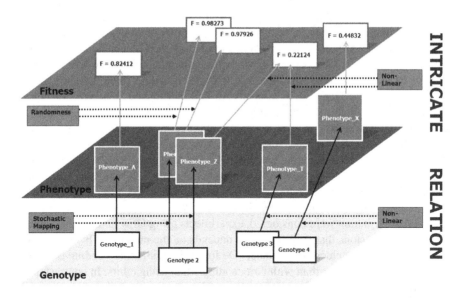

Fig. 5.3 The highly complex, non-linear and stochastic relationship taking place across the mapping from genotype to phenotype and then from phenotype to fitness.

i.e. $-0.15 \leq$ FDC ≤ 0.15, would suggest that more nuisances, perhaps including scatter plots analyses, of the fitness versus distance to the optimum should be done and, in general the problem is categorized as difficult. The formula for computing the FDC is shown in Equation 5.2, where n is the number of samples, f_i is the fitness of sample i with distance to the known optimum d_i, $\overline{f}$ and S_F are the mean and standard deviation of the fitness values, and $\overline{d}$ and S_D are the mean and standard deviation of the distances.

$$FDC = \frac{(1/n)\sum_{i=1}^{n}(f_i - \overline{f})(d_i - \overline{d})}{S_F S_D}$$

$$\overline{f} = \frac{1}{n}\sum_{i=1}^{n} f_i$$

$$\overline{d} = \frac{1}{n}\sum_{i=1}^{n} d_i$$

$$S_F = \sqrt{\frac{1}{n-1}\sum_{i=1}^{n}(f_i - \overline{f})^2}$$

$$S_D = \sqrt{\frac{1}{n-1}\sum_{i=1}^{n}(d_i - \overline{d})^2} \tag{5.2}$$

A study focused on whether FDC predicts the GA behaviour, and whether it detects differences in encoding and representation for a number of well-studied minimisation problems has been given in [3]. When predicting the GA behaviour, the FDC confirmed that Deb & Goldberg's 6-bit fully deceptive function and Whitley's 4-bit (F2 and F3) fully deceptive functions are indeed misleading since the correlation values were 0.30, 0.51 and 0.36 respectively, and the fitnesses tended to increase with the distance from the global optimum. In addition, FDC also confirmed that problems like Ackley's One Max, Two Max, Porcupine and NK landscape problems for $K \leq 3$ are easy since the correlation values resulted in -0.83, -0.55, -0.88 and -0.35 respectively. Nevertheless, the FDC indicated that NK(12,11) landscape, Walsh polynomials on 32 bits with 32 terms each of order 8, Royal Road functions R1 and R2, as well as some of the De Jong's functions like F2(12) are difficult since the resulting correlation values were close to 0.0. When the differences in encoding and representation were considered, experiments using Gray and binary coding led to the conclusions that the superiority depends on the number of bytes used to encode the numeric values. For instance, De Jong's F2 with binary coding is likely to make the search easier than with Gray coding when using 8 bits. In contrast, the correlation value of F12 indicated that Gray coding works better than binary when using 12 or 24 bits. Despite its successful application on a wide benchmarking set of problems, FDC is still not considered to be a very accurate predictor in some other problems. For instance, a case where FDC failed as a difficulty predictor has been presented when studying a GA maximising *Ridge functions* [8].

In summary, although FDC cannot be expected to be a perfect predictor of performance, previous work reported in [4] [1] [13] [12] [14] suggests that it is indeed a good *indicator* of performance. Our goal is then to assess how effectively the fitness of an individual correlates to its genotype when using Minkowski functionals as fitness function for the GA presented in [11].

5.3   Correlating the Self-assembly Wang Tiles Design

Since different set of tiles may self-assemble in aggregates similar in shape to the target structure, it is of our interest to study here how effectively the fitness of an individual correlates to its genotypic distance to a known optimum (see Fig. 5.4). In the rest of this section, we carry out FDC analysis in order to study Minkowski functionals effectiveness as fitness function for the evaluation of the achieved self-assembled aggregates.

5.3.1   *Experiments and Results*

In order to perform a FDC analysis, we first choose the best individual found among the four simulation environments. In this case, the best individual belongs to the results achieved when using probabilistic criteria and no rotation simulator. Next, a

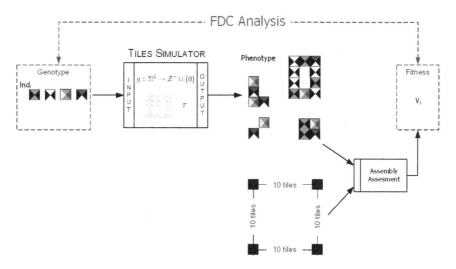

Fig. 5.4 Diagram of mappings from genotype onto phenotype and from phenotype onto numerical fitness value, and relationship to the Fitness Distance Correlation.

data set comprising 500 individuals at different Hamming distances from the best individual is created. In particular, given two individuals Ind_i and Ind_j of same length, their Hamming distance H is defined as in Equation 5.3.

$$H(Ind_i, Ind_j) = \sum_{k=1}^{n} diff(T_k^i, T_k^j)$$

$$diff(T_i, T_j) = \sum_{l=0}^{3} c_l^i \ominus c_l^j$$

$$c_a \ominus c_b = \begin{cases} 1 \text{ if } c_a \neq c_b \\ 0 \text{ otherwise} \end{cases}$$

$$c_a, c_b \in \Sigma \qquad\qquad (5.3)$$

Thus, this 500 individuals data set comprises all the possible chromosomes at Hamming distance of 1 plus some other randomly generated individuals at greater distance, all of these systematically generated following the pseudocode described in Algorithm 5.1 where $DuplicateReplacing(T_k, c_l, c_{new})$ duplicates tile T_k replacing glue type c_l with c_{new}, $DuplicateReplacing(Ind_i, T_k, T_{new})$ duplicates individual Ind_i replacing tile T_k with T_{new}, $TileAt(Ind_i, k)$ returns the tile at position k of an individual, and $Replace(T_k, c_l, c_{new})$ replaces glue type c_l in tile T_k with glue type c_{new}.

Algorithm 5.1. GenerateIndividuals

Require: *Ind* an individual
Ensure: *S* a set of individuals
  1. **for all** tiles T_k in Ind_i **do**
  2.   **for all** glue types c_l in T_k **do**
  3.   **for all** glue type $c_{new} \in \Sigma$ **do**
  4. $T_{new} \leftarrow DuplicateReplacing(T_k, c_l, c_{new})$
  5. $Ind_{new} \leftarrow DuplicateReplacing(Ind_i, T_k, T_{new})$
  6. $Insert(S, Ind_{new})$
  7.   **end for**
  8.   **end for**
  9. **end for**
10. **while** $|S| < 500$ **do**
11. $Ind_{new} \leftarrow Duplicate$(randomly chosen $Ind_i \in S$)
12. $n \leftarrow Random(0, |Ind_{new}|)$
13.   **for all** k to n **do**
14. $T_k \leftarrow TileAt(Ind_{new}, k)$
15. $m \leftarrow Random(0, 3)$
16.   **for all** l to m **do**
17. $c_{new} \leftarrow Random(\Sigma \setminus c_l)$
18. $Replace(T_k, c_l, c_{new})$
19. $Insert(S, Ind_{new})$
20.   **end for**
21.   **end for**
22. **end while**

In total, each of the generated individuals was simulated 5 times giving as a result a group with equal number of final configurations. Thus, a configuration in turn was considered as a target ($Conf_T$) against which the remaining configurations of all the groups ($Conf_i$) were evaluated on fitness (f_i) and on distance (d_i) among their associate genotypes (see Equation 5.4).

$$d_i = H(ind_i, ind_T)$$

$$f_i = f(Conf_i) = Eval(Conf_i, Conf_T) = \sqrt{(\Delta \mathscr{A})^2 + (\Delta \mathscr{P})^2 + (\Delta \mathscr{N})^2}$$

$$\Delta \mathscr{A} = max\{A_1^T, \ldots, A_m^T\} - max\{A_1^i, \ldots, A_n^i\}$$

$$\Delta \mathscr{P} = \sum_{k=1}^{m} P_k^T - \sum_{k=1}^{n} P_k^i$$

$$\Delta \mathscr{N} = m - n \qquad (5.4)$$

Since a configuration comprises a collection of aggregates, a way is needed to perform an evaluation involving all its aggregations collectively. For this reason, considering a target configuration $Conf_T$ and an arbitrary one $Conf_i$ with aggregates $\{A_1^T, \ldots, A_m^T\}$ and $\{A_1^i, \ldots, A_n^i\}$ respectively, $Conf_i$ will be evaluated upon $Conf_T$ in

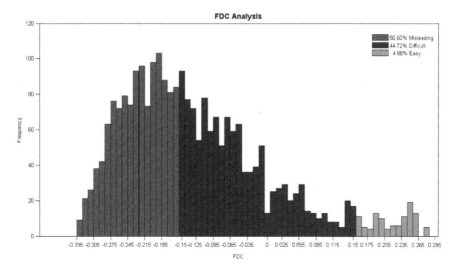

Fig. 5.5 Proportion of FDC values falling into difficult, misleading and easy to solve categories. From the 2500 analyses performed over 500 individuals, only a 4.68% reveals that a GA may successfully treat the problem.

terms of the difference in areas, perimeters and number of achieved aggregations as shown in Equation 5.4.

After performing the calculations, the findings show that the FDC values range from -0.331444 to 0.281457. Since Equation 5.4 defines a minimisation, 50.60% of the FDC values indicate that using a GA may not be effective, 44.72% that the problem is difficult to solve and a 4.68% that the GA may successfully treat the problem (see Fig. 5.5).

In particular, visual inspections over scatter plots obtained from the values captured into the smallest percentage depict good correlation on some individuals. A representative sample of these is depicted in Fig. 5.6 but at http://www.cs.nott.ac.uk/~gzt/fdcminkowski we provide the rest of the experiments. For each plot, ind_ij identifies the j-simulation of individual i, where $j \in \{a, b, c, d, e\}$ and $i \in \{1, \ldots, 500\}$. Hence, from the sampling of 500 individuals and 2500 simulations subject to FDC analyses, it emerges that employing Minkowski functionals as fitness function offers a relatively satisfactory correlation upon the relationship genotype-fitness for half of the putative samples.

Contrary to the interpretations given by some of the FDC figures seen in this section, the results reported in [11] reveal that using Minkowsi functionals as evaluation method of a GA has positively addressed the self-assembly Wang tiles design problem. Henceforth, we consider that studying and analysing the phenotype-fitness relationship may shed light on the reasons for which such evolutionary approach has been effective.

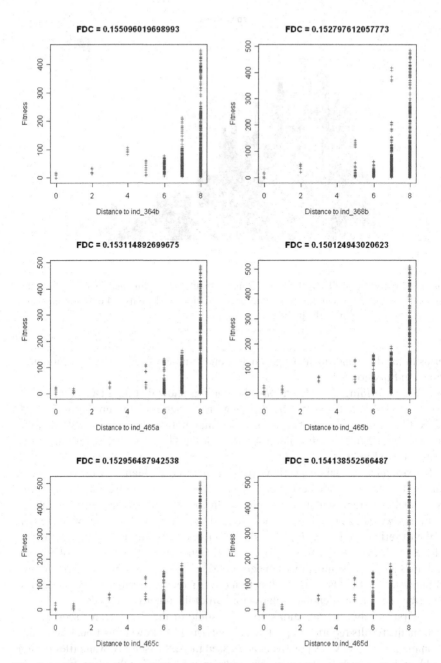

Fig. 5.6 Graphics of the resultant scatter plots and correlation coefficients for the self-assembly Wang tiles model showing that the Minkowski functionals has a relatively satisfactory correlation with the genotype for some of the self-assembly Wang tile families.

5.4   Conclusions

This paper has presented an assessment protocol to study the effectiveness of Minkowski functionals as fitness function of our GA employed for the design optimisation of self-assembly Wang tiles.

The results obtained with the morphological image analyses in [11] supports the use of Minkowski functionals as fitness function. However, from the systematically obtained individuals, only 5% of the FDC values have revealed that our GA may successfully solve the problem and 44.72% that the problem could be difficult to solve. Such is the complexity of the genotype-phenotype-fitness mapping, that clearly FDC cannot, alone, be guaranteed to give a completely accurate picture. Indeed, the objective function itself is also only an approximation of two individuals' phenotypic similarity. For these reasons, we conclude that relying on only FDC to validate complex problems would not be adequate. Therefore, we see here the necessity of combining FDC with another method to show with better accuracy whether a given fitness function is a suitable evaluation mechanism for the evolutionary design problem addressed in [11].

As a general conclusion, the application of the methodology shown in Section 5.3 reveals that employing a fitness function in terms of Minkowski functionals for the evolutionary design optimisation of self-assembly Wang tiles results in a complex mechanism of evaluation where, although its success as phenotype evaluator seems to be appropriate, a more robust analysis is needed for an assessment of how effectively an individual correlates to its genotypic distance to a known optimum. Therefore, the application of such methodology before starting long and expensive evolutionary runs should be considered in any problem where the genotype-phenotype-fitness mapping is complex, stochastic, many-to-many and computationally expensive.

Acknowledgements. The research reported here is funded by EPSRC grant EP/H010432/1 Evolutionary Optimisation of Self Assembling Nano-Designs (ExIStENcE).

References

[1] Altenberg, L.: Fitness Distance Correlation Analysis: An Instructive Counter example. In: 7th International Conference on Genetic Algorithms, pp. 57–64. Morgan Kaufmann, San Francisco (1997)

[2] Jones, T.: Evolutionary algorithms, fitness landscapes and search. PhD thesis, University of New Mexico (1995)

[3] Jones, T., Forrest, S.: Fitness Distance Correlation as a Measure of Problem Difficulty for Genetic Algorithms. In: 6th International Conference on Genetic Algorithms, pp. 184–192. Morgan Kaufmann Publishers Inc., San Francisco (1995)

[4] Koljonen, J.: On fitness distance distributions and correlations, GA performance, and population size of fitness functions with translated optima. In: Honkela, T., Kortela, J., Raiko, T., Valpola, H. (eds.) 9th Scandinavian Conference on Artificial Intelligence, Finnish Artificial Intelligence Society, Espoo, Finland, pp. 68–74 (2006)

[5] Li, L., Siepmann, P., Smaldon, J., Terrazas, G., Krasnogor, N.: Automated Self-Assembling Programming. In: Krasnogor, N., Gustafson, S., Pelta, D., Verdegay, J.L. (eds.) Systems Self-Assembly: Multidisciplinary Snapshots. Elsevier, Amsterdam (2008)

[6] Michielsen, K., Raedt, H.D.: Morphological image analysis. Computer Physics Communications 1, 94–103 (2000)

[7] Michielsen, K., Raedt, H.D.: Integral-geometry morphological image analysis. Physics Reports 347, 461–538 (2001), doi:10.1016/S0370-1573(00)00106-X

[8] Quick, R.J., Rayward-Smith, V.J., Smith, G.D.: Fitness distance correlation and ridge functions. In: Eiben, A.E., Bäck, T., Schoenauer, M., Schwefel, H.-P. (eds.) PPSN 1998. LNCS, vol. 1498, pp. 77–86. Springer, Heidelberg (1998)

[9] Rothemund, P.W.K., Winfree, E.: The program-size complexity of self-assembled squares (extended abstract). In: 32nd ACM Symposium on Theory of computing, pp. 459–468. ACM, New York (2000),
doi: http://doi.acm.org/10.1145/335305.335358

[10] Terrazas, G., Krasnogor, N., Kendall, G., Gheorghe, M.: Automated Tile Design for Self-Assembly Conformations. In: IEEE Congress on Evolutionary Computation, vol. 2, pp. 1808–1814. IEEE Press, Los Alamitos (2005)

[11] Terrazas, G., Gheorghe, M., Kendall, G., Krasnogor, N.: Evolving Tiles for Automated Self-Assembly Design. In: IEEE Congress on Evolutionary Computation, pp. 2001–2008. IEEE Press, Los Alamitos (2007)

[12] Tomassini, M., Vanneschi, L., Collard, P., Clergue, M.: A Study of Fitness Distance Correlation as a Difficulty Measure in Genetic Programming. Evolutionary Computation 13(2), 213–239 (2005), doi: http://dx.doi.org/10.1162/1063656054088549

[13] Vanneschi, L., Tomassini, M.: Pros and Cons of Fitness Distance Correlation in Genetic Programming. In: Barry, A.M. (ed.) Bird of a Feather Workshops, Genetic and Evolutionary Computation Conference, pp. 284–287. AAAI, Chigaco (2003)

[14] Vanneschi, L., Tomassini, M., Collard, P., Clergue, M.: Fitness Distance Correlation in Structural Mutation Genetic Programming. In: Ryan, C., Soule, T., Keijzer, M., Tsang, E.P.K., Poli, R., Costa, E. (eds.) EuroGP 2003. LNCS, vol. 2610, pp. 455–464. Springer, Heidelberg (2003)

Chapter 6
Model Order Reduction of Single Input Single Output Systems Using Artificial Bee Colony Optimization Algorithm

Jagdish Chand Bansal, Harish Sharma, and K.V. Arya

Abstract. In many practical situations a fairly complex and high order model is obtained in modeling different components/subsystems of a system. The analysis of such high order system is not only tedious but also cost ineffective for online implementation. Therefore, deriving reduced order models of high-order linear time invariant systems attracted researchers to develop new methods for this purpose. Artificial Bee Colony (ABC) optimization algorithm is an effective and recent addition to swarm based optimization algorithm for optimization in continuous search space. In this paper, Artificial Bee Colony optimization algorithm is applied to solve Model Order Reduction of Single Input Single Output (SISO) Systems. The results obtained by ABC are compared with two most popular deterministic approaches namely Pade and Routh approximation method. The results reported are encouraging and shows that this technique is comparable in quality with existing conventional methods.

Jagdish Chand Bansal
ABV-Indian Institute of Information Technology and Management, Gwalior
Tel.: +0751-2449819,
Fax: + 0751-2449813
e-mail: jcbansal@gmail.com

Harish Sharma
ABV-Indian Institute of Information Technology and Management, Gwalior
Tel.: +09479810157,
Fax: + 0751-2449813
e-mail: harish.sharma0107@gmail.com

K.V. Arya
ABV-Indian Institute of Information Technology and Management, Gwalior
Tel.: +0751-2449819,
Fax: + 0751-2449813
e-mail: kvarya@gmail.com

D.A. Pelta et al. (Eds.): NICSO 2011, SCI 387, pp. 85–100, 2011.
springerlink.com   © Springer-Verlag Berlin Heidelberg 2011

6.1 Introduction

The complexity of physical systems makes their exact analysis difficult and possibly
a non-desirable task, mainly due to the difficult economic and computational con-
siderations involved. To deal with the physical systems, the first step is to develop
a mathematical model. This quest for mathematical models is justified because the
use of such models facilitates the analysis of the systems they describe. In many
practical situations a fairly complex and high order model is obtained in modeling different components/subsystems of a system. The analysis of such high order
system is not only tedious but also not cost effective for online implementation.

In many contexts of control and system theory, it is advisable or even necessary to
refer to simplified models of complex processes both for simulation and for control
system synthesis. Many reduction methods have been introduced in last decades
for high order state space models or high degree transfer function/transfer function
matrices of large scale linear time invariant single input single output systems. So far
no universally accepted model order reduction method has been developed which
can be applied to all systems. Each method is best applied in a specific situation
and has its own advantages and disadvantages. One method may produce a model
that approximates low frequency behavior well, whereas other may produce good
approximation to impulse or step responses. The choice of the particular technique
to be used depends on the specific problem at hand. In general, the retention of
the stability of the original system and a "good" approximation of its response are
considered to be essential features of a reduction method.

A wide variety of model order reduction methods have been proposed by several authors such as Continued Fraction Expansion Method [3], Moment Matching Method [30], Pade Approximation Method [27], Routh Approximation Method
[11], Routh Hurwitz Array Method [15], Stability Equation Method [4], Differen-
tiation Method [26], Truncation Method [7], Dominant Pole Retention Method [5],
Factor Division Method [17], Least Square Method [34] etc.

Although many conventional approaches for model order reduction guarantee
the stability of the reduced order model but sometimes the model may turn out to
be nonminimum phase. Therefore to obtain better reduced order models, the use of
some kind of optimization is necessary by itself and in combination with other techniques. Error minimization is one of the popular techniques for model order reduction of continuous time systems. In this technique, lower order model is obtained by
minimizing an error function constructed form the time responses (or alternatively
frequency responses) of the system and reduced order model. Various error criteria
are available for minimization such as Integral Square Error (ISE), Integral Time
Square Errors (ITSE), Integral of Absolute Error (IAE) and Integral Time Absolute
Errors (ITAE). Out of these ISE is used most frequently. Most of the error minimization methods minimize the integral of the squared error between the unit (impulse
or step) response of the high order original system and the reduced order model.
This technique has been used by several researchers for obtaining the reduced order
models of a given high order system in time domain [8, 9, 22] and in frequency
domain [9, 12, 16, 23, 36, 37, 39]. Yang and Luss [40] obtained reduced models by

minimizing output response errors. Luss [19] used numerical optimization to minimize the deviation between the frequency response of the high order and low order models. Sanathanan and Quinn [32] used simplex method of minimization to obtain reduced models. Xiheng [10] combined the Pade approximation technique with the minimization of frequency deviations. Quang et al. [20] introduced a method which utilizes both pole retention and frequency response matching. Sivanandam and Deepa [38] used Particle Swarm Optimization (PSO) and Genetic Algorithm (GA) for reduction of linear time invariant discrete systems. *ISE* is used as an indicator for selecting the lower order model. Satakshi et al. [24] applied GA for the reduction of linear discrete systems.

The methods of error minimization based on classical techniques suffer from many limitations like dependency on the initial guess for the convergence to an optimal solution; lack of robustness i.e. can only be applied to restricted set of problems and often too time consuming or sometimes unable to solve real world problems. When solving the nonlinear optimization problems having many local optimal solutions, if traditional methods get stuck in any of locally optimal solutions, there is no escape from it and hence is a major limitation of classical techniques. Therefore, stochastic optimization techniques may be more useful. These techniques are applicable to wider set of problems and objective function need not be convex, continuous or explicitly defined. One of their most attractive characteristics, besides being naturally parallel and robust, is that they do not require the computation of derivatives.

The main contribution of this paper is the application of ABC algorithm to solve model order reduction problem for single input single output systems. Many other meta-heuristic methods have already been applied to solve this problem. All these methods have considered to minimize Integral Squared Error (ISE) between the transient responses of original higher order model and the reduced order model pertaining to a unit step input. But in this paper, minimization is carried out based on both Integral Squared Error (*ISE*) and Impulse Response Energy (*IRE*). The obtained results are compared with existing conventional methods and the results reported are encouraging and show that this technique is comparable in quality with existing conventional methods.

Rest of the paper is organized as follows: Section 2 describes brief overview of Artificial Bee Colony Algorithm. In Section 3, Model Order Reduction(MOR) problem for Single Input Single Output (SISO) Systems is defined. Numerical Examples are explained in Section 4. Finally, in Section 5, paper is concluded.

6.2   Brief Overview of Artificial Bee Colony Algorithm

Swarm based optimization algorithms find solution by collaborative trial and error. Peer to peer learning behavior of social colonies is the main driving force behind the development of many efficient swarm based optimization algorithms. Artificial Bee Colony (ABC)optimization algorithm is a recent addition in this category. ABC is developed in 2005 by Dervis Karaboga [13]. Like any other population based optimization algorithm, ABC consists of a population of potential solutions. With reference to ABC, the potential solutions are food sources of honey bees. The fitness is determined in terms of the quality (nectar amount) of the food source. The total number of bees in the colony are divided into three groups: Onlooker Bees, Employed Bees and Scout Bees . Number of employed bees or onlooker bees are equal to the food sources. Employed bees are associated with food sources while onlooker bees are those bees that stay in the hive and use the information gathered from employed bees to decide the food source. Scout bee searches the new food sources randomly.

Working of ABC may be described as follows:

Initially, ABC generates a uniformly distributed initial population of SN solutions where each solution $x_i (i = 1, 2, ...; SN)$ is a D-dimensional vector. Here D is the number of variables in the optimization problem. ABC process requires cycles of three phases: Employed bees phase, Onlooker bees phase and Scout bee phase. In employed bee phase, employed bees modify the current solution based on the information of individual experience and the fitness value of the new solution (nectar amount). If the fitness value of the new source is higher than that of the old solution, the bee updates her position with the new one and discards the old one. The position update equation for i^{th} candidate in this phase is

$$v_{ij} = x_{ij} + \phi_{ij}(x_{ij} - x_{kj}) \tag{6.1}$$

where $k \in \{1, 2, ..., SN\}$ and $j \in \{1, 2, ..., D\}$ are randomly chosen indices. k must be different from i. ϕ_{ij} is a random number between [-1, 1].

After completion of the employed bees phase, the onlooker bees phase is started. In onlooker bees phase, all the employed bees share the new fitness information (nectar) of the new solutions (food sources) and their position information with the onlooker bees in the hive. Onlooker bees analyse the available information and select a solution with a probability, p_i, related to its fitness. The probability p_i may be calculated using following expression (there may be some other but must be a function of fitness):

$$p_i = \frac{fit_i}{\sum_{i=1}^{SN} fit_i} \tag{6.2}$$

where fit_i is the fitness value of the solution i. As in the case of the employed bee, she produces a modification on the position in her memory and checks the fitness of the candidate source. If the fitness is higher than that of the previous one, the bee memorizes the new position and forgets the old one.

If the position of a food source is not updated for predetermined number of cycles, then the food source is assumed to be abandoned and scout bees phase is started. In this phase the bee associated with the abandoned food source becomes scout bee and the food source is replaced by the randomly chosen food source within the search space. In ABC, predetermined number of cycles is a crucial control parameter which is called *limit* for abandonment.

Assume that the abandoned source is x_i and $j \in \{1, 2, ..., D\}$ then the scout bee replaces this food source with x_i. This operation can be defined as follows

$$x_i^j = x_{min}^j + rand[0, 1](x_{max}^j - x_{min}^j) \tag{6.3}$$

where x_{min}^j and x_{max}^j are bounds of x_i in j^{th} direction.

It is clear from the above discussion that there are three control parameters in ABC search process: The number of food sources SN (equal to number of onlooker or employed bees), the value of *limit* and the maximum number of cycles MCN.

In the ABC algorithm,the exploitation process is carried out by onlookers and employed bees and exploration process is carried out by scout bees in the search space. Following is the pseudo-code of the ABC algorithm [14]:

Algorithm 6.1. Artificial Bee Colony Algorithm

Initialize the population of solutions, $x_i (i = 1, 2, ...; SN)$;
cycle = 1;
while cycle $<>$ MCN **do**
    Produce new solutions v_i for the employed bees and evaluate them;
    Apply the greedy selection process for the employed bees
    Calculate the probability values P_i for the solutions x_i;
    Produce the new solutions v_i for the onlookers from the solutions x_i selected depending on p_i and evaluate them;
    Apply the greedy selection process for the onlookers;
    Determine the abandoned solution for the scout, if exists, and replace it with a new randomly produced solution x_i;
    Memorize the best solution achieved so far;
    cycle = cycle + 1;
end while

6.3   Model Order Reduction(MOR)

Model Order Reduction (MOR) problem is studied in the branch of systems and control theory. In real world situation, usually we get a system of very high order which is inappropriate for representing some properties that are important for effective use of the system. Model Order Reduction (MOR) problem deals with reduction of complexity of a dynamical system, while preserving their input-output behavior. This problem applies both to continuous-time and discrete-time systems. MOR techniques are used to create a possibly smallest model while preserving system's behavior.

6.3.1 MOR as an Optimization Problem

Consider an nth order linear time invariant dynamic SISO system given by following expression:

$$G(s) = \frac{N(s)}{D(s)} = \frac{\sum\limits_{i=0}^{n-1} a_i s^i}{\sum\limits_{i=0}^{n} b_i s^i} \qquad (6.4)$$

where a_i and b_i are known constants.

The problem is to find a rth order reduced model in the transfer function form $R(s)$, where $r < n$ represented by equation (6.5), such that the reduced model retains the important characteristics of the original system and approximates its step response as closely as possible for the same type of inputs with minimum Integral Square Error.

$$R(s) = \frac{N_r(s)}{D_r(s)} = \frac{\sum\limits_{i=0}^{r-1} a_i' s^i}{\sum\limits_{i=0}^{r} b_i' s^i} \qquad (6.5)$$

where a_i' and b_i' are unknown constants.

Mathematically, the Integral Square Error of step responses of original and reduced system can be expressed by error index J as [6],

$$J = \int_0^\infty [y(t) - y_r(t)]^2 \, dt. \qquad (6.6)$$

where $y(t)$ is the unit step response of original system and $y_r(t)$ is the unit step response of reduced system. This error index J is the function of unknown coefficients a_i' and b_i'. The aim is to determine the coefficients a_i' and b_i' of reduced order model so that the error index J is minimized.

6.3.2 Modified Objective Function for MOR

As explained in Section 6.3.1, the aim of the optimization techniques, reported earlier in literature for MOR, is to minimize the ISE of the reduced order model. But in this paper, minimization is carried out based on both ISE and IRE. The low order model is obtained by minimizing an error function, constructed from minimization of the Integral Square Error (ISE) between the transient responses of original higher order model and the reduced low order model pertaining to a unit step input as well as minimization of the difference between the high order model's impulse response energy (IRE) and the reduced low order IRE.

The impulse response energy (IRE) is calculated for original and various reduced order models, which is given by :

$$IRE = \int_0^\infty g(t)^2 dt. \tag{6.7}$$

where, $g(t)$ is the impulse response of the system.

Therefore, in this paper, both, ISE and IRE, are used to construct the objective function for minimizing the ISE and difference between IRE of high order model and reduced order model. The following modified objective function is constructed to carry out the results.

$$objective\_value = |ISE| + \frac{|IRE_R - IRE_O|}{IRE_R + IRE_O} \tag{6.8}$$

where ISE is an integral squared error of difference between the responses given by the equation (6.6), IRE_O is the impulse response energy of the original high order model and IRE_R is the impulse response energy of the reduced order model. The advantage of this modified objective function is that it minimizes ISE as well as the differences of IRE of both the models (high order and reduced order).

6.4   Experimental Results and Numerical Examples

In the present study, following values of ABC Parameters are taken into consideration for doing all experiments.

- Colony size (employed bees+onlooker bees) $NP=20$;
- The number of food sources equals the half of the colony size $FoodNumber = NP/2$;
- A food source which could not be improved through $limit$ trials is abandoned by its employed bee $limit = 100$;
- The number of runs $= 30$;
- The number of cycles for foraging a stopping criteria $maxCycle = 100$;
- The number of parameters of the problem to be optimized $D = 3$;
- The probability values are calculated by using fitness values and normalized by dividing maximum fitness value as shown by equation (6.2);

Altogether four examples are given in this section. The results in the form of step responses and impulse responses are shown and the comparison among ABC other probabilistic and deterministic approaches is carried out. It should be noted that all other approaches minimizes only Integral Square Error (ISE).

Example - 1

Let us consider the system described by the transfer function due to Shamash [33],

$$G_1(s) = (18s^7 + 514s^6 + 5982s^5 + 36380s^4 + 122664s^3 + \\ 222088s^2 + 185760s + 40320)/(s^8 + 36s^7 + 546s^6 + 4536s^5 + 22449s^4 + \\ 67284s^3 + 118124s^2 + 109584s + 40320)$$

A second order reduced model $R_1(s)$ is desired. The second order reduced system $R_1(s)$ obtained using ABC whose transfer function is given by

$$R_1(s) = \frac{17.387s + 5.3743}{s^2 + 7.091s + 5.3743}$$

A comparison of ABC with some other existing methods for this example is given in Table 1. It may be seen that in Table 1; *ISE* and absolute difference between *IRE* of original system and that of $R_1(s)$ (let us call it *IREabs*), calculated by ABC,

Table 6.1 Comparison of the Methods for example 1

Method of order reduction	Reduced Models; $R_1(s)$	ISE	IRE
Original	$G_1(s)$	-	21.740
ABC	$\dfrac{17.387s+5.3743}{s^2+7.091s+5.3743}$	0.00085	21.696
[32]]	$\dfrac{24.114s+8}{s^2+9s+8}$	1.792	32.749
Pade Approximation	$\dfrac{15.1s+4.821}{s^2+5.993s+4.821}$	1.6177	19.426
Routh Approximation	$\dfrac{1.99s+0.4318}{s^2+1.174s+0.4318}$	1.9313	1.8705
[26]	$\dfrac{4[133747200s+203212800]}{85049280s^2+552303360s+812851200}$	8.8160	4.3426
[11]	$\dfrac{1.98955s+0.43184}{s^2+1.17368s+0.43184}$	18.3848	1.9868
[15]	$\dfrac{155658.6152s+40320}{65520s^2+75600s+40320}$	17.5345	2.8871
[1]	$\dfrac{7.0908s+1.9906}{s^2+3s+2}$	6.9159	9.7906
[23]	$\dfrac{7.0903s+1.9907}{s^2+3s+2}$	6.9165	9.7893
[25]	$\dfrac{11.3909s+4.4357}{s^2+4.2122s+4.4357}$	2.1629	18.1060
[28]	$\dfrac{151776.576s+40320}{65520s^2+75600s+40320}$	17.6566	2.7581
[31]	$\dfrac{17.98561s+500}{s^2+13.24571s+500}$	18.4299	34.1223
[33]	$\dfrac{6.7786s+2}{s^2+3s+2}$	7.3183	8.9823

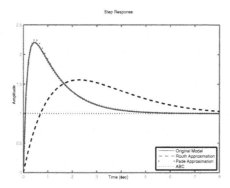

Fig. 6.1 Comparison of step responses for example 1.

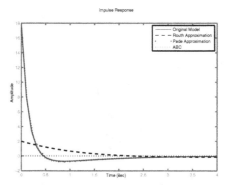

Fig. 6.2 Comparison of impulse responses for example 1.

is least in comparison of the other existing methods. The unit step responses and impulse responses of original and reduced system using ABC, Pade Approximation and Routh Approximation are shown in Figure 1 and Figure 2 respectively. It may be seen that the steady state responses of original and reduced order models by ABC are exactly matching while the transient response matching is also very close as compare to other methods. Thus the example establishes the superiority of ABC over other methods for this problem.

Example - 2

Consider the example taken from [18] with transfer function

$$G_2(s) = \frac{8169.13s^3 + 50664.97s^2 + 9984.32s + 500}{100s^4 + 10520s^3 + 52101s^2 + 10105s + 500}$$

The reduced second order model using ABC is

$$R_2(s) = \frac{485s + 50000}{s^2 + 4187s + 50000}$$

A comparison of ABC with some other reduced models of this example available in the literature is given in Table 2. It may be seen that in Table 2, ABC provides least value of the *ISE* except [21] and [35] whereas *IRE abs* is least than any other method. The unit step responses and impulse responses of original and reduced system using ABC, Pade and Routh approximation are shown in Figure 3 and Figure 4 respectively. It may be seen that the steady state responses of original and reduced order models by ABC are exactly matching while the transient response matching is also very close as compare to other methods. Thus the example shows that ABC is better then other methods for this problem.

Table 6.2 Comparison of the Methods for example 2

Method of order reduction	Reduced Models; $R_2(s)$	ISE	IRE
Original	$G_2(s)$	-	34.069
ABC	$\dfrac{485s+50000}{s^2+4187s+50000}$	0.011624954	34.065841
[35]	$\dfrac{93.7562s+1}{s^2+100.10s+10}$	0.008964	43.957
Pade Approximation	$\dfrac{23.18s+2.36}{s^2+23.75s+2.36}$	0.0046005	11.362
Routh Approximation	$\dfrac{0.1936s+0.009694}{s^2+0.1959s+0.009694}$	2.3808	0.12041
[26]	$\dfrac{0.19163s+0.00959}{s^2+0.19395s+0.00959}$	2.4056	0.11939
[4]	$\dfrac{0.38201s+0.05758}{s^2+0.58185s+0.05758}$	1.2934	0.17488
[21]	$\dfrac{83.3333s+499.9998}{s^2+105s+500}$	0.00193	35.450

Example - 3

This example is taken from [29]. The transfer function is

$$G_3(s) = \frac{s+4}{s^4 + 19s^3 + 113s^2 + 245s + 150}$$

The reduced second order model using ABC is

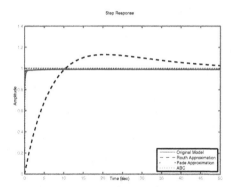

Fig. 6.3 Comparison of step responses for example 2.

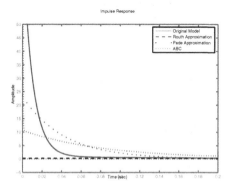

Fig. 6.4 Comparison of impulse responses for example 2.

$$R_3(s) = \frac{0.0318s + 4.0074}{s^2 + 13.7409s + 150.2775}$$

Table 3 provides comparison of ABC with some other existing methods for this example. *ISE* and *IREabs* of $R_3(s)$ are least among all methods while that of Pade and Routh approximation method is coming out to be infinity because of the steady state error between original and reduced system. The unit step responses and impulse responses of original and reduced system $R_3(s)$ using ABC, Pade Approximation and Routh Approximation are shown in Figure 4 and Figure 5 respectively. It may be seen that the steady state responses of original and reduced order models by ABC are exactly matching while the transient response matching is also very close to the original system. Thus the example establishes the superiority of ABC over other methods for this problem.

Table 6.3 Comparison of the Methods for example 3

Method of order reduction	Reduced Models; $R_3(s)$	ISE	IRE
Original	$G_3(s)$	-	0.26938×10^{-3}
ABC	$\dfrac{0.0318s+4.0074}{s^2+13.7409s+150.2775}$	0.0005	0.0039
[35]	$\dfrac{-494.596s+405.48}{150s^2+2487s+15205.5}$	0.2856×10^{-2}	0.2476×10^{-3}
Pade Approximation	$\dfrac{-0.005017s+0.08247}{s^2+4.09s+3.093}$	∞	0.27192×10^{-3}
Routh Approximation	$\dfrac{0.009865s+0.03946}{s^2+2.417s+1.48}$	∞	0.23777×10^{-3}

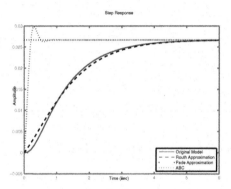

Fig. 6.5 Comparison of step responses for example 3.

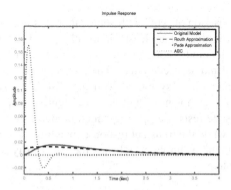

Fig. 6.6 Comparison of impulse responses for example 3.

Example - 4

Consider a fourth order practical system taken from [2]. It is the transfer function of a fuel control system of an actual boiler, which integrates a real power plant.

$$G_4(s) = \frac{4.269s^3 + 5.10s^2 + 3.9672s + 0.9567}{4.3992s^4 + 9.0635s^3 + 8.021s^2 + 5.362s + 1}$$

The reduced second order model using ABC is

$$R_4(s) = \frac{0.2034s + 8.994}{s^2 + 7.9249s + 9.4008}$$

A comparison of ABC with some other existing methods for this example is shown in Table 4. It may be seen that in Table 5 ABC provides least value of the *ISE* and *IREabs*in comparison of the other existing methods. The unit step responses and impulse responses of original and reduced system using ABC and some other existing methods are shown in Figure 7 and Figure 8 respectively. It may be seen that the unit step responses and impulse responses of original and reduced order model by ABC are very close as compare to other methods.

Table 6.4 Comparison of the Methods for example 4

Method of order reduction	Reduced Models; $R_4(s)$	ISE	IRE
Original	$G_4(s)$	-	.54536
ABC	$\dfrac{0.2034s+8.994}{s^2+7.9249s+9.4008}$	0.03501	0.54552
[35]	$\dfrac{4.0056s+0.9567}{8.021s^2+5.362s+1}$	0.22372	0.27187
Pade Approximation	$\dfrac{1.869s+0.5585}{s^2+2.663s+0.5838}$	∞	.75619
Routh Approximation	$\dfrac{0.6267s+0.1511}{s^2+0.847s+0.158}$	∞	.31715

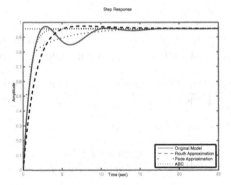

Fig. 6.7 Comparison of step responses for example 4.

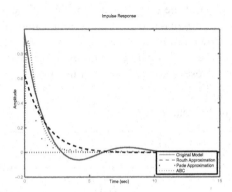

Fig. 6.8 Comparison of impulse responses for example 4.

6.5   Conclusion

Artificial Bee Colony optimization algorithm is applied for the model order reduction problem. Integral Square Error (ISE) and impulse response energy ($IREabs$) are minimized using ABC. Different order systems are taken into consideration and results are obtained. It is shown that different types of systems when reduced using the ABC, it gives better results in terms of ISE and $IREabs$ when compared with the results of other researchers. Based on our experiments it is suggested that ABC could be a better choice for model order reduction problem.

Acknowledgements. Author: Harish Sharma, acknowledges ABV- Indian Institute of Information Technology and Management Gwalior, for providing research grant to carry out this work.

References

[1] Mittal, A.K., Prasad, R., Sharma, S.P.: Reduction of linear dynamic systems using an error minimization technique. Journal of Institution of Engineers India, IE(I) Journal EL 84, 201–206 (2004)

[2] Aguirre, L.: The least squares padé method for model reduction. International Journal of Systems Science 23(10), 1559–1570 (1992)

[3] Chen, C., Shieh, L.: A novel approach to linear model simplification. International Journal of Control 8(6), 561–570 (1968)

[4] Chen, T., Chang, C., Han, K.: Reduction of transfer functions by the stability-equation method. Journal of the Franklin Institute 308(4), 389–404 (1979)

[5] Davison, J.E.: A method for simplifying linear dynamic systems. IEEE Trans. Automatic Control AC-11(1), 93–101 (1966)

[6] Gopal, M.: Control systems: principles and design. Tata McGraw-Hill, New York (2002)

[7] Gustafson, R.D.: A paper and pencil control system design. Trans. ASME J. Basic Engg., 329–336 (1966)

[8] Hickin, J., Sinha, N.: Canonical forms for aggregated models. International Journal of Control 27(3), 473–485 (1978)

[9] Hickin, J., Sinha, N.K.: Reduction of linear system by canonical forms. Electronics Letters 12(21), 551–553 (1976)

[10] Hu, X.: Ff-padé method of model reduction in frequency domain. IEEE Transactions on Automatic Control 32(3), 243–246 (1987)

[11] Hutton, M., Friedland, B.: Routh approximations for reducing order of linear, time-invariant systems. IEEE Transactions on Automatic Control 20(3), 329–337 (1975)

[12] Hwang, C.: Mixed method of routh and ise criterion approaches for reduced-order modeling of continuous-time systems. Journal of dynamic systems, measurement, and control 106, 353 (1984)

[13] Karaboga, D.: An idea based on honey bee swarm for numerical optimization. Techn. Rep. TR06, Erciyes Univ Press, Erciyes (2005)

[14] Karaboga, D., Akay, B.: A comparative study of artificial bee colony algorithm. Applied Mathematics and Computation 214(1), 108–132 (2009)

[15] Krishnamurthy, V., Seshadri, V.: Model reduction using the routh stability criterion. IEEE Transactions on Automatic Control 23(4), 729–731 (1978)

[16] Lamba, S., Gorez, R., Bandyopadhyay, B.: New reduction technique by step error minimization for multivariable systems. International Journal of Systems Science 19(6), 999–1009 (1988)

[17] Lucas, T.: Factor division: A useful algorithm in model reduction. IEE Proceedings D, IET, Control Theory and Applications 130, 362–364 (1983)

[18] Lucas, T.: Continued-fraction expansion about two or more points: a flexible approach to linear system reduction. Journal of the Franklin Institute 321(1), 49–60 (1986)

[19] Luus, R.: Optimal reduction of linear systems. Journal of the Franklin Institute 336(3), 523–532 (1999)

[20] Ouyang, M., Liaw, C.M., Pan, C.T.: Model reduction by power decomposition and frequency response matching. IEEE Trans. Auto Control AC 32, 59–62 (1987)

[21] Marshall, S.: Comments on viability of methods for generating stable reduced order models. IEEE Transactions on Automatic Control 28(5), 630–631 (1983)

[22] Mishra, R., Wilson, D.: A new algorithm for optimal reduction of multivariable systems. International Journal of Control 31(3), 443–466 (1980)

[23] Mukherjee, S., Mishra, R.: Order reduction of linear systems using an error minimization technique. Journal of the Franklin Institute 323(1), 23–32 (1987)

[24] Mukherjee, S., Mittal, R.: Order reduction of linear discrete systems using a genetic algorithm. Applied Mathematical Modelling 29(6), 565–578 (2005)

[25] Mukherjee, S., et al.: Model order reduction using response-matching technique. Journal of the Franklin Institute 342(5), 503–519 (2005)

[26] Gutman, P.O., Mannerfelt, C.F., Molander, P.: Contributions to the model reduction problem. IEEE Trans Automatic Control AC-27(2), 454–455 (1982)

[27] Pade, H.: Sur la representation approaches dune function par des fraction rationelles. Annales Scientifiques de l'Ecole Normale Supieure 9, 1–93 (1892)

[28] Pal, J.: Stable reduced-order padã â© approximants using the routh-hurwitz array. Electronics letters 15(8), 225–226 (1979)

[29] Pal, J.: An algorithmic method for the simplification of linear dynamic scalar systems. International Journal of Control 43(1), 257–269 (1986)

[30] Paynter, H.M., Takahashi, Y.: A new method of evaluation dynamic response of counter flow and parallel flow heat exchangers. Trans. ASME J. Dynam. Syst. Meas. Control 27, 749–753 (1956)

[31] Prasad, R., Pal, J.: Stable reduction of linear systems by continued fractions. Journal-Institution of Engineers India Part El Electrical Engineering Division 72, 113–113 (1991)

[32] Sanathanan Stanley, B., et al.: A comprehensive methodology for siso system order reduction. Journal of the Franklin Institute 323(1), 1–21 (1987)

[33] Shamash, Y.: Linear system reduction using pade approximation to allow retention of dominant modes. International Journal of Control 21(2), 257–272 (1975)

[34] Shoji, K., et al.: Model reduction for a class of linear dynamic systems. Journal of the Franklin Institute 319(6), 549–558 (1985)

[35] Singh, N.: Reduced order modeling and controller design. PhD thesis, Indian Institute of Technology Roorkee, India (2007)

[36] Sinha, N., Bereznai, G.: Optimum approximation of high-order systems by low-order models. International Journal of Control 14(5), 951–959 (1971)

[37] Sinha, N., Pille, W.: A new method for reduction of dynamic systems. International Journal of Control 14(1), 111–118 (1971)

[38] Sivanandam, S., Deepa, S.: A genetic algorithm and particle swarm optimization approach for lower order modelling of linear time invariant discrete systems. In: IC-CIMA, pp. 443–447. IEEE Computer Society, Los Alamitos (2007)

[39] Vilbe, P., Calvez, L.: On order reduction of linear systems using an error minimization technique. Journal of the Franklin Institute 327(3), 513–514 (1990)

[40] Yang, S., Luus, R.: Optimisation in linear system reduction. Electronics Letters 19(16), 635–637 (1983)

Chapter 7
Lamps: A Test Problem for Cooperative Coevolution

Alberto Tonda, Evelyne Lutton, and Giovanni Squillero

Abstract. We present an analysis of the behaviour of Cooperative Co-evolution algorithms (CCEAs) on a simple test problem, that is the optimal placement of a set of lamps in a square room, for various problems sizes. Cooperative Co-evolution makes it possible to exploit more efficiently the artificial Darwinism scheme, as soon as it is possible to turn the optimisation problem into a co-evolution of inter-dependent sub-parts of the searched solution. We show here how two cooperative strategies, Group Evolution (GE) and Parisian Evolution (PE) can be built for the lamps problem. An experimental analysis then compares a classical evolution to GE and PE, and analyses their behaviour with respect to scale.

7.1 Introduction

Cooperative co-evolution algorithms (CCEAs) share common characteristics with standard artificial Darwinism-based methods, i.e. Evolutionary Algorithms (EAs), but with additional components that aim at implementing collective capabilities. For optimisation purpose, CCEAs are based on a specific formulation of the problem where various inter- or intra-population interaction mechanisms occur. Usually,

Alberto Tonda
ISC-PIF, CNRS CREA, UMR 7656, 57-59 rue Lhomond, Paris France
e-mail: Alberto.Tonda@gmail.com

Evelyne Lutton
AVIZ Team, INRIA Saclay - Ile-de-France, Bat 490, Université Paris-Sud, 91405 ORSAY Cedex, France
e-mail: Evelyne.Lutton@inria.fr

Giovanni Squillero
Politecnico di Torino - Dip. Automatica e Informatica, C.so Duca degli Abruzzi 24 - 10129 Torino - Italy
e-mail: Giovanni.Squillero@polito.it

D.A. Pelta et al. (Eds.): NICSO 2011, SCI 387, pp. 101–120, 2011.
springerlink.com

these techniques are efficient as optimiser when the problem can be split into smaller interdependent subproblems. The computational effort is then distributed onto the evolution of smaller elements of similar or different nature, that aggregates to build a global solution.

Cooperative co-evolution is increasingly becoming the basis of successful applications [1, 6, 8, 15, 19], including learning problems, see for instance [3]. These approaches can be shared into two main categories: co-evolution process that happens between a fixed number of separate populations [5, 13, 14] or within a single population[7, 10, 18].

The design and fine tuning of such algorithms remain however difficult and strongly problem dependent. A critical question is the design of simple test problem for CCEAs, for benchmarking purpose. A first test-problem based on Royal Road Functions has been proposed in [12]. We propose here another simple problem, the Lamps problem, for which various instances of increasing complexity can be generated, according to a single ratio parameter. We show below how two CCEAs can be designed and compared against a classical approach, whith a special focus on scalability.

The paper is organised as follows: the Lamps problem is described in Section 7.2, then the design of two cooperative co-evolution strategies, Parisian Evolution and Group Evolution, is detailed in Sections 7.3 and 7.4. The experimental setup is described in Section 7.5: three strategies are tested, a classical genetic programing approach, (CE for Classical Evolution), the Group Evolution (GE) and the Parisian Evolution (PE). All methods are implemented using the μGP toolkit [16]. Results are presented and analysed in Section 7.6, and conclusions and future work are given in Section 7.7.

7.2   The Lamps problem

The optimisation problem chosen to test cooperative coevolution algorithms requires to find the best placement for a set of lamps, so that a target area is fully brightened with light. The minimal number of lamps needed is unknown, and heavily depends on the topology of the area. All lamps are alike, modeled as circles, and each one may be evaluated separately with respect to the final goal. In the example, the optimal solution requires 4 lamps (Figure 7.1, left): interestingly, when examined independently, all lamps in the solution waste a certain amount of light outside the target area. However, if one of the lamps is positioned to avoid this undesired effect, it becomes impossible to lighten the remaining area with the three lamps left (Figure 7.1, right). Since lamps are simply modeled as circles, the problem may also be seen as using the circles to completely cover the underlying area, as efficiently as possible.

This apparently simple benchmark exemplifies a common situation in real-world applications: many problems have an optimal solution composed of a set of homogeneous elements, whose individual contribution to the main solution can be evaluated

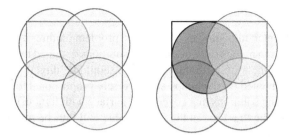

Fig. 7.1 Placement of a set of lamps. The aim is to enlighten all the square area. It is interesting to notice how a solution where some of the light of each lamp is wasted outside the area (left) overall performs better than a solution where the grayed lamp maximises its own performance (right).

separately. Note that, in this context, *homogeneous* is used to label elements sharing the same base structure.

A similar toy problem has been sketched in [17], but the structure of the benchmark has been improved and parametrised, and the fitness function has been modified, to increase the complexity and the number of local optima on the fitness landscape.

7.2.1   Size of the Problem

It is intuitive that the task of enlightening a room with a set of lamps can be more or less difficult, depending on the size of the room and the cone of light of each lamp. If small sources of light are used to brighten a large room, surely a greater number of lamps will be required, and the number of possible combinations will increase dramatically.

With these premises, the complexity of the problem can thus be expressed by the ratio between the surface to be enlightened and the maximum area enlightened by a single lamp:

$$problem\_size = \frac{area\_room}{area\_lamp}$$

as this ratio increases, finding an optimal solution for the problem will become harder.

It is interesting to notice how variations in the shape of the room could also influence the complexity of the task: rooms with an irregular architecture may require more intricate lamp placements. However, finding a dependency between the shape of the room and the difficulty of the problem is not trivial, and results might be less intuitive to analyze. For all these reasons, the following experiments will feature square rooms only.

7.2.2 Fitness Value

Comparing different methodologies on the same problem requires a common fitness function, to be able to numerically evaluate the solutions obtained by each approach.

Intuitively, the fitness of a candidate solution should be directly proportional to the area fully brightened by the lamps and inversely proportional to the number of lamps used, favoring solutions that cover more surface with light using the minimal number of lamps. The first term in the fitness value will thus be proportional to the ratio of the area enlightened by the lamps,

$$\frac{area\_enlightened}{total\_area}$$

To increase the complexity of the fitness landscape, a further contribution is added: the area brightened by more than one lamp is to be minimised, in order to have as little overlapping as possible. The second term will be then proportional to:

$$-\frac{area\_overlap}{total\_area}$$

It is interesting to note that minimising the overlap also implies an optimisation of the number of lamps used, since using a greater number would lead to more overlapping areas.

The final fitness function will then be:

$$fitness = \frac{area\_enlightened}{total\_area} - W \cdot \frac{area\_overlap}{total\_area} =$$

$$= \frac{area\_enlightened - W \cdot area\_overlap}{total\_area}$$

where W is a weight associated to the relative importance of the overlapping, set by the user.

Using this function with $W = 1$, fitness values will range between $(0, 1)$, but it is intuitive that it is impossible to reach the maximum value: by problem construction, overlapping and enlightenment are inversely correlated, and even good solutions will feature overlapping areas and/or parts not brightened. This problem is actually multi-objective, the fitness function we propose corresponds to a compromise between the two objectives.

7.3 Parisian Evolution

Initially designed to address the inverse problem for Iterated Function System (IFS), a problem related to fractal image compression[7], this scheme has been successfully applied in various real world applications: in stereovision [4], in photogrammetry [9, 11], in medical imaging [18], for Bayesian Network Structure learning [2], in data retrieval[10].

Parisian Evolution (PE) is based on a two-level representation of the optimisation problem, meaning that an individual in a Parisian population represents only a part of the solution. An aggregation of multiple individuals must be built to complete a meaningful solution to the problem. This way, the co-evolution of the whole population (or of a major part of it) is favoured over the emergence of a single best individual, as in classical evolutionary schemes.

This scheme distributes the workload of solution evaluations at two levels. Light computations (e.g. existence conditions, partial or approximate fitness) can be done at the individual's level (local fitness), while the complete calculation (i.e. global fitness) is performed at the population level. The global fitness is then distributed as a bonus to individuals who participate the global solution. A Parisian scheme has all the features of a classical EA (see figure 7.2) with the following additional components:

- A grouping stage at each generation, that selects individuals that are allowed to participate to the global solution.
- A redistribution step that rewards the individuals who participate to the global solution : their bonus is proportional to the global fitness.
- A sharing scheme, that avoids degenerate solutions where all individuals are identical.

Efficient implementations of the Parisian scheme are often based on partial redundancies between local and global fitness, as well as clever exploitation of computational shortcuts. The motivation is to make a more efficient use of the evolution of the population, and reduce the computational cost. Successful applications of such a scheme usually rely on a lower cost evaluation of the partial solutions (i.e. the individuals of the population), while computing the full evaluation only once at each generation or at specified intervals.

7.3.1   Implementation of the Lamps Problem

For the lamps problem, the PE has been implemented as follows. An individual represents a lamp, its genome is its (x, y) position, plus a third element, e, that can assume values 0 or 1 (on/off switch). Lamps with $e = 1$ are "on" (expressed) and contribute to the global solution, while lamps with $e = 0$ do not.

Global fitness is computed as described in subsection 7.2.2. In generation 0 the global solution is computed simply considering the individuals with $e = 1$ among the μ initial ones. Then, at each step, λ individuals are generated. For each new individual with $e = 1$, its potential contribution to the global solution is computed. Before evaluation, a random choice ($p = 0.5$) is performed: new individuals are either considered in addition to or in replacement of the existing ones.

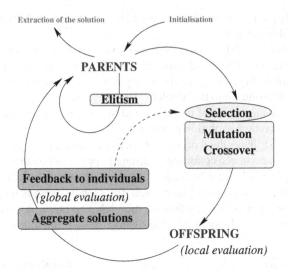

Fig. 7.2 A Parisian EA: a monopopulation cooperative-coevolution. Partial evaluation (local fitness) is applied to each individual, while global evaluation is performed once a generation.

If individuals are considered for addition, the contribution to the global solution of each one is computed. Otherwise, the less performing among the old individuals is removed, and only then the contribution to the global solution of each new individual is evaluated. If the addition or replacement of the new individuals leads to an improvement over the previous global fitness, the new individual selected is rewarded with a high local fitness value (*local_fitness* = 2), together with all the old individuals still contributing to the global solution. New expressed individuals ($e = 1$) that are not selected for the global solution are assigned a low fitness value (*local_fitness* = 0). Non-expressed individuals ($e = 0$) have an intermediate fitness value (*local_fitness* = 1).

Sharing follows the simple formula

$$fitness\_sharing(I_k) = \frac{local\_fitness(I_k)}{\sum_{i=0}^{individuals} sharing(I_k, I_{i \neq k})}$$

with

$$sharing(I_1, I_2) = \begin{cases} 1 - \frac{d(I_1, I_2)}{2 \cdot lamp\_radius} & d(I_1, I_2) < 2 \cdot lamp\_radius \\ 0 & d(I_1, I_2) \geq 2 \cdot lamp\_radius \end{cases}$$

Lamps that have a relatively low number of neighbours will be preferred for selection over lamps with a bigger number of neighbours. In this implementation, sharing is computed only for expressed lamps ($e = 1$) and used only when selecting the less performing individual to be removed from the population.

7.4   Group Evolution

Group Evolution (GE) is a novel generational cooperative coevolution concept presented in [17]. The approach uses a population of partial solutions, and exploits non-fixed sets of individuals called *groups*. GE acts on individuals and groups, managing both in parallel. During the evolution, individuals are optimised as in a common EA, but concurrently groups are also evolved. The main peculiarity of GE is the absence of *a priori* information about the grouping of individuals.

At the beginning of the evolutionary process, an initial population of individuals is randomly created on the basis of a high-level description of a solution for the given problem. Groups at this stage are randomly determined, so that each individual can be included in any number of different groups, but all individuals are part of at least one group.

Population size $\mu_{individuals}$ is the maximum number of individuals in the population, and it is set by the user before the evolution starts. The number of groups μ_{groups}, the minimum and maximum size of the groups are set by the user as well. Figure 7.3 (left) shows a sample population where minimum group size is 2, and maximum group size is 4.

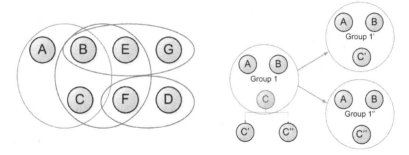

Fig. 7.3 Individuals and Groups in a sample population of 8 individuals (left). While individual A is part of only one group, Individual B is part of 3 different groups. On the right, the effect of a Individual Genetic Operator, applied to individual C. Since individual C is part of Group 1, two groups are created and added to the population.

7.4.1   Generation of New Individuals and Groups

GE exploits a generational approach: at each evolutionary step, a number of genetic operators is applied to the population. Genetic operators can act on both individuals and groups, and produce a corresponding offspring, in form of individuals and groups.

The offspring creation phase comprehends two different actions at each generation step (see Figure 7.4):

1. Application of *group genetic operators*;
2. Application of *individual genetic operators*.

Each time a genetic operator is applied to the population, parents are chosen and offspring is generated. The children are added to the population, while the original parents are unmodified. Offspring is then evaluated, while it is not compulsory to reconsider the fitness value of the parents again. It is important to notice that the number of children produced at each evolutionary step is not fixed: each genetic operator can have any number of parents as input and produce in output any number of new individuals and groups. The number and type of genetic operators applied at each step can be set by the user.

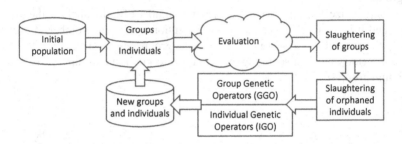

Fig. 7.4 Schema of Group Evolution algorithm.

7.4.1.1   Group Genetic Operators

Group Genetic Operators (GGOs) work on the set of groups. Each operator needs a certain number of groups as parents and produces a certain number of groups as offspring that will be added to the population. GGOs implemented in our approach are:

1. **crossover:** generates offspring by selecting two individuals, one from parent group A and one from parent group B. Those individuals are switched, creating two new groups;
2. **adding-mutation:** generates offspring by selecting one or more individuals from the population and a group. Chosen individuals are added (if possible) to the parent group, creating a single new group;
3. **removal-mutation:** generates offspring by selecting a group and one or more individuals inside it. Individuals are removed from the parent group.
4. **replacement-mutation:** generates offspring by selecting a group and one or more individuals inside it. Individuals are removed from the parent group, and replaced by other individuals selected from the population.

Parent groups are chosen via tournament selection.

7.4.1.2   Individual Genetic Operators

Individual Genetic Operators (IGOs) operate on the population of individuals, very much like they are exploited in usual GA. The novelty of GE is that for each individual produced as offspring, new groups are added to the group population. For each group the parent individual was part of, a copy is created, with the offspring taking the place of the parent.

This approach, however, could lead to an exponential increase in the number of groups, as the best individuals are selected by both GGOs and IGOs. To keep the number of groups under a strict control, we choose to create a copy only of the highest-fitness groups the individual was part of.

IGOs select individuals by a tournament selection in two parts: first, a group is picked out through a tournament selection with moderate selective pressure; then an individual in the group is chosen with low selective pressure. The actual group and the highest-fitness groups the individual is part of are cloned once for each child individual created: in each clone group the parent individual is replaced with a child. An example is given in Figure 7.3 (right): an IGO, selects individual C as a parent. The chosen individual is part of only one group, Group 1. The IGO produces two children individuals: since the parent was part of a group, a new group is created for each new individual generated. The new groups (Group 1' and Group 1") are identical to Group 1, except that individual C is replaced with one of its children, C' in Group 1' and C" in Group 1" respectively.

The aim of this process is to select individuals from well-performing groups to create new groups with a slightly changed individual, in order to explore the a near area in the solution space.

7.4.2   Evaluation

During the evaluation phase, a fitness value is associated to each group: the fitness value is a number that measures the goodness of the candidate solutions with respect to the given problem. When a group is evaluated, a fitness value is also assigned to all the individuals composing it. Those values reflect the goodness of the solution represented by the single individual and have the purpose to help discriminate during tournament selection for both IGOs and GGOs.

An important strength of the approach resides in the evaluation step: if there is already a fitness value for an individual that is part of a new group, it is possible to take it into account instead of re-evaluating all the individuals in the group. This feature can be exceptionally practical when facing a problem where the evaluation of a single individual can last several minutes and the fitness of a group can be computed without examining simultaneously the performance of the individuals composing it. In that case, the time-wise cost of both IGOs and GGOs becomes very small.

7.4.3 Slaughtering

After each generation step, the group population is resized. The groups are ordered
fitness-wise and the worst one is deleted until the desired population size is reached.
Every individual keeps track of all the groups it belongs to in a set of references.
Each time a group ceases to exist, all its individuals remove it from their set of
references. At the end of the group slaughtering step, each individual that has an
empty set of references, and is therefore not included in any group, is deleted as
well.

7.4.4 Implementation of the Lamps Problem

For the lamps problem, GE has been implemented as follows. One individual represents a lamps, its genome is the (x, y) position. The fitness of a single individual,
which must be independent from all groups it is part of, is simply the area of the
room it enlightens. The group fitness is computed as described in Subsection 7.2.2.

7.5 Experimental Setup

Before starting the experiments on the cooperative coevolution algorithms, a series
of 10 runs for each *problem_size* of a classical evolutionary algorithm is performed,
to better understand the characteristics of the problem and to set parameters leading
to a fair comparison. The genome of a single individual is a set of lamps, modeled
as an array of N couples of values $((x_1, y_1), (x_2, y_2), ..., (x_N, y_N))$, where each (x_i, y_i)
describes the position of individual i in the room. The algorithm uses a classical
$(\mu + \lambda)$ evolutionary paradigm, with $\mu = 20$ and $\lambda = 10$, probability of crossover
0.2 and probability of mutation 0.8. Each run lasts 100 generations.

By examining these first experimental results, it is possible to reach the following
conclusions: good solutions for the problem use a number of lamps in the range
$(problem\_size, 3 \cdot problem\_size)$; and, as expected, as the ratio grows, the fitness
value of the best individual at the end of the evolution tends to be lower.

In order to perform a comparison with GE and PE as fair as possible, the stop
condition for each algorithm will be set as the average number of single lamp evaluations performed by the classical evolutionary algorithm for each *problem_size*. In
this way, even if the algorithms involved have significantly different structures, the
computational effort is commensurable. Table 7.1 summarises the results.

Table 7.1 Average evaluations for each run of the classical evolutionary algorithm.

Problem size	Evaluations	Average best fitness
3	3,500	0.861
5	5,000	0.8216
10	11,000	0.7802
20	22,000	0.6804
100	120,000	0.558

7.5.1   PE Setup

Due to the observation of the CE runs, $\mu = 3 \cdot problem\_size$, while $\lambda = \mu/2$. The probability of mutation is 0.8 and the probability of crossover is 0.2.

7.5.2   GE Setup

The number of groups in the population is fixed, $\mu_{groups} = 20$, as is the number of genetic operators selected at each step $\lambda = 10$. The number of individuals in each group is set to vary in the range $(problem\_size, 3 \cdot problem\_size)$. Since the number of individuals in each group will grow according to the problem size, the number of individuals in the population will be $\mu_{individuals} = 3 \cdot \mu_{groups} \cdot problem\_size$. The probability of mutation is 0.8, while the probability of crossover is 0.2, for both individuals and groups.

7.5.3   Implementation in μGP

The two CCEAs used in the experience have been implemented using μGP [16], an evolutionary toolkit developed by CAD group of Politecnico di Torino. Exploiting μGP's flexible structure, with the fitness evaluator completely independent from the evolutionary core, the great number of adjustable parameters, and the modular framework composed of clearly separated C++ classes, it is possible to obtain the behavior of very different EAs.

In particular, to implement PE, it is sufficient to operate on the fitness evaluator, setting the environment to evaluate the whole population and the offspring at each step. Obtaining GE behavior is slightly more complex, and requires the addition of new classes to manage groups. CE is simply μGP standard operation mode.

7.6   Results and Analysis

In a series of experiments, 100 runs of each evolutionary approach are executed, for a set of meaningful values of *problem_size*. To exploit a further measurement of comparison, the first occurrence of an acceptable solution also appears in the results: here an *acceptable solution* is defined as a global solution with at least 80% of the final fitness value obtained by the CE. Table 7.2 summarises the results for significant values of *problem_size*. For each evolutionary algorithm are reported the results reached at the end of the evolution: average fitness value, average enlightenment percentage of the room, average number of lamps, and average number of lamps evaluated before finding an acceptable solution (along with the standard deviation for each value).

Table 7.2 Average results for 100 runs, for each evolutionary approach. *Due to the times involved, data for problem size 100 is computed on 10 runs.

Problem size	Evolution	Avg. fitness	Std dev	Avg. enlight.	Std dev	Avg. lamps	Std dev	Avg. overlap	Std dev	Avg. lamps before acceptable	Std dev
3	Classical Evolution	0.861	0.0149	0.8933	0.0212	4	0	0.0323	0.0168	313.2	126
	Group Evolution	0.8764	0.0498	0.8963	0.0533	3.75	0.435	0.0198	0.0103	267.32	138.1
	Parisian Evolution	0.8355	0.064	0.8945	0.0439	4.02	0.3344	0.059	0.0502	316.9	262.2
5	Classical Evolution	0.7802	0.023	0.8574	0.04	6.2	0.64	0.0772	0.0278	572.7	215.38
	Group Evolution	0.8136	0.0241	0.8537	0.0349	6.03	0.7372	0.0401	0.0166	741.36	221.86
	Parisian Evolution	0.7825	0.03	0.8803	0.0335	6.96	0.6936	0.0978	0.0395	511.35	373.85
10	Classical Evolution	0.7487	0.0149	0.834	0.0235	11.3	0.62	0.0853	0.0216	1,779.8	407.4
	Group Evolution	0.7532	0.0178	0.8132	0.0255	10.66	0.6336	0.0599	0.0215	1,836.87	412.08
	Parisian Evolution	0.7791	0.0221	0.8847	0.0207	12.84	0.8184	0.1055	0.0274	1,018.47	546.16
20	Classical Evolution	0.6804	0.0117	0.7749	0.0148	20.6	0.72	0.0946	0.0123	3,934.7	702.24
	Group Evolution	0.697	0.0127	0.7837	0.0147	20.49	0.5978	0.0867	0.0142	4602.1	1156.5
	Parisian Evolution	0.7624	0.0147	0.8762	0.0168	23.57	1.053	0.1138	0.0177	2,759.28	965.45
100*	Classical Evolution	0.558	0.0049	0.7334	0.0062	102.9	1.88	0.1755	0.0093	29,567.3	4,782.96
	Group Evolution	0.5647	0.0057	0.7309	0.0073	101.5	1.7	0.1662	0.0072	30,048	8,876.8
	Parisian Evolution	0.6867	0.0073	0.8708	0.0078	117.2	3.2	0.1841	0.0134	5,318.9	332.72

For each *problem_size*, a box plot is provided in Figure 7.5. It is noticeable how the performance of each evolutionary algorithm is very close for small values of *problem_size*, while GE and PE gain the upper hand when the complexity increases.

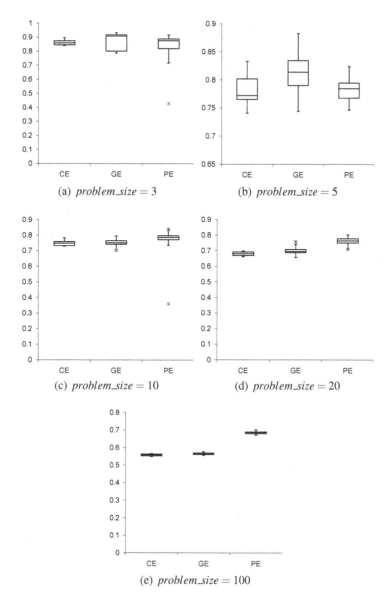

Fig. 7.5 Box plots of fitness values for each *problem_size*.

In particular, PE obtains the best performance from *problem_size* = 10 onwards. The extra information inserted allows the approach to obtain high enlightenment percentages even when the complexity of the task increases, as shown in Figure 7.6.

On the other hand, GE obtains enlightenment percentages close to CE, but on the average it uses a lower number of lamps, that leads to a lower overlap, as it is noticeable in Figure 7.8 and 7.7.

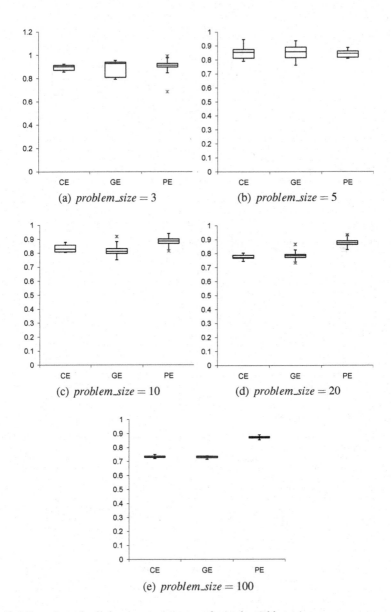

Fig. 7.6 Box plots of enlightenment percentage for each *problem_size*.

When dealing with the number of lamp evaluations needed before reaching what is defined an *acceptable* solution, PE is again in the lead, see Figure 7.9.

In Figure 7.10, a profile of the best run for *problem_size* = 100 for each algorithm is reported. PE enjoys a rapid growth in the first stages of the evolution, thanks to

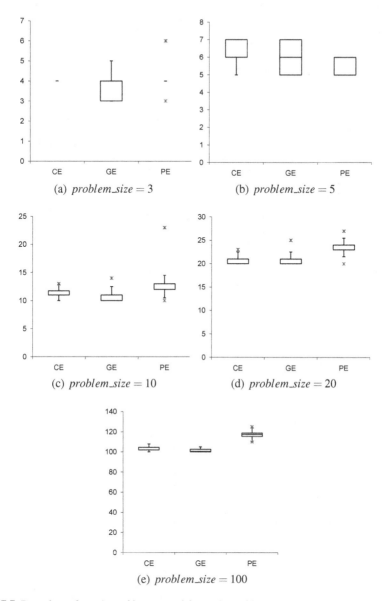

Fig. 7.7 Box plots of number of lamps used for each *problem_size*.

the extra information it can make use of, while GE proves more efficient than CE in the last part of the evolution, where exploitation becomes more prevalent.

As it is noticeable in Figure 7.11, while the number of lamps evaluated before reaching an acceptable solution grows more than linearly for GE and CE, PE shows a less steep growth. On the other hand, GE presents the lowest overlap for all values of *problem_size* (Figure 7.12).

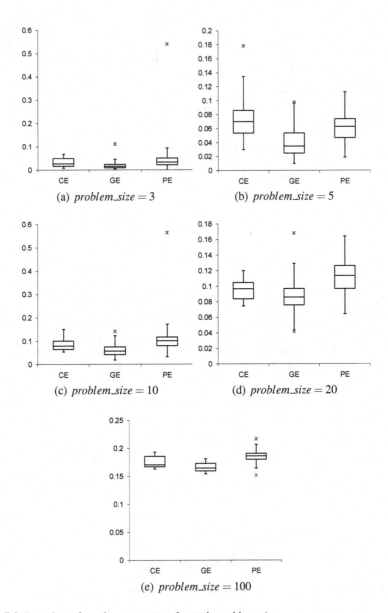

Fig. 7.8 Box plots of overlap percentage for each *problem_size*.

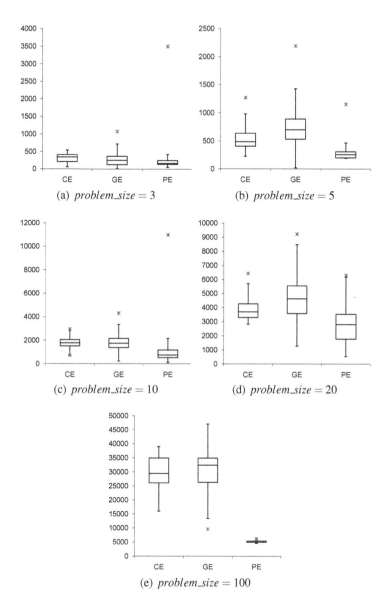

Fig. 7.9 Box plots of number of lamps evaluated before reaching an acceptable solution, for each *problem_size*.

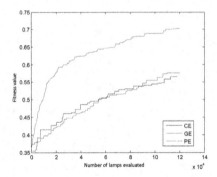

Fig. 7.10 Profile of the best run for each evolutionary algorithm at *problem_size* = 100.

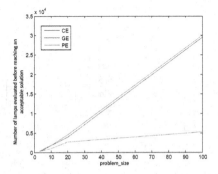

Fig. 7.11 Average number of lamps evaluated before reaching an acceptable solution, for different values of *problem_size*.

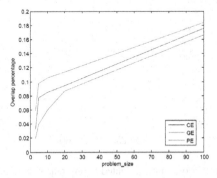

Fig. 7.12 Average overlap, for different values of *problem_size*.

7.7   Conclusion and Future Work

The Lamps benchmark has the major advantage to provide a set of toy problems that are simple, and for which the scale can be characterised with a single real value (the surface ratio between room and lamp sizes). This formulation is very convenient to get some insight on the behaviour of algorithms with respect to scale.

The intuition that guided the development of Group Evolution and Parisian Evolution has been confronted to experimental analysis, to yield the following conclusions: Parisian Evolution is the most efficient approach in terms of computational expense, and scalability; Group Evolution yields better and more precise results, in a computational time that is similar to Classical Evolution.

These differences can be explained by the nature of *a priori* information that has been injected into the algorithms. Parisian Evolution relies on a deterministic algorithm for selecting the group of lamps that are used as the current global solution at each generation, while Group Evolution does not make any assumption on it and let evolution decide which lamps can be grouped to build various solutions.

In some sense Parisian Evolution is making a strong algorithmic choice for the grouping stage, that acts as a constraint on the evolution of the population of lamps. It has the major advantage to reduce the complexity of the problem by providing solutions from the evolution of simpler structures (lamps instead of groups of lamps for Group Evolution or for Classical Evolution). It may be considered as a "quick and dirty" approach.

Future work on this topic will investigate hybridisation between Parisian and Group Evolution, i.e. running a population of elements in the Parisian mode to rapidly get a good rough solution, and using the Group scheme to refine it. A more realistic modelisation of the lamp's light could also be used, taking into account the gradual fading from the source; this approach would shift the problem from enlightening the room to ensure that each point of it receives at least a certain amount of light.

References

[1] Amaya, J.E., Cotta, C., Leiva, A.J.F.: A memetic cooperative optimization schema and its application to the tool switching problem. In: Schaefer, R., Cotta, C., Kołodziej, J., Rudolph, G. (eds.) PPSN XI. LNCS, vol. 6238, pp. 445–454. Springer, Heidelberg (2010)

[2] Barrire, O., Lutton, E., Wuillemin, P.H.: Bayesian network structure learning using cooperative coevolution. In: Genetic and Evolutionary Computation Conference, GECCO 2009 (2009)

[3] Bongard, J., Lipson, H.: Active coevolutionary learning of deterministic finite automata. Journal of Machine Learning Research 6, 1651–1678 (2005)

[4] Boumaza, A.M., Louchet, J.: Dynamic flies: Using real-time parisian evolution in robotics. In: Boers, E.J.W., Gottlieb, J., Lanzi, P.L., Smith, R.E., Cagnoni, S., Hart, E., Raidl, G.R., Tijink, H. (eds.) EvoIASP 2001, EvoWorkshops 2001, EvoFlight 2001, EvoSTIM 2001, EvoCOP 2001, and EvoLearn 2001. LNCS, vol. 2037, pp. 288–297. Springer, Heidelberg (2001)

[5] Bucci, A., Pollacj, J.B.: On identifying global optima in cooperative coevolution. In: Proceedings of the 2005 Conference on Genetic and Evolutionary, Washington DC, USA (2005)

[6] Chen, W., Weise, T., Yang, Z., Tang, K.: Large-scale global optimization using cooperative coevolution with variable interaction learning. In: Schaefer, R., Cotta, C., Kołodziej, J., Rudolph, G. (eds.) PPSN XI. LNCS, vol. 6239, pp. 300–309. Springer, Heidelberg (2010)

[7] Collet, P., Lutton, E., Raynal, F., Schoenauer, M.: Polar ifs + parisian genetic programming = efficient ifs inverse problem solving. Genetic Programming and Evolvable Machines Journal 1(4), 339–361 (2000)

[8] De Jong, E.D., Stanley, K.O., Wiegand, R.P.: Introductory tutorial on coevolution. In: Proceedings of the 2007 GECCO Conference Companion on Genetic and Evolutionary Computation, London, United Kingdom (2007)

[9] Dunn, E., Olague, G., Lutton, E.: Automated photogrammetric network design using the parisian approach. In: EvoIASP 2005, Lausanne, nominated for the best paper Award (2005)

[10] Landrin-Schweitzer, Y., Collet, P., Lutton, E.: Introducing lateral thinking in search engines. GPEM, Genetic Programming an Evolvable Hardware Journal 1(7), 9–31 (2006); Banzhaf, W., et al (eds.)

[11] Lutton, E., Olague, G.: Parisian camera placement for vision metrology. Pattern Recognition Letters 27(11), 1209–1219 (2006)

[12] Ochoa, G., Lutton, E., Burke, E.K.: Cooperative royal road functions. In: Evolution Artificielle, Tours, France, October 29-31 (2007)

[13] Panait, L., Luke, S., Harrison, J.F.: Archive-based cooperative coevolutionary algorithms. In: Proceedings of the 8th Annual Conference on Genetic and Evolutionary Computation, Seattle, Washington, USA (2006)

[14] Popovici, E., De Jong, K.: The effects of interaction frequency on the optimization performance of cooperative coevolution. In: Proceedings of the 8th Annual Conference on Genetic and Evolutionary Computation, Seattle, Washington, USA (2006)

[15] Potter, M.A., Couldrey, C.: A cooperative coevolutionary approach to partitional clustering. In: Schaefer, R., Cotta, C., Kołodziej, J., Rudolph, G. (eds.) PPSN XI. LNCS, vol. 6238, pp. 374–383. Springer, Heidelberg (2010)

[16] Sanchez, E., Schillaci, M., Squillero, G.: Evolutionary Optimization: the μGP toolkit, 1st edn. Springer, Heidelberg (2011)

[17] Sanchez, E., Squillero, G., Tonda, A.: Group evolution: Emerging synergy through a coordinated effort. In: Proceedings of the 2011 IEEE Congress of Evolutionary Computation, CEC (2011)

[18] Vidal, F.P., Louchet, J., Rocchisani, J.-M., Lutton, É.: New genetic operators in the fly algorithm: Application to medical PET image reconstruction. In: Di Chio, C., Cagnoni, S., Cotta, C., Ebner, M., Ekárt, A., Esparcia-Alcazar, A.I., Goh, C.-K., Merelo, J.J., Neri, F., Preuß, M., Togelius, J., Yannakakis, G.N. (eds.) EvoApplicatons 2010. LNCS, vol. 6024, pp. 292–301. Springer, Heidelberg (2010)

[19] Wiegand, R.P., Potter, M.A.: Robustness in cooperative coevolution. In: Proceedings of the 8th Annual Conference on Genetic and Evolutionary Computation, Seattle, Washington, USA (2006)

Chapter 8
A Game Theoretic Approach to Community Detection in Social Networks

Rodica Ioana Lung, Anca Gog, and Camelia Chira

Abstract. The problem of detecting community structures in social networks is a complex problem of great importance in sociology, biology and computer science. Communities are characterized by dense intra-connections and comparatively sparse inter-cluster connections. The community detection problem is empirically formulated from a game theoretic point of view and solved using a Crowding based Differential Evolution algorithm adapted for detecting Nash equilibria of noncooperative games. Numerical results indicate the potential of this approach.

8.1 Introduction

The science of networks has recently become a field by itself. There was always a great interest in studying networks but different fields like mathematics, neuroscience, biology, epidemiology, sociology, social-psychology, economy and many others have studied networks independently. The new science of networks aroused starting with the papers [1, 24].

The complexity of real networks is given by non-trivial topological features (skewed degree distribution, high clustering coefficient, hierarchical structure) and also refers to the fact that local interactions between simple components bring forth a complex global behaviour in a non-trivial manner. A great review on the field of complexity and network thinking can be found in [11].

Rodica Ioana Lung
Babeş-Bolyai University of Cluj Napoca
e-mail: rodica.lung@econ.ubbcluj.ro

Anca Gog
Babeş-Bolyai University of Cluj Napoca
e-mail: anca@cs.ubbcluj.ro

Camelia Chira
Babeş-Bolyai University of Cluj Napoca
e-mail: cchira@cs.ubbcluj.ro

D.A. Pelta et al. (Eds.): NICSO 2011, SCI 387, pp. 121–131, 2011.
springerlink.com   © Springer-Verlag Berlin Heidelberg 2011

The detection of community structure in complex networks is a challenging problem which attracted a lot of attention in recent years. A community in a network is a group of nodes densely connected but sparsely connected with the nodes belonging to other communities. Communities unfold an organization of the network that might not be revealed from empirical tests and allow the deduction of certain relationships between nodes. Identifying community structure can lead to better understanding and exploiting real networks, as many complex networks in nature, society and technology exhibit community structures: identifying locations for dedicated mirror servers in order to increase the performance of the WWW; creation of recommendation systems by identifying groups of customers with similar interests; preventing crime by identifying hidden communities on the WWW; parallel computing; vaccination of hubs in the case of developing epidemics and limited vaccinating resources; identifying groups of similar items in social, biochemical and neural networks could simplify the functional analysis of the networks. Generally, a better understanding and visualization of network structure can be facilitated by efficient community structure detection.

The problem of detecting community structure has been recognized to be NP-hard [3] and exact methods are not suitable for approaching it. An extensive description of existing methods for community detection can be found in [3]. These methods can be classified as traditional, divisive, modularity-based, spectral, based on statistical inference or alternative. Some of the most recently proposed approaches include stochastic blockmodels [4], message passing [5], combining defensive/offensive label propagation [23]. The drawback of these techniques is the computational complexity that makes them unsuitable for large networks.

Bio-inspired techniques provide promising algorithms for solving NP-hard problems. Several evolutionary approaches to the problem of detecting community structures in complex networks have been proposed in the literature [2, 15, 21]. In [21], the authors propose a genetic algorithm based on the network modularity [13] as fitness function. Individuals are represented as arrays of integers, where each position corresponds to a node in the network and each value in the array represents the id of the corresponding community. In [15], a genetic algorithm called *GA-Net* is proposed to discover communities in networks. Individuals are represented as array of integers of size N (number of nodes in the network) where each position $i, i = 1...N$ has a value $j, j = 1...N$ with the meaning that nodes i and j will be placed in the same cluster. The concept of community score - a quality measure of a partitioning favouring highly intra-connected and sparsely inter-connected communities - is engaged as fitness function. In [2] a collaborative evolutionary algorithm based on the representation and fitness function from [15] is proposed. The main features of the proposed evolutionary approach refer to collaborative selection and recombination which sustain a balanced search process. Each individual has information about its best ancestor and the global optimal and worst solutions already detected. Selection of parents considers individuals which are not genetically related while the recombination operator takes into account the intermediary global best and worst solutions.

In this paper a new game theoretic approach to the community detection problem is presented. The problem is reconsidered as a game in which each node is a player.

Players choose to belong to a community according to a fitness/payoff that has to be maximized. The assertion is that the an equilibrium of this game represents a community structure of the network. The paper is organized as follows: Section 2 contains some prerequisites for game theory; the proposed approach for community detection is presented in Section 3; the evolutionary approach of Nash equilibria detection is briefly described in Section 4; numerical results in Section 5 and conclusions and directions for future research in Section 6.

8.2   Game Theory - Prerequisites

A finite strategic game Γ [14] is defined by the set of players involved, their possible actions and corresponding payoffs, $\Gamma = ((N, S_i, u_i), i = 1, n)$ where:

- N represents the set of players, $N = \{1, ..., n\}$, n is the number of players;
- for each player $i \in N$, S_i represents the set of actions available to him, $S_i = \{s_{i_1}, s_{i_2}, ..., s_{i_{m_i}}\}$ where m_i represents the number of strategies available to player i and $S = S_1 \times S_2 \times ... \times S_N$ is the set of all possible situations of the game;
- for each player $i \in N$, $u_i : S \to \mathbb{R}$ represents the payoff function.

The most common concept of solution for a non cooperative game is the concept of Nash equilibrium (NE) [10, 12]. A collective strategy $s \in S$ for the game Γ represents a Nash equilibrium if no player has anything to gain by changing only his own strategy. More formal, the strategy s^* is a Nash equilibrium if and only if the inequality

$$u_i(s_i, s^*_{-i}) - u_i(s^*) \le 0, \ \forall s_i \in S_i, \forall i \in N$$

holds, where by (s_{i_j}, s^*_{-i}) the strategy profile obtained from s^* by replacing the strategy of player i with s_{i_j} i.e.

$$(s_{i_j}, s^*_{-i}) = (s^*_1, s^*_2, ..., s^*_{i-1}, s_{i_j}, s^*_{i+1}, ..., s^*_n).$$

is denoted.

Consider two strategy profiles x and y from S. An operator $k : S \times S \to \mathbb{N}$ that associates the cardinality of the set composed by the players i that would benefit if - given the strategy profile x - would change their strategy from x_i to y_i is defined.

Let $x, y \in S$. We say the strategy profile x *Nash ascends* the strategy profile y and we write $x \prec y$ if the inequality

$$k(x, y) < k(y, x)$$

holds.

Thus a strategy profile x ascends strategy profile y if there are less players that can increase their payoffs by switching their strategy from x_i to y_i than vice-versa. It can be said that strategy profile x is more stable (closer to equilibrium) then strategy y.

The strategy profile $s^* \in S$ is called non-ascended in Nash sense (NAS) if

$$\nexists s \in S, s \neq s^* \text{ such that } s \prec s^*.$$

In [8] it is shown that all non-ascended strategies are NE and also all NE are non-ascended strategies. Thus the Nash ascendancy relation can be used to characterize the equilibria of a game. Considering this result we can regard the Nash ascendancy relation as a generative relation for the NEs of a game.

8.3   The Proposed Community Detection Game

A network can be represented as a graph and a community is a group of nodes (a subgraph) densely interconnected and sparsely connected with nodes outside the community. There are many available community definitions in the literature but there is no general agreement on the concept of connection density [17, 18].

Let A be the adjacency matrix of size $N \times N$ for a network with N nodes. The value a_{ij} of A at position (i, j) is 1 if there is an edge from i to j and 0 otherwise. The degree k_i of a node i is defined as:

$$k_i = \sum_{j=1}^{N} a_{ij}.$$

Considering a subgraph C, the total degree of a node i belonging to C can be split in two contributions:

$$k_i = k_i^{in}(C) + k_i^{out}(C).$$

The internal degree $k_i^{in}(C)$ is defined as the number of edges connecting i to other nodes in C:

$$k_i^{in}(C) = \sum_{j \in C} a_{ij}.$$

The external degree $k_i^{out}(C)$ of a node $i \in C$ is defined as the number of edges connecting i to all nodes outside community C:

$$k_i^{out}(C) = \sum_{j \notin C} a_{ij}.$$

In [18], the subgraph C is considered to be a community in the strong sense if each node has more connections within the community than with the rest of the network, that is:

$$k_i^{in}(C) > k_i^{out}(C), \forall i \in C.$$

The subgraph C is a community in the weak sense if the sum of all degrees within C is higher than the sum of all degrees towards other communities, that is:

$$\sum_{i \in C} k_i^{in}(C) > \sum_{i \in C} k_i^{out}(C).$$

In this paper, we adopt the community definition proposed by Lancichinetti in [6]. A community is a subgraph identified by the maximization of the following fitness [6]:

$$f_C = \frac{k^{in}(C)}{[k^{in}(C) + k^{out}(C)]^\alpha},$$ (8.1)

where

- $k^{in}(C)$ is the total internal degree of nodes in community C and equals the double of the number of internal links of that community.
- $k^{out}(C)$ is the total external degree of nodes in community C and can be computed as the number of links joining each member of the module with the rest of the graph.
- α is a positive real-valued parameter, controlling the size of the communities.

Let us consider now the community detection problem from a game theoretic point of view. In order to do this we have to define a game by describing the set of players, their possible actions and their corresponding payoff functions. There are several ways this may be accomplished. In the following we propose the following empirical approach:

- **The players:** Consider each node of the network as a player; the number of nodes gives the number of players. Let N be the number of nodes. The players will be denoted by i, $i = 1,...,N$;
- **The strategies:** Each player may choose a certain community. The number of players represents also the maximum number of possible communities $C_1,...,C_N$. A situation of the game would be defined as a network cover in which each node belongs to a community:

$$P = (C_{i_1},...,C_{i_n}),$$

  where C_{i_k} represents the community chosen by player k;
- **The payoffs** The considered payoff of each player will be the score of the community the player has chosen as defined by Lancichinetti in [6]. This score is computed as the difference between the 'quality' of the community containing that player and the 'quality' of the same community but without that player. The 'quality' of a community is given by f_C as defined in Equation 8.1 given above. The payoff of player i is thus computed:

$$u_i(P) = f_{C_i} - f_{C_i - i}$$

  where C_i represents the community chosen by player i and $C_i - i$ denotes the community chosen by player i without node i.

Thus each player seeks to maximize its payoff by choosing the community that has to gain mostly from its belonging to it, or has more to lose for not having him as a member.

The Nash equilibrium of this game may be such a situation in which no player (no node) can change its strategy (can switch community) to improve its payoff when all other keeps theirs unchanged. The existence of a Nash equilibrium in pure strategies of such a game cannot be guaranteed and it obviously depends on the choice of payoff function. The mathematical instrument behind the existence of this equilibrium is not in the scope of this paper. From a computational point of view, an approximation of the NE of this game can be computed using the Nash ascendancy relation presented in Section 8.2.

Thus, the Nash ascendancy relation can be rephrased as: having two situations P and P' of the game, P is better in Nash sense than P' if there are less nodes i that can improve their payoffs by individually switching from C_i to C_i' than the players j that improve their payoffs from switching from C_j' to C_j. Moreover, this relation can be embedded within an evolutionary algorithm in order to direct its search towards solutions that are nondominated with respect to this relation.

Fig. 8.1 A simple network with a natural community structure

Fig. 8.2 The node 6 does not belong to the right community

Example:

A simple example to illustrate the construction of the game and payoff function considered is presented. Thus, consider the network depicted in Figure 8.1 containing the two communities C_1 (circles) and C_2 (squares). The corresponding game can be described as follos:

1. the **players** are the 9 nodes of the network;
2. **Strategies**: each player can choose a community out of a maximum number of 9 - $C_1, C_2, ..., C_9$;
3. The **Payoff** functions based on the Lacichinetti score as described above. For example consider the following situation of the game

$$P_1 = (C_1, C_1, C_1, C_1, C_2, C_1, C_2, C_2, C_2)$$

which represents the community structure illustrated in Figure 8.2. The payoff for player 6 can be written $u_6(P_1) = f_{C_1} - f_{C_1-6}$ because in situation P_1 player 6 has chosen community C_1. Thus

$$u_6(P_1) = \frac{14}{18} - \frac{12}{14} = -0.0793.$$

By switching from community C_1 to C_2 a new situation of the game is obtained

$$P_2 = (C_1,C_1,C_1,C_1,C_2,C_2,C_2,C_2,C_2)$$

which is also corresponds to the correct community structure of the network and the payoff of player 6 becomes

$$u_6(P_2) = f_{C_2} - f_{C_2-6} = \frac{16}{18} - \frac{10}{14} = 0.174.$$

Obviously, the strategy of player 6 would be to choose situation P2 of the game as $u_6(P_2) > u_6(P_1)$ as the objective of each player is to maximize its payoff.

8.4 Evolutionary Detection of Nash Equilibria

Crowding Differential Evolution (CrDE) [22] which extends the Differential Evolution (DE) algorithm [19] with a crowding scheme has already been used for Nash equilibria detection [7].

Individuals evolved by CrDE represent in this case situations of the game. A situation of the game is composed by strategies chosen by each player. Thus an individual in the current CrDE population is represented by an n-dimensional vector

$$x = (C_{i_1},...,C_{i_n}),$$

where C_{i_k} represents the community chosen by player k.

In CrDE the only modification to the conventional DE is made regarding the individual (parent) being replaced. Usually, the parent producing the offspring is substituted, whereas in CrDE the offspring replaces the most similar individual among the population if it is better. In the case of the community detection game the offspring replaces the most similar of the parents if it **Nash ascends** it. A *DE/rand/1/exp* [20] scheme is used to modify individuals in the population. As a similarity measure the Euclidean distance in $\mathbb{R}^n$ is used.

The set of nondominated solutions with respect to the Nash ascendancy relation in the final population represents the output of the algorithm.

8.5 Numerical Examples

The quality of the obtained results is evaluated using the Normalized Mutual Information (NMI) proposed in [6]. NMI is a real number between 0 and 1 that represents a similarity measure between two distributions of nodes into overlapping communities: a value of 1 means that the two structures that are being compared are identical and 0 means that the two structures are completely different. Since its proposal in [6], NMI is being increasingly used by researchers for evaluating the quality of overlapping communities structures. Of course, this similarity measure can only be applied for this purpose when the real solution of the problem

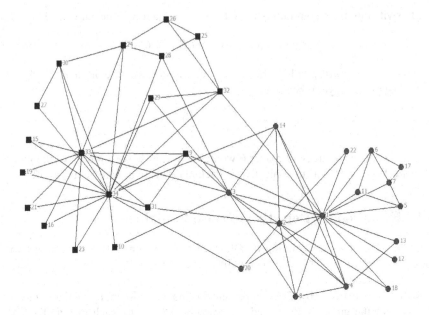

Fig. 8.3 Zachary's karate club network

is known. For computing the NMI in our experiments we have used the source code that the authors have made available and can be freely downloaded from http://sites.google.com/site/andrealancichinetti/mutual.

For each considered network 10 independent runs of CrDE were performed with the following parameters:

- Population size: 50;
- Crowding factor: 50;
- Crossover probability: 0.9;
- $F = 0.5$;
- maximum number of individual payoff function evaluations; $2 \cdot 10^7$.
- $\alpha = 1$

The average values of the NMI and standard deviations over 10 runs are presented. From the NAS individuals in the final population provided by CrDE the one having the best NMI is considered.

The first network considered is the one presented in Section 4. For this simple network CrDE detected the correct community structure in all ten runs ($Avg_{NMI} = 1$).

The next network analyzed is the Zachary's karate club network [25] which describes the friendship of 34 members of a karate club over two years. The network has 34 nodes (representing the members) and 78 edges (representing the social interactions between members). The accepted solution for this network illustrated in Figure 8.3 is detected by CrDE in all 10 out of the 10 runs also $Avg_{NMI} = 1$. This result matches that obtained by the multi-objective approach of Pizzuti in [16].

The social network of 62 bottlenose dolphins [9] living in Doubtful Sound, New Zealand reflects seven years of dolphins behavior. A link between two dolphins was established by their statistically significant frequent association, with a total number of 159 links. The network split naturally into two large groups. The average best NMI obtained by CrDE for this network was $Avg_{NMI} = 0.629$ with standard deviation 0.04. For the dolphins database Pizzuti reported an average best NMI of 1 in [16].

Although not better than others on all networks tested, these numerical results illustrate the potential of this approach.

8.6   Conclusions and Further Work

A new game theoretic approach to the community structure detection in social network problem is presented. Within this approach each node of the network is considered as a player in a non-cooperative game. The strategy of a player is to choose an appropriate community. The payoff for each player is computed by evaluating how much its presence in his chosen community improves the score of that community.

The empirical assertion that for the game thus defined the Nash equilibrium represents an acceptable community structure over the network is supported by preliminary numerical experiments presented in Section 5. A deeper theoretical investigation of this game is however necessary to prove this assertion as well as more numerical experiments. Different methods to compute the payoff functions can also be considered.

Acknowledgements. This research is supported by Grant PN II TE 320, Emergence, auto-organization and evolution: New computational models in the study of complex systems, funded by CNCS Romania.

References

[1] Barabási, A.L.: Linked: The New Science of Networks, 1st edn. Basic Books, New York (2002)

[2] Chira, C., Gog, A.: Collaborative community detection in complex networks. In: Corchado, E., Kurzyński, M., Woźniak, M. (eds.) HAIS 2011, Part I. LNCS, vol. 6678, pp. 380–387. Springer, Heidelberg (2011)

[3] Fortunato, S.: Community detection in graphs (2010),
http://arxiv.org/abs/0906.0612

[4] Karrer, B., Newman, M.E.J.: Stochastic blockmodels and community structure in networks. Physical Review E 83(1), 016, 107+ (2011),
http://dx.doi.org/10.1103/PhysRevE.83.016107,
doi:10.1103/PhysRevE.83.016107

[5] Lai, D., Nardini, C., Lu, H.: Partitioning networks into communities by message passing. Physical Review E 83(1), 016, 115+ (2011),
http://dx.doi.org/10.1103/PhysRevE.83.016115,
doi:10.1103/PhysRevE.83.016115.

[6] Lancichinetti, A., Fortunato, S., Kertész, J.: Detecting the overlapping and hierarchical community structure in complex networks. New Journal of Physics 11(3), 033, 015+ (2009), http://dx.doi.org/10.1088/1367-2630/11/3/033015,
doi:10.1088/1367-2630/11/3/033015

[7] Lung, R., Mihoc, T., Dumitrescu, D.: Nash equilibria detection for multi-player games. In: 2010 IEEE Congress on Evolutionary Computation (CEC), pp. 1–5 (2010), doi:10.1109/CEC.2010.5586174

[8] Lung, R.I., Dumitrescu, D.: Computing nash equilibria by means of evolutionary computation. Int. J. of Computers, Communications & Control III(suppl. issue), 364–368 (2008)

[9] Lusseau, D., Newman, M.E.J.: Identifying the role that individual animals play in their social network (2004)

[10] McKelvey, R.D., McLennan, A.: Computation of equilibria in finite games. In: Amman, H.M., Kendrick, D.A., Rust, J. (eds.) Handbook of Computational Economics, vol. 1, ch. 2, pp. 87–142. Elsevier, Amsterdam (1996)

[11] Mitchell, M.: Complexity: A Guided Tour. Oxford University Press, USA (2009)

[12] Nash, J.F.: Non-cooperative games. Annals of Mathematics 54, 286–295 (1951)

[13] Newman, M.E.J., Girvan, M.: Finding and evaluating community structure in networks. Physical Review E 69(2), 026, 113+ (2004),
http://dx.doi.org/10.1103/PhysRevE.69.026113,
doi: 10.1103/PhysRevE.69.026113

[14] Osborne, M.J.: Introduction to Game Theory: International Edition. No. 9780195322484 in OUP Catalogue. Oxford University Press, Oxford (2009)

[15] Pizzuti, C.: Ga-net: A genetic algorithm for community detection in social networks. In: Rudolph, G., Jansen, T., Lucas, S., Poloni, C., Beume, N. (eds.) PPSN 2008. LNCS, vol. 5199, pp. 1081–1090. Springer, Heidelberg (2008)

[16] Pizzuti, C.: A multi-objective genetic algorithm for community detection in networks. In: IEEE International Conference on Tools with Artificial Intelligence, vol. 0, pp. 379–386 (2009), doi: http://doi.ieeecomputersociety.org/
10.1109/ICTAI.2009.58

[17] Pizzuti, C.: Overlapped community detection in complex networks. In: GECCO, pp. 859–866 (2009)

[18] Radicchi, F., Castellano, C., Cecconi, F., Loreto, V., Parisi, D.: Defining and identifying communities in networks. In: Proc. Natl. Acad. Sci. USA (PNAS 2004), vol. 101, pp. 2658–2663 (2004)

[19] Storn, R., Price, K.: Differential evolution - a simple and efficient adaptive scheme for global optimization over continuous spaces. Tech. Rep. TR-95-012, Berkeley, CA (1995), http://citeseer.ist.psu.edu/article/
storn95differential.html

[20] Storn, R., Price, K.: Differential evolution a simple evolution strategy for fast optimization. Dr. Dobb's Journal of Software Tools 22(4), 18–24 (1997)

[21] Tasgin, M., Bingol, H.: Community detection in complex networks using genetic algorithm (2006), http://arxiv.org/abs/cond-mat/0604419

[22] Thomsen, R.: Multimodal optimization using crowding-based differential evolution. In: Proceedings of the 2004 IEEE Congress on Evolutionary Computation, pp. 1382–1389. IEEE Press, Portland (2004)

[23] Šubelj, L., Bajec, M.: Unfolding communities in large complex networks: Combining defensive and offensive label propagation for core extraction. Physical Review E 83(3), 036, 103+ (2011), http://dx.doi.org/10.1103/PhysRevE.83.036103, doi: 10.1103/PhysRevE.83.036103

[24] Watts, D.J., Strogatz, S.H.: Collective dynamics of 'small-world' networks. Nature 393(6684), 440–442 (1998), http://dx.doi.org/10.1038/30918, doi:10.1038/30918

[25] Zachary, W.W.: An information flow model for conflict and fission in small groups. Journal of Anthropological Research 33(4) (1977), http://dx.doi.org/10.2307/3629752, doi:10.2307/3629752

Chapter 9
Simplified Social Impact Theory Based Optimizer in Feature Subset Selection

Martin Macaš and Lenka Lhotská

Abstract. This chapter proposes a simplification of the original Social Impact Theory based Optimizer (oSITO). Based on the experiments with seven benchmark datasets it is shown that the novel method called simplified Social Impact Theory based Optimizer (sSITO) does not degrade the optimization abilities and even leads to smaller testing error and better dimensionality reduction. From these points of view, it also outperforms another well known social optimizer - the binary Particle Swarm Optimization algorithm. The main advantages of the method are the simple implementation and the small number of parameters (two). Additionally, it is empirically shown that the sSITO method even outperforms the nearest neighbor margin based SIMBA algorithm.

9.1 Introduction

The feature subset selection is widely used dimensionality reduction technique and one of the most important parts of classifier design cycle. It reduces computational cost by reducing dimensionality and can increase classifiers efficiency by removing noisy and irrelevant features. The main task is to select a subset of inputs for a classifier that maximizes an optimization criterion. In wrapper approach, an estimate of classifiers error rate is used as the criterion. One of the most popular criteria is the cross validation estimate that is often used in combination with 1-Nearest Neighbor (1NN) classifier.

In [1], the original version of the Social Impact Theory based Optimizer (oSITO) was applied on feature subset selection. It is relatively novel optimization algorithm

Martin Macaš
CTU in Prague, Technicka 2, Prague
e-mail: `macas.martin@fel.cvut.cz`

Lenka Lhotská
CTU in Prague, Technicka 2, Prague
e-mail: `lhotska@fel.cvut.cz`

D.A. Pelta et al. (Eds.): NICSO 2011, SCI 387, pp. 133–147, 2011.
springerlink.com   © Springer-Verlag Berlin Heidelberg 2011

inspired by socio-psychological models of human social interactions. It was compared to well known binary Particle Swarm Optimization algorithm (bPSO) that is inspired by animal social interactions but can be understood as a model of decision making processes in the group. The two methods were compared on four datasets from UCI machine learning repository. It was concluded that both methods are able to significantly reduce input dimensionality and simultaneously keep up the generalization ability. On the benchmark datasets, the oSITO algorithm was powerful and comparable to the bPSO. However, the methods were compared only from the final cost value and final testing accuracy obtained after a predefined number of iterations. One never knows, what is a proper choice of the optimization length and it is therefore important to compare the whole dynamics of the two methods. This paper shows the whole evolution of the cost function and testing error values and compares the corresponding curves.

Another important point is that although the oSITO algorithm used in [1] is simple, it can be even simplified. It uses a 2D rectangular grid as the social topology and distance measures. In this paper, we propose a simplified version of the SITO algorithm (sSITO) that defines the neighborhood as a number of randomly selected individuals and that does not use any distance measure. Moreover, we revisited the strength computation and the strength is now dependent on both the influencing and influenced individual. Such changes show better behavior from the testing error point of view. The sSITO algorithm needs to preset only two parameters - the number of individuals and the neighborhood size. It is much smaller number compared to the number of parameters of common evolutionary algorithms and to bPSO.

The paper is organized as follows. The feature subset selection, classifier, and non-wrapper based algorithm SIMBA are described in Sect. 9.2. The social interactions based search methods are summarized in Sect. 9.3. The complete experimental work and its results can be found in Sect. 9.4. Finally, in Sect. 9.5, some more discussion is presented and the paper is concluded.

9.2   Feature Subset Selection

Let $\mathscr{D}$ be a dataset $\mathscr{D} = \{(\mathbf{x}, y)\}$ with N data instances, where $\mathbf{x}$ is D-dimensional feature vector and y represents the target classification of an object described by the feature vector $\mathbf{x}$. Let $\gamma : \mathscr{R}^D \to \{\omega_1, \omega_2\}$ be a classifier that assigns the vector $\mathbf{x}$ into one of the classes. Let $\mathbf{s}$ be a binary vector $(s^1, \ldots, s^D)$, where $s^m \in \{-1, 1\}$. The vector represents one feature subset. If $s^m = 1$ or $s^m = -1$, the mth feature is included in or excluded from the feature subset, respectively. The general feature selection process tries to reach $\mathbf{s}^*$ that minimizes a cost function $f(\mathbf{s})$:

$$\mathbf{s}^* = arg \min f(\mathbf{s}).$$

For wrappers, the cost function is usually an estimate of the true error of classifier trained on available data with selected features. A very simple method is the hold-out estimate. The dataset $\mathscr{D}$ is split into two disjunctive subsets the training set $\mathscr{T}$ and the testing set $\mathscr{D} \setminus \mathscr{T}$. The estimate is the empirical error of classifier trained on

$\mathscr{T}$ measured on $\mathscr{D} \setminus \mathscr{T}$. Here, the hold out method is used only for testing purposes for larger datasets (described below).

Further, we choose the most popular error estimation technique - the cross-validation (CV). It is often used and has been recommended in [2]. In the K-fold CV, the dataset $\mathscr{D}$ is partitioned into K disjunctive folds of similar size. At each ith of K iterations, ith fold is used for testing and remaining $K-1$ folds are used for training. The CV error estimate is the testing error averaged over all folds (the empirical error of classifier trained on $\mathscr{D} \setminus \mathscr{D}_i$ measured on $\mathscr{D}_i$). The K-fold CV estimate is unbiased as an estimator of the expected value of error arising from design of sample of size $|\mathscr{D}| - |\mathscr{D}|/K$. *In this paper we use cross-validation with K=10 applied on $\mathscr{D}$ with selected features* s *as the optimization criterion f for our social metaheuristics.* (Moreover, 10-fold CV is also used for evaluation of selection algorithms on smaller datasets).

9.2.1   1-Nearest Neighbor Classifier

The nearest neighbor classifiers use all available features in their distance computation and are therefore quite sensitive on certain disturbing features. Thus it is extremely important to perform the feature selection for these classifiers. The central classification model of this study is 1-nearest neighbor classifier (1NN). It has been widely used in wrapper feature selection studies. The reason is its simplicity, zero training time and high performance especially in the large sample limit It simply finds the training data instance $(\mathbf{x}, y)$ that is most similar (we use the Euclidean distance for dissimilarity quantification) to the testing instance and classifies the instance into class y. Obviously, the error of 1NN classifier can be used as the selection criterion for other classifiers that are more complex and their use in the optimization loop would be thus impractical.

9.2.2   Iterative Search Margin Based Algorithm (SIMBA)

For the comparison purposes we use also one well known nearest neighbor based technique called SIMBA. We choose this method, because it has been proven effective for feature selection and it outperformed very popular RELIEF algorithm. SIMBA [8] is the gradient based optimization of the nearest neighbor margin based feature selection criterion. It assigns weight to each feature. Thus, the s vector contains real components $s^m \in \mathscr{R}$ instead of the binary ones. The feature subset of a given size can be extracted by selection of the features with the highest weights. The margin $\theta^{\mathbf{s}}_{\mathscr{D} \setminus \mathbf{x}}(\mathbf{x})$ of one data instance $\mathbf{x}$ with respect to weights s is defined as:

$$\theta^{\mathbf{s}}_{\mathscr{D} \setminus \mathbf{x}}(\mathbf{x}) = \frac{1}{2}(||\mathbf{x} - nearmiss(\mathbf{x})||_{\mathbf{s}} - ||\mathbf{x} - nearhit(\mathbf{x})||_{\mathbf{s}}), \qquad (9.1)$$

where $nearmiss(\mathbf{x})$ and $nearhit(\mathbf{x})$ is the nearest instance to $\mathbf{x}$ from $\mathscr{D} \setminus \mathbf{x}$ from the same and different class, respectively. The norm $||z||_{\mathbf{s}}$ with respect to weights s is defined as $||\mathbf{z}||_{\mathbf{s}} = \sqrt{\sum_m (s^m)^2 (z^m)^2}$.

The margin based criterion that is minimized (which corresponds to the maximization of sum of particual margins) is

$$f(\mathbf{s}) = - \sum_{\mathbf{x} \in \mathscr{D}} \theta^{\mathbf{s}}_{\mathscr{D} \setminus \mathbf{x}}(\mathbf{x}). \qquad (9.2)$$

The SIMBA algorithm starts with $s^m = 1$ for all m and in each iteration it adds $-\frac{\delta f(\mathbf{s})}{\delta s^m}$ to each component of $\mathbf{s}$. Such a gradient based minimization leads to a local minimum of (9.2). Here, the number of iterations is $5N$. The detailed description of the SIMBA algorithm can be found in [8]. SIMBA is quite different to our social optimization methods. First, it is not wrapper approach because it does not directly use the classifier's error. Second, its search is gradient based. Third, its output is the weight vector, the binary vector must be created by thresholding. Next, we describe the social optimizers that use different, binary representation (i.e. $s^m \in \{-1, 1\}$).

9.3   SITO – The Social Impact Based Optimizer

The SITO algorithm [3] is related to the social psychology issue. It is inspired by opinion formation models [4]. Particularly, its precursor is a computer simulation modeling the processes of social influence investigated by social psychology [5]. It has been successfully applied to feature selection in [1], where the original SITO (oSITO) algorithm with spatially distributed population was used. First, we propose a general pseudocode for a SITO algorithm. Both the oSITO and sSITO algorithm are the particular cases of this framework. The algorithm maintains a population of individuals. Each individual i consists of candidate solution represented by a binary vector $\mathbf{s}_i = (s_i^1, \dots, s_i^d)$, where $s_i^k \in \{-1, 1\}$ and -1 or 1 means the feature is or is not included in the corresponding feature subset, respectively. The binary vector $\mathbf{s}_i$ is evaluated by the cost function $f_i = f(s_i^1, \dots, s_i^d)$ in terms of binary optimization problem, i.e. the minimization of f.

Each candidate solution of an optimization problem is associated with one vertex in a graph that represents a social topology. The topological graph is further used for definition of neighborhoods and distances. Each vector $\mathbf{s}_i$ has assigned its cost value $f_i = f(\mathbf{s}_i)$. Furthermore, each candidate solution $\mathbf{s}_i$ has assigned its strength value q_i, which is generally derived from the cost value of solution $\mathbf{s}_i$ and from the cost values of all the other solutions $\{\mathbf{s}_j\}_{j \neq i}$ by transformation $q_i = q(f_i, \{f_j\}_{j \neq i})$ called strength function.

Many opinion formation models combine the social information using the notion of social impact function that numerically characterizes the total influence of social neighborhood of a particular individual [4]. The new opinion is further computed according to the value of impact function. We use the analogy with the social impact based opinion formation models. A candidate solution is influenced by its social neighborhood in the following manner: the value of impact function is computed that depends on the examined component and strength value of its social neighbors. The neighbors with higher strength value have higher influence on the impact value. The negative impact value leads to preference of component inversion. Contrary, the positive value have supportive character and leads to preference of keeping the component value. The pseudocode of a general SITO algorithm is described in algorithm 9.1.

Algorithm 9.1. General social optimization algorithm

$t = 0$
initialize $s_i(t)$ randomly
while stop condition not met **do**
    for all i **do**
        evaluate $f_i(t) = f(s_i(t))$
        compute strength $q_i(t) = q(f_i(t), \{f_j(t)\}_{j \neq i})$
    end for
    for all i, m **do**
        compute total impact $I_j^m(t)$ according to $\{s_j^m(t)\}_{j \in \mathcal{N}_i(t)}$ and $\{q_j(t)\}_{j \in \mathcal{N}_i(t)}$
    end for
    for all i, m **do**
        compute $s_i^m(t+1)$ according to $I_j^m(t)$
    end for
    $t = t + 1$
end while

9.3.1   Original SITO (oSITO)

In the original SITO algorithm (oSITO), the individuals were placed on cells forming a 2D grid. The topology enabled the definition of distances d_{ij} between two individuals i and j. Usually, the squared neighborhood with a predefined radius and Euclidean distances were considered. Each individual belonged to its own neighborhood and the distance $\delta = d_{ij}$ was called self distance parameter and represented a relative weight of self confidence.

For minimizer, the strength value was inversely proportional to the cost value:

$$q_i = \frac{f_{max} - f_i}{f_{max} - f_{min}}, \tag{9.3}$$

where f_{max} and f_{min} are maximum and minimum values of the cost function (achieved by the worst and best individual).

Further, for each dimension m of each individual j, the neighbors of j are divided into two disjoint subsets, persuaders $\mathcal{P}_j$ with mth bit opposite to s_j^m and supporters $\mathcal{S}_j$ with the same value of mth bit. Further the two impacts are computed:

$$P_j^m = \frac{1}{|\mathcal{P}_j|} \sum_{i \in \mathcal{P}_j} \frac{q_i}{d_{ij}^2}, S_j^m = \frac{1}{|\mathcal{S}_j|} \sum_{i \in \mathcal{S}_j} \frac{q_i}{d_{ij}^2}, I_j^m(t) = P_j^m(t) - S_j^m(t) \tag{9.4}$$

where P_j^m and S_j^m are persuasive and supportive impacts respectively.

Further, all candidate solutions are updated. Each bit is inverted if the persuasive impact is greater than the supportive one:

$$\tilde{s}_j^m(t+1) = \begin{cases} -s_j^m(t) \cdot sgn(I_j^m(t)) & \text{if } (I_j^m(t)) \neq 0, \\ s_j^m(t) & \text{otherwise.} \end{cases} \tag{9.5}$$

After this deterministic update, $\mathbf{s}_j(t+1)$ is generated by random inversion applied with probability $1/d$ on each bit of $\tilde{\mathbf{s}}_j(t+1)$. It has the same function as mutation known from evolutionary computation — improves the explorative capability and prevents a loss of diversity.

9.3.2   Simplified SITO (sSITO)

Here we propose a simplified version of SITO algorithm that uses only random topology. It means that there is no spatial grid and no distance definition and the neighborhood is the set of randomly chosen individuals that is re-initialized at each iteration. Thus, an individual can but need not to be the neighbor of itself (depends on the instantaneous random social graph). Further, the way of strength computation is different. Consider one particular individual i and its influence on an another individual j. We will assume the minimization of cost function and the social strength q_{ij} by which the individual i affects the individual j depends on their cost values according to the following formula:

$$q_{ij} = \max(f_j - f_i, 0), \tag{9.6}$$

where f_i and f_j are the cost values of the individual i and individual j, respectively. This equation means that fitter individual have a non-zero influence on less fitter individual and is not influenced by it. During each iteration, the whole social impact I_j^m affecting each mth bit of jth candidate solution is computed from the strength values of the other individuals from j's predefined random neighborhood $\mathcal{N}_j$ of a given size:

$$P_j^m = \sum_{i \in \mathscr{P}_j} q_{ij}, S_j^m = \sum_{i \in \mathscr{S}_j} q_{ij}, I_j^m(t) = P_j^m(t) - S_j^m(t) \tag{9.7}$$

The neighborhood is a set of randomly chosen individuals that is re-initialized at each iteration. Each mth bit of jth candidate solution is further updated according to the formula 9.5 into which we substitute 9.7 and obtain the following simple equation:

$$\tilde{s}_j^m(t+1) = \begin{cases} sgn\left[\sum_{i \in \mathcal{N}_j} q_{ij}^m s_i^m(t)\right] & \text{if } \sum_{i \in \mathcal{N}_j} q_{ij}^m s_i^m(t) \neq 0, \\ s_j^m(t) & \text{otherwise.} \end{cases} \tag{9.8}$$

Note that each individual works similarly to perceptron and the whole society works as a perceptron network with recurrent connections. First, all individuals in the society initialize their binary candidate solutions randomly according to uniform distribution ($P(-1) = P(+1) = 0.5$)). At each iteration, individuals evaluate their candidate solutions using the cost function and compute their strength using (9.6). Next, each individual, for each dimension, considering its predefined neighborhood, changes the corresponding bit according to the update formula (9.8). Finally, the random mutation is applied (with probability $1/D$) on each $\tilde{s}_j^m(t+1)$ which gives the new $s_j^m(t+1)$. Further the next step is evaluated and the algorithm runs until a stopping condition is satisfied. Here, the stopping condition is the reach of a predefined number of iterations.

9.3.3   Binary Particle Swarm Optimization (bPSO)

The second nature-inspired method based on social influences is the binary Particle Swarm Optimization (bPSO) ([6], [7]). Each candidate solution (particle) consists of a set of binary parameters and represents a position in a binary multi-dimensional space of attitudes. Each particle i is represented as a binary D-dimensional position vector $s_i(t)$ with a corresponding real-valued velocity vector $v_i(t)$.

Two kinds of information are available to the particles. The first is their own experience - they have tried the choices and know which state has been best so far and how good it was. The other information is social knowledge - the particles know how the other individuals in their neighborhood have performed. The *local best PSO* algorithm is used in underlying experiments, where the neighborhood of each particle is a part of the entire swarm.

Furthermore, it remembers its individual best cost function value and position p_i (vector of attitudes) which has resulted in that value. During each iteration t, the velocity update rule is applied to each particle in the swarm:

$$v_i(t) = wv_i(t-1) + \varphi_1 R_1(p_i - s_i(t-1)) + \varphi_2 R_2(p_l - s_i(t-1)). \qquad (9.9)$$

The p_l is the position of the best particle in the predefined neighborhood of particle i and represents the social knowledge. The ring lattice sociometry was used, where every particle is connected to the particle on either side of it in the population array. The neighborhood is defined by r particles on each side, where r is the neighborhood radius. The parameter w is called inertia weight and its behavior determines the character of search. The symbols R_1 and R_2 represent the diagonal matrices with random diagonal elements drawn from a uniform distribution between 0 and 1, $U(0,1)$. The parameters φ_1 and φ_2 are scalar constants that weight influence of particles' own experience and the social knowledge and were set as $\varphi_1 = \varphi_2 = 2$. If any component of v_i is less than $-V_{max}$ or greater than $+V_{max}$, the corresponding value is replaced by $-V_{max}$ or $+V_{max}$, respectively.

Next, the position update rule is applied:

$$s_{i,j}(t) = \begin{cases} 1 \text{ if } R_3 < \frac{1}{1+e^{-v_{i,j}(t)}} \\ 0 \text{ otherwise,} \end{cases} \qquad (9.10)$$

where R_3 is random number drawn from $U(0,1)$. Thus, the position update rule is based on probability produced by normalization of velocity components using sigmoid function.

9.4 Experimental Results

In this section, the experimental comparison of the feature selection methods is described. First, the datasets are summarized in Sect. 9.4.1. Second, the testing methodology is described in Sect. 9.4.2 and some variables that are used for comparisons are defined in Sect. 9.4.3. Finally, the results are shown in Sect. 9.4.4.

9.4.1 Datasets

Majority of datasets were taken from the UCI Machine Learning Repository [9]. We choose only 2-class datasets with the number of features D between 30 and 166. The datasets are described in Table 9.1. Moreover, we focused on the small sample size case (hundreds of instances), where the variance of error estimators is high and where the cross validation is usually being used. Furthermore, we considered only binary classification case. For completeness it must be pointed out that for each outer-loop testing (described below), normalization of features into zero mean and unitary variance was applied to the training set and the pre-computed normalization transform was further applied to the testing set. This is much fairer approach than preliminary normalization of the whole dataset.

The BASEHOCK dataset can be obtained from [10] and only 100 best of 4862 features were preselected before to obtain suitable benchmark dataset (individual rank and class separability criterion used). The CTG dataset is described in [11] and has been used with kind permission of authors.

Table 9.1 Datasets, the full numbers of features, sample sizes and class frequencies and the informations about the true error estimation used for testing

Name	D	N	Test method/$|\mathscr{D}_{tr}|$/$|\mathscr{D}_{ts}|$
MUSK	166	2034 (1017/1017)	HO/204/1830
BASEHOCK	100	1993 (994/999)	HO/199/1794
SPAMBASE	57	4601 (2788/1813)	HO/230/4371
CTG	64	217 (90/127)	CV/195/22
IONOSPHERE	34	351 (126/225)	CV/315/36
SONAR	61	208 (111/97)	CV/187/21
WDBC	30	569 (357/212)	CV/512/57

9.4.2   Testing Error

One of the goals of the paper is to evaluate and compare some feature selection approaches, thus it would be appropriate to measure the true error of classifiers trained with selected features. Because the true error is usually unavailable, one needs independent data that are not used during classifier design process. Such data are further used for estimation of the true performance of feature selection methods and benefits of feature selection. Two main estimates of the true performance were used corresponding to two dataset types. For larger datasets (MUSK, BASEHOCK and SPAMBASE) an outer loop of hold-out error estimate was used (testing set 10 times bigger than the testing set). The available training data were modeled using a very small part $\mathcal{D}_{tr}$ of the whole dataset $\mathcal{D}$. The remaining data $\mathcal{D}_{ts} = \mathcal{D} \setminus \mathcal{D}_{tr}$ were used for testing. For small datasets (CTG,IONOSPHERE,SONAR and WDBC) an outer loop of 10-fold cross-validation method was used. The respective sizes of the training and testing sets are described in the last column of Table 9.1. Obviously, these error estimates can be quite inaccurate and depends on the particular partition of instances into HO splits or CV folds. This problem is partly reduced by repeating the estimate 50 times and averaging the results. To reduce the variability of results caused by differences in the content of training and testing set over the evaluation runs, the distribution of data instances into the training and the testing sets was the same for all examined methods in one individual run. This emphasizes the real difference between the particular methods and suppresses the random component.

9.4.3   Observed Variables

Further, we summarize all measures that will be used for experimental analysis of the algorithms. At each iteration of a search algorithm, we measure the selection criterion value f of all candidate solutions (feature subsets) The criterion value is to be minimized with respect to the feature subset by the search algorithm. We also measure the selection criterion value of the instantaneous best candidate solution (feature subset). During the search, the best-so-far candidate solution and its cost value are memorized and become output when the search is stopped. The first important variable is $f_{BSF}(t)$, the best-so-far criterion value averaged over 50 runs. Its final value is f_{BSF}.

Although for wrappers, the selection criterion should be a true error estimate of the classifier, it does not correspond to the true classification error and cannot be used for evaluating the feature selection benefits. The main reasons are the limited size of testing data and the feature selection bias. The best so far criterion value only measures the optimization abilities of the search methods. To measure the real potential of a particular feature subset, we must estimate the true error using a suitable estimation method running in outer loop.

As it was described above, for larger datasets (upper part of Table 9.1), we estimate the true error of the final best-so-far candidate solution using hold-out method. The whole dataset is split into two disjoint parts - the second part is much greater and is used for testing the classifier trained with the first part (see the last column of Table 9.1 for particular splits). For small datasets (lower part of Table 9.1), we estimate the true error of the best candidate solution using 10-fold cross-validation method in outer loop. During the optimization, we measure the feature selection benefits using these outer-loop-estimates of true error of the best-so-far candidate solution (the one with minimum criterion value) and call it the best-so-far true error estimate. Its value averaged over 50 runs is denoted $e_{BSF}(t)$.

Finally, we will observe so called relative dimensionality D_{BSF}/D, which is the number of features selected by the final best-so-far candidate solution divided by the original dimensionality. The smaller is the relative dimensionality, the more features have been eliminated by the feature selection.

9.4.4 Results

First, we follow [1] and compare the final values of the criterion and testing error for the bPSO, oSITO and the sSITO algorithm. The Table 9.3 show the values after the predefined number of iterations. One can see that all algorithms are comparable from the final criterion value perspective. On the other hand, the sSITO algorithm leads to the best final e_{BSF} value on 6 of 7 datasets and thus, it is better feature selector. This shows that the simplification of the original SITO algorithm does not lead to any performance degradation.

For all experiments, the same parameter settings were used (see the Table 9.2). It is crucial to note that the parameters for the bPSO and oSITO algorithm were tuned in [1] and we used the parameters. On the other hand, the sSITO algorithm has only two parameters - the number of individuals and the neighborhood size. We did not performed any tuning and used the same population size as for bPSO and the neighborhood size 6 which is the same as in bPSO with neighborhood radius $r = 3$.

Further, we leave the oSITO algorithm and compare the sSITO algorithm to the well established and commonly used bPSO algorithm. For the examined method, it is impossible to preset the number of iterations. Moreover, the relative efficiency of the methods can change through time (as will be demonstrated on the figures) and thus a small change in the number of iterations can cause a significant change in the

Table 9.2 The parameter settings

bPSO Parameter	Value	oSITO Parameter	Value	sSITO Parameter	Value
Population size	25	Population size	25	Population size	25
Neighborhood size	6	Neighborhood size	8	Neighborhood size	6
Inertia weight w	1	Self distance δ	1		
Individual weight φ_1	2				
Social weight φ_2	2				
Max. velocity V_{max}	5				

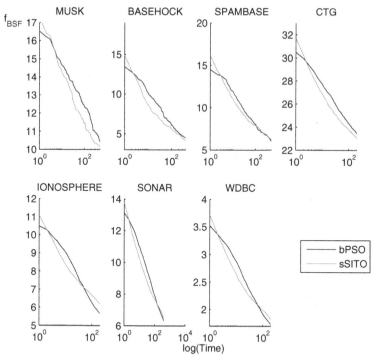

Fig. 9.1 The evolution of best so far criterion value f_{BSF} [%]. The logarithmic x-axis was used to emphasize the difference.

Table 9.3 The final values of f_{BSF} and e_{BSF}. The bolded text emphasizes the best of the three methods

Dataset	Final f_{BSF} [%]			Final e_{BSF} [%]		
	bPSO	oSITO	sSITO	bPSO	oSITO	sSITO
MUSK	10.44	10.16	**10.13**	17.89	17.88	**17.8**
BASEHOCK	4.53	4.25	**4.24**	11.06	10.72	**11.05**
SPAMBASE	**6.11**	6.20	6.14	15.4	15.5	**15.1**
CTG	23.4	23.2	**23.0**	**34.5**	34.9	34.7
IONOSPHERE	5.68	**5.80**	6.18	10.28	10.20	**9.70**
SONAR	6.33	**6.30**	6.50	13.82	13.16	**13.15**
WDBC	**1.73**	1.83	1.82	4.36	4.32	**4.22**

relative order of the two methods. Thus, we performed the testing (computation of e_{BSF}) in every iteration of the optimization progress) and compare the methods from the dynamical point of view using the temporal evolution of the observed variables. This is much more predicative approach compared to [1], where only evaluation of final values was performed.

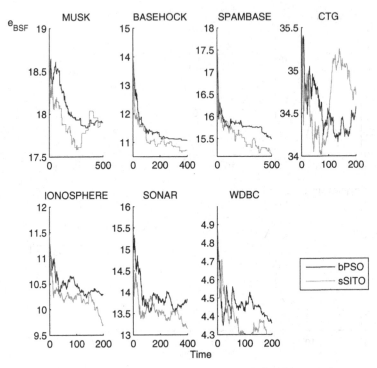

Fig. 9.2 The evolution of the best so far testing error e_{BSF} [%] estimated using hold out or 10 fold cross validation method

The first point of view is the evolution of the cost function value. In Fig. 9.1, the curve for $f_{BSF}(t)$ is depicted. To point out the differences between the two curves, the x-axis is logarithmic. One can observe that at the very early stage of the optimization (few iterations) the bPSO finds smaller cost values. However, after less than ten time steps, the situation changes and sSITO algorithm performs better. For IONOSPHERE and WDBC data, the sSITO does not remain better until the end of optimization and is again outperformed by the bPSO algorithm at the end of optimization. Thus, for our benchmark datasets, for number of iteration between 10 and 100, the SITO finds feature subsets with smaller criterion value for all datasets.

Although the f_{BSF} value evaluates the optimization abilities of the search methods, it says nothing about the feature selection benefits. For such a purpose, the Fig. 9.2 shows the testing error. One can see that the testing performance corresponds poorly to the criterion value except for its decreasing trend which is evident for both the criterion value and the testing error. For all datasets, the sSITO method leads to smaller testing error. Thus, although for some datasets the sSITO algorithm optimizes worse than the bPSO, it always leads to feature subsets with smaller testing error. The only exception is the result for CTG data, where the sSITO possibly leads to a strong over fitting.

Next, the Fig. 9.3 shows the behavior of the algorithms from the dimensionality reduction point of view. The sSITO algorithm selects smaller feature subsets for all datasets and during the whole optimization process. There are only non-significant short-term exceptions. The random initialization method was used, which means that $-1/1$ are initialized with probability 0.5 and the cost function does not take the number of selected features into account. These two facts cause that the relative dimensionality decreases from about 0.5 after the first iterations to a smaller value that is however still close to 0.5. In case of a need for smaller number of features, it would be promising to use a different initialization probabilities or to penalize the number of $+1$s in the encoded candidate solution.

Finally we show how the proposed sSITO algorithm performs in comparison with the SIMBA method. To get a feature subset from SIMBA, one must decide how many features will be selected. The sSITO algorithm chooses the output dimensionality implicitly. Thus, it is difficult to compare the two algorithms. Fig. 9.4 presents the dependence of testing error obtained by SIMBA on the number of selected features. The cross represents the result of sSITO algorithm. It is clear that for the particular number of features selected by the sSITO, SIMBA is always outperformed by sSITO with the only exception – the MUSK dataset. Moreover, the sSITO reaches often the testing error that is smaller than the minimum of the curve

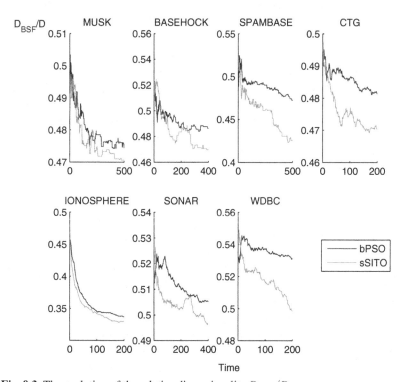

Fig. 9.3 The evolution of the relative dimensionality D_{BSF}/D

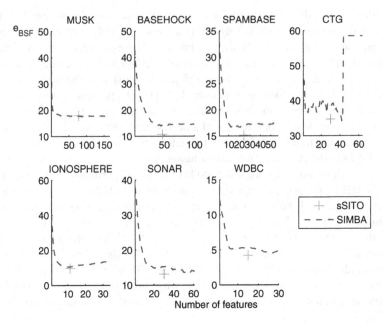

Fig. 9.4 The testing error e_{BSF} [%] obtained by SIMBA for different number of features (averaged over multiple runs). The cross depicts the particular number of features and the testing error obtained by sSITO method

for SIMBA (for BASEHOCK, SPAMBASE, IONOSPHERE, SONAR and WDBC). The user of SIMBA however does not never know how many features to select and thus does not usually find the minimum point. One can conclude for our datasets that sSITO performs better than the SIMBA algorithm from the testing error point of view.

9.5 Conclusions

In this paper, the simplified Social Impact Theory based Optimizer was introduced and applied on wrapper feature subset selection. The novel method is much easier to implement than the original Social Impact Theory based Optimizer and does not need any computation of distances. Moreover, it needs only two parameters. The results obtained for seven randomly selected "small sample/moderate number of features" datasets shown that although the sSITO algorithm leads to slightly better optimization abilities than bPSO, it finds feature subsets that corresponds to permanently smaller testing errors in connection with 1-nearest neighbor classifier. Moreover, the sSITO algorithm reduces the dimensionality better than bPSO. For future, the behavior of the method in connection with other classifiers should be tested and the algorithm should be also applied on different real world problems. Recently, it is being tested on classifier ensemble optimization or feature extraction optimization.

The SITO methods were developed and implemented in Matlab. The codes are used at Czech Technical University in "Advanced Artificial Intelligence Methods" course for teaching the social intelligence principles and can be downloaded from our websites[1]. The code for SIMBA method was downloaded from the websites of its creators[2].

Acknowledgements. The research is supported by The University Development Foundation grant No. FRVS 1885/2011 and by the research program No. MSM6840770012 Transdisciplinary Research in the Area of Biomedical Engineering II of the CTU in Prague, sponsored by the Ministry of Education, Youth and Sports of the Czech Republic.

References

[1] Macaš, M., Lhotská, L.: Social Impact based Approach to Feature Subset Selection. In: Nature Inspired Cooperative Strategies for Optimization (NICSO 2007), pp. 239–248 (2008)

[2] Kohavi, R.: A Study of Cross-Validation and Bootstrap for Accuracy Estimation and Model Selection. In: IJCAI, pp. 1137–1145 (1995)

[3] Macaš, M., Lhotská, L.: Social Impact and Optimization. International Journal of Computational Intelligence Research 4(2), 129–136 (2008)

[4] Bordogna, C.M., Albano, E.V.: Statistical methods applied to the study of opinion formation models: a brief overview and results of a numerical study of a model based on the social impact theory. Journal of Physics: Condensed Matter 19(6), 065144 (2007)

[5] Nowak, A.J., Szamrej, J., Latané, B.: From private attitude to public opinion - A dynamic theory of social impact. Psychological Review 97(3), 362–376 (1990)

[6] Kennedy, J., Eberhart, R.C.: A discrete binary version of the particle swarm algorithm. In: Proceedings of the World Multiconference on Systemics, Cybernetics and Informatics, pp. 4104–4109 (1997)

[7] Kennedy, J., Eberhart, R.C., Shi, Y.: Swarm intelligence. The Morgan Kaufmann Series in Evolutionary Computation. Morgan Kaufmann Publishers, San Francisco (2001)

[8] Gilad-Bachrach, R., Navot, A., Tishby, N.: Margin Based Feature Selection - Theory and Algorithms. In: Proceedings of International Conference on Machine Learning (ICML 2004), pp. 43–50 (2004)

[9] Frank, A., Asuncion, A.: UCI Machine Learning Repository (2010)

[10] National Center for Biotechnology Information,
http://featureselection.asu.edu/datasets.php

[11] Chudáček, V., Spilka, J., Huptych, M., Georgoulas, D., Lhotská, L., Stylios, C., Koucký, M., Janků, P.: Linear and Non-linear Features for Intrapartum Cardiotocography Evaluation. In: Computing in Cardiology, vol. 37, pp. 999–1002 (2010)

---

[1] http://bio.felk.cvut.cz
[2] http://www.cs.huji.ac.il/labs/learning/Papers/
Simba_04_11_2004.zip

Chapter 10
Evolutionary Detection of Berge and Nash Equilibria

Noémi Gaskó, D. Dumitrescu, and Rodica Ioana Lung

Abstract. The problem of equilibria detection in many-player games is computationally untractable by standard techniques. Generative relations represent an useful tool for equilibria characterization and evolutionary equilibria detection. The generative relation for k-Berge-Zhukovskii equilibrium is introduced. An evolutionary technique based on differential evolution capable to cope with hundred players is proposed. Experimental results performed on a multi-player version of Prisoner's Dilemma indicate the effectiveness of the approach.

10.1 Introduction

Within Game Theory (GT) [2, 8] players utilities depend on the actions of all other players. GT represents a basis for neo-classical microeconomic theory. Moreover some decision making processes can be described as non-cooperative games.

Some NP-hard combinatorial optimization problems may also be formulated as decision problems in a game theoretical framework. The problem of detecting game equilibria is a computationally difficult one, particularly in multi-player games.

Berge equilibrium [10] has been proposed as an alternative solution in situations where a game presents none or more than one Nash equilibria. Based on the notion of equilibrium of a coalition P relatively to a coalition K of Berge, Zhukovskii has introduced the Berge equilibrium as an alternative solution to Nash equilibrium for

Noémi Gaskó
Babes-Bolyai University, Cluj-Napoca
e-mail: gaskonomi@cs.ubbcluj.ro

D. Dumitrescu
Babes-Bolyai University, Cluj-Napoca
e-mail: ddumitr@cs.ubbcluj.ro

Rodica Ioana Lung
Babes-Bolyai University, Cluj-Napoca
e-mail: rodica.lung@econ.ubbcluj.ro

D.A. Pelta et al. (Eds.): NICSO 2011, SCI 387, pp. 149–158, 2011.
springerlink.com   © Springer-Verlag Berlin Heidelberg 2011

normal form non-cooperative games. The essential advantage of this equilibrium is that it does not require negotiation of any player with the remaining players, which is not the case when a game has more than one Nash equilibria. Moreover Berge-Zhukovskii equilibrium describes a situation of reciprocation altruism. Therefore it may be useful as a robust solution concept.

An algebraic characterization of k-Berge-Zhukovskii equilibrium by a generative relation is proposed. The generative relation is the base of our evolutionary technique as it induces a rank-based fitness assignment. The n-Prisoner's Dilemma is used for numerical experiments to illustrate the effectiveness of the proposed method.

10.2 Game Theoretic Prerequisites

A finite strategic game is a system $G = ((N, S_i, u_i), i = 1, ..., n)$, where:

- N represents a set of players, and n is the number of players;
- for each player $i \in N$, S_i is the set of available actions,

$$S = S_1 \times S_2 \times ... \times S_n$$

is the set of all possible situations of the game and $s \in S$ is a strategy (or strategy profile) of the game;
- for each player $i \in N$, $u_i : S \to R$ represents the payoff function (utility) of the player i.

10.2.1 Nash Equilibrium

One of the most important solution concept in GT is the Nash equilibrium [6]. The Nash equilibrium can be described as a state, such that no player can change unilaterally her strategy to increase the payoff.

This notion can described mathematically as follows:

Definition 10.1. A strategy profile $s^* \in S$ is a Nash equilibrium if the inequality holds:

$$u_i(s_i, s^*_{-i}) \geq u_i(s^*), \forall i = 1, ..., n, \forall s_i \in S_i,$$

where (s_i, s^*_{-i}) denotes the strategy profile obtained from s^* by replacing the strategy of player i with s_i.

10.2.2 Berge-Zhukovskii Equilibrium

The Berge equilibrium [10] represents an alternative to Nash equilibria for games in which this is not relevant. In contrast to the Nash equilibrium, where the players are self-regarding, the Berge equilibrium determines cooperation in a non-cooperative framework.

The strategy s^* is a *Berge equilibrium (in the sense of Zhukovskii)*, if, when at least one of the players of the coalition $N - \{i\}$ deviates from her equilibrium strategy, the payoff of the player i in the resulting strategy profile would be at most equal to her payoff $u_i(s^*)$ in the equilibrium strategy.

Formally we can write:

Definition 10.2. A strategy profile $s^* \in S$ is a simple Berge (or Berge-Zhukovskii) equilibrium if the inequality

$$u_i(s^*) \geq u_i(s_i^*, s_{N-i})$$

holds for each player $i = 1, ..., n$, and $s_{N-i} \in S_{N-i}$.

10.2.3   k-Berge-Zhukovskii Equilibrium

The k-Berge-Zhukovskii equilibrium is a subset of the Berge-Zhukovskii equilibrium. In this case the size of the coalitions of players is equal to k. The strategy s^* is a *k-Berge-Zhukovskii equilibrium*, if, when at least one of the players of the coalition K deviates from her equilibrium strategy, the payoff of the player i in the resulting strategy profile would be at most equal to her payoff $u_i(s^*)$ in the equilibrium strategy.

Formally:

Definition 10.3. A strategy profile $s^* \in S$ is a k-Berge-Zhukovskii equilibrium if the inequality

$$u_i(s^*) \geq u_i(s_i^*, s_K)$$

holds for each player $i = 1, ..., n$, $i \in N - K$, and $s_K \in S_K$, where $K \subset N$, $card(K) = k$.

Example 10.1. Let us consider the two-person continuous game G_1 [7], having the following payoff functions:

$$u_1(x_1, x_2) = -x_1^2 - x_1 + x_2,$$

$$u_2(x_1, x_2) = 2x_1^2 + 3x_1 - x_2^2 - 3x_2, x_1, x_2 \in [-2, 1].$$

Detected Berge-Zhukovskii equilibrium is $(1; 1)$ having the payoff $(-1; 1)$. This Berge-Zhukovskii equilibrium corresponds to the 1-Berge-Zhukovskii equilibrium (the size of the coalition equal to 1). The Nash equilibrium of the game is the strategy pair $(-0.5; 1.5)$ with the corresponding payoff $(-1.25; 1.25)$.

10.3 Generative Relations

Generative relations represent relations defined over the set of all possible situations of the game, relations that induce a certain type of equilibria. Thus, a generative relation for Nash equilibrium called the *Nash ascendancy relation* has been introduced in [5].

In this section the generative relation for the k-Berge-Zhukovskii equilibrium is introduced.

Consider two strategy profiles x and y from S. Denote by $b_k(x,y)$ the number of players who lose by remaining to the initial strategy x, while the other players are allowed to play the corresponding strategies from y.

We may express $b_k(x,y)$ as:

$$b_k(x,y) = card\{i \in N - K, K \subset N, u_i(x) < u_i(x_i, y_K), card(K) = k\}.$$

Definition 10.4. Let $x,y \in S$. We say the strategy x is better than strategy y with respect to k-Berge-Zhukovskii equilibrium, and we write $x \prec_{B_k} y$, if and only if the inequality

$$b_k(x,y) < b_k(y,x),$$

holds.

Definition 10.5. The strategy profile $y \in S$ is a Berge-Zhukovskii non-dominated strategy, if and only if there is no strategy $x \in S, x \neq y$ such that x dominates y with respect to $\prec_{B_k}$ i.e.

$$x \prec_{B_k} y.$$

We may consider relation $\prec_{B_k}$ as the generative relation of the k-Berge-Zhukovskii equilibrium. This means the set of the non-dominated strategies with respect to the relation $\prec_{B_k}$ equals the set of k-Berge-Zhukovskii equilibria.

Example 10.2. Let us consider game G_1 from the example 10.1, and two strategy profiles: $x = (0.5, 0.25)$ and $y = (0;0)$, and $k = 1$ (each coalition of size 1). In this case strategy x is better then strategy y: $x \prec B_1 y$.

10.4 Evolutionary Equilibria Detection

To the best of our knowledge there is no computationally effective technique for detecting Berge equilibria. However, using the Berge-Zhukovskii any search operator may be adapted to detect these equilibria. In this paper a simple differential evolution algorithm proposed and used to detect both Nash and Berge-Zhukovskii equilibria for the multi-player version of the prisoner's dilemma.

10.4.1 Crowding Based Differential Evolution

Crowding Differential Evolution (CrDE) [9] extends the Differential Evolution (DE) algorithm with a crowding scheme. The only modification to the conventional DE is made regarding the individual (parent) being replaced. Usually, the parent producing the offspring is substituted, whereas in CrDE the offspring replaces the most similar individual among the population if it Nash ascends or if it is better than it with respect to the k-Berge-Zhukovskii equilibrium respectively. A *DE/rand/1/exp* scheme is used.

10.4.2   Nondominated Sorting Genetic Algorithm

NSGA-II [3] is a genetic algorithm for multi-objective optimization problem. Instead of computing the Pareto front, we compute the Nash and k-Berge-Zhukovskii equilibria (using the generative relations).

An initial population is generated randomly. Real polynomial mutation [3] and simulated binary crossover (SBX) [4] operators are used. The generative relation is used for rank-based fitness assignment.

10.5   The Prisoner's Dilemma

A classical form of the prisoner's dilemma is the following: two suspects are arrested. The police has insufficient information to condemn them. They decide to separate them and to make for each of them an offer: if one confesses and the other remains silent, the betrayer will be free, and the other receives ten years sentence. If both remain silent, the punishment for them will be two years. If both confess they receive six year prison.

The payoffs can be summarized in Table 10.1.

Table 10.1 The payoff functions of the two players in Prisoner's Dilemma

		Player 2	
		Cooperate (Stay silent)	Defect (Confess)
Player 1	Cooperate (Stay silent)	(2, 2)	(10, 0)
	Defect (Confess)	(0, 10)	(6, 6)

The payoffs represented by preferences are described in Table 10.2. The game has a pure Nash equilibrium *(Defect, Defect)*, which is not the best solution. The paradox is that both agents act rationally, but producing an apparently irrational result.

The Berge-Zhukovskii equilibrium of the PD game is the *(Cooperate, Cooperate)*. This equilibrium is a better solution, because the payoff for each agent is higher in this case.

Table 10.2 The payoff functions (preferences) of the two players in Prisoner's Dilemma

	Player 2 Cooperate	Defect
Player 1 Cooperate	(2, 2)	(0, 3)
Defect	(3, 0)	(1, 1)

Let us consider the n-person version of the Prisoner's Dilemma (PD). The payoff function is expressed as:

$$u_i(s) = \begin{cases} 2\sum_{j \neq i} s_j + 1 & \text{if } s_i = 0; \\ 2\sum_{j \neq i} s_j & \text{if } s_i = 1. \end{cases}$$

Remark 10.1. For $n = 2$ we get the two person version of the game.

The PD is a classical example that if all players choose a strategy rationally (they play Nash) the result will not be the best for all of them while Berge-Zhukovskii equilibrium can be a better choice.

10.6 Numerical Experiments

CrDE is used to compute k-Berge-Zhukovskii and Nash equilibrium for seven instances of the prisoner's dilemma considering 2, 10, 20, 50 and 100 players respectively. Parameters used for CrDE are presented in Table 10.3. Average and standard deviation of distances to the Berge-Zhukovskii and Nash equilibria respectively over 30 runs are computed. Results are presented in Table 10.4.

Table 10.3 Parameter settings for CrDE

Parameter	2 10 20 50 100
Pop size	50
Max no evaluations	5×10^5 5×10^6
CF	50
F	0.1
Crossover rate	0.9

Table 10.4 Average and standard deviation of distances to $(n-1)$-Berge-Zhukovskii (n is the number of players) and Nash equilibria over 30 runs using CrDE

No players	Nash	$(n-1)$-Berge
2	0 ± 0	0 ± 0
10	0 ± 0	0 ± 0
20	0 ± 0	0 ± 0
50	0 ± 0	0 ± 0
100	0 ± 0	0 ± 0

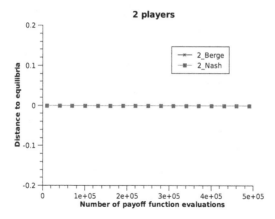

Fig. 10.1 Distance to equilibria versus no. of function evaluations for 2 player PD for CrDE

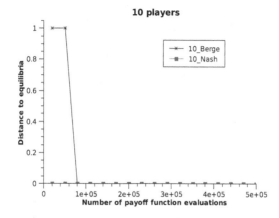

Fig. 10.2 Distance to equilibria versus no. of function evaluations for 10 player PD for CrDE

Although these results might indicate that the problem may be trivial, this is not the case. Consider that a 100-player game may be assimilated with the corresponding many-objective optimization problem of 100 objectives to be maximized. The only difference between the two is represented by the solution concept to be searched - Nash or Berge-Zhukovskii equilibria in games and Pareto optimal solutions in many-objective optimization.

In order to illustrate the evolution of CrDE on the six instances of PD the average distance to each type of equilibria versus the number of payoff function evaluations is presented in Figures 10.1-10.5. A common observation is that the Nash

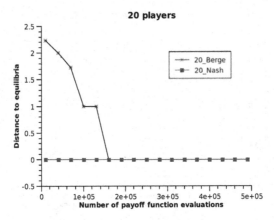

Fig. 10.3 Distance to equilibria versus no. of function evaluations for 20 player PD for CrDE

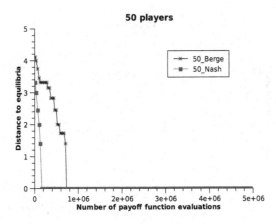

Fig. 10.4 Distance to equilibria versus no. of function evaluations for 50 player PD for CrDE

equilibrium is detected earlier in all the instances. The number of fitness function evaluations necessary to reach each type of equilibria is related to the size of the search space. In all cases the equilibria are detected long before the maximum number of payoff function evaluations is reached, indicating that the CrDE was not only capable of detecting the equilibria but also to maintain their position throughout the search.

For the second part of the experiments NSGA-II is used for the same PD problem, considering 2, 10, 20, 50 and 100 players respectively. Parameters used for NSGA-II are presented in Table 10.5. Average distances to the $(n-1)$-Berge-Zhukovskii and Nash equilibria respectively over 30 runs are computed. Results are presented in Table 10.6. The detection method is based on [3] and [5].

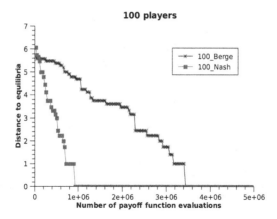

Fig. 10.5 Distance to equilibria versus no. of function evaluations for 100 player PD for CrDE

Table 10.5 Parameter settings for CrDE

Parameter	2 10 20 50 100
Pop size	50
Max no evaluations	5×10^5 5×10^6
prob. of crossover	0.9
prob. of mutation	0.5

Table 10.6 Average and standard deviation distances from the Nash equilibrium strategy and from the $(n-1)$-Berge-Zhukovskii (n is the number of players) equilibrium strategy using NSGA-II

No players	Nash equilibrium	$(n-1)$-Berge equilibrium
2	0 ± 0	0 ± 0
10	0 ± 0	0 ± 0
20	0 ± 0	0 ± 0
50	0 ± 0	3.93 ± 0.08
100	1.2 ± 0.29	4.02 ± 0.26

10.7   Conclusions

An evolutionary approach for detecting k-Berge-Zhukovskii equilibria in many-player games is presented. Two well known algorithms, Crowding Differential Evolution and Non-dominated Sorting Genetic Algorithm II have been adapted for detecting the Nash equilibrium and k-Berge-Zhukovskii equilibria by using appropriate generative relations. To the best of our knowledge this is the first effective attempt to address the problem of detecting k-Berge-Zhukovskii equilibrium computationally. The generative relation for the k-Berge-Zhukovskii equilibrium is introduced. The proposed computational method is used for solving several instances of the Prisoner's Dilemma involving up to 100 players.

The numerical results indicate the effectiveness of the technique for detecting Nash and k-Berge-Zhukovskii equilibria for the multi-player Prisoner's Dilemma.

Acknowledgements. The first author wishes to thank for the financial support provided from programs co-financed by The SECTORAL OPERATIONAL PROGRAMME HUMAN RESOURCES DEVELOPMENT, Contract POSDRU 6/1.5/S/3 - "Doctoral studies: through science towards society". This publication was made possible through the support of a grant from the John Templeton Foundation. The opinions expressed in this publication are those of the authors and do not necessarily reflect the views of the John Templeton Foundation. This research is also supported by Grant PN II TE 320, Emergence, auto-organization and evolution: New computational models in the study of complex systems, funded by CNCS Romania.

References

[1] Aumann, R.: Acceptable Points in General Cooperative n Person Games. In: Contributions to the Theory of Games. Annals of Mathematics Studies 40, vol. IV, pp. 287–324 (1959)

[2] Aumann, R., Hart, S. (eds.): Handbook of Game Theory. Handbooks in Economics (11), vol. 1. North-Holland, Amsterdam (1992)

[3] Deb, K., Agrawal, S., Pratab, A., Meyarivan, T.: A Fast Elitist Non-Dominated Sorting Genetic Algorithm for Multi-Objective Optimization: NSGA-II. In: Schoenauer, M., Deb, K., Rudolph, G., Yao, X., Lutton, E., Merelo, J.J., Schwefel, H.-P. (eds.) Proceedings of the Parallel Problem Solving from Nature VI Conference, Paris, France (2000)

[4] Deb, K., Beyer, H.: Self-adaptive genetic algorithms with simulated binary crossover. Complex Systems 9, 431–454 (1995)

[5] Lung, R.I., Dumitrescu, D.: Computing Nash Equilibria by Means of Evolutionary Computation. Int. J. of Computers, Communications & Control, 364–368 (2008)

[6] Nash, J.F.: Non-cooperative games. Annals of Mathematics 54, 286–295 (1951)

[7] Nessah, R., Larbani, M., Tazdait, T.: A note on Berge equilibrium. Applied Mathematics Letters 20, 926–932 (2007)

[8] Osborne, M.: An Introduction to Game Theory. Oxford University Press, New York (2004)

[9] Thomsen, R.: Multimodal optimization using crowding-based differential evolution. In: Proceedings of the 2004 IEEE Congress on Evolutionary Computation, June 20-23, pp. 1382–1389. IEEE Press, Portland (2004)

[10] Zhukovskii, V.I.: Linear Quadratic Differential Games. Naukova Doumka, Kiev (1994)

Chapter 11
Gravitational Swarm Approach for Graph Coloring

Israel Rebollo Ruiz and Manuel Graña Romay

Abstract. We introduce a new nature inspired algorithm to solve the Graph Coloring Problem (GCP): the Gravitational Swarm. The Swarm is composed of agents that act individually, but that can solve complex computational problems when viewed as a whole. We formulate the agent's behavior to solve the GCP. Agents move as particles in the gravitational field defined by some target objects corresponding to graph node colors. Knowledge of the graph to be colored is encoded in the agents as friend-or-foe information. We discuss the convergence of the algorithm and test it over well-known benchmarking graphs, achieving good results in a reasonable time.

11.1 Introduction

The Graph Coloring Problem (GCP) is a classical NP-hard problem which has been widely studied [10, 17, 19, 21, 22]. There are a lot of algorithms like dealing with it [2, 6, 8], some of them using Ant Colony Optimization (ACO) [11] and Particle Swarm Optimization (PSO) [15]. Up to our knowledge, Swarm Intelligence approaches have not been applied to this problem. The GCP consist in assigning a color to the vertices of a graph with the limitation that any pair of vertices linked by an edge cannot have the same color.

We introduce a new nature inspired strategy to solve this problem following a Swarm Intelligence (SI) [26] approach. The bees [1], ants [13] and flocking birds [9, 24, 25] form swarms that can be interpreted as working in a cooperative way. In SI models, the emergent collective behavior is the outcome of a process of

Israel Rebollo Ruiz
Informática 68 Investigación y Desarrollo S.L., Computational Intelligence
Group- University of the Basque Country
e-mail: beca98@gmail.com

Manuel Graña Romay
Computational Intelligence Group-University of the Basque Country
e-mail: ccpgrrom@gmail.com

D.A. Pelta et al. (Eds.): NICSO 2011, SCI 387, pp. 159–168, 2011.
springerlink.com   © Springer-Verlag Berlin Heidelberg 2011

self-organization, where the agents evolve autonomously following a set of internal rules for its motion and interaction with the environment and the other agents. Intelligent complex collective behavior emerges from simple individual behaviors. An important feature of SI is that there is no leader agent or central control. One of its biggest advantage is that it allows a high level of scalarity, because the problem to be solved is naturally divided into small problems, one for each agent. In real life, if some ants of a colony (also valid for bees, birds or other swarms) fail in its task, that would not alter too much the behavior of the overall system, , so we expect algorithms based on SI to be robust against individual failure.

A technique very close to SI is Particle Swarm Optimization (PSO) [7]. The PSO agents, the particles, have complete knowledge of the problem statement and incorporate a specific solution each, having memory of the best position visited in the search space, so they move in the neighboring of that position. Therefore, PSO performs a random population-based search in the solution space using swarm coherence behavior rules.

The work on GCP presented in this paper follows the approach published in [3, 12]. As in the previous works, SI agents correspondend to graph nodes, and the SI agents try to approach specific space places (goals) where the corresponding graph node acquires a color. We establish a correspondence of the graph edges with antagonist relations between agents. However, in this paper we model the agent's behavior by a simplified dynamic model, removing some unnecesary complexity.

We place the SI agents in a torus shaped space, moving towards the color goals. The only behavior rule that we use is the attraction of the goals exerted on the SI agents. We model the attraction to a color goal as a gravitational field, extended to the entire environment space, which is a fundamental departure from the Reynolds model. When the SI agent reaches a goal, it remains there. The SI agents know the friend/foe relation between them. When a SI agent approaches a goal, the antagonistic agents may be expelled from goal if the approaching agent has enough discomfort pressure. This discomfort reaction allows the system to escape local minima that are not solutions of the GCP. We will demonstrate that with this simple individual behavior, our algorithm can solve the GCP with good performance.

The rest of the paper is organized as follow: Section 2 presents our Gravitational Swarm Intelligence algorithm, and three ways of application of the algorithm, supervised, unsupervised and a mix of both. In Section 3 we discuss the convergence of the algorithm. In Section 4 we show experimental results comparing our algorithm performance with other methods using well-known graphs. Finally, Section 5 gives some conclusions and lines for future work.

11.2   Gravitational Swarm Intelligence

The agent's world is a toric surface where the SI agents move attracted by the color goals. If all the SI agents are in a color goal the algorithm stops and the problem is solved, if not, they have to continue moving through the world. When a SI agent reaches a goal it must be sure that there are no enemy SI agents in that goal. We call enemies the SI agents connected by an arc in the underlying graph. Let be $G = (V,E)$ a graph with V vertices and E edges. We define B as a group of SI agents $B = \{b_1, b_2, ..., b_n\}$ where n is the number of vertices. Each SI agent b_i is represented by a position $p_i = (x_i, y_i)$ in the search space and a speed vector $\vec{v_i}$. Let $C = \{1, 2, ..., k\}$ be the set of color goals that we propose to solve the problem, with attraction speed $\vec{v_{gc}}$ moving each agent to one color goal.

We model the problem as a tuple $F = \{B, \vec{v}, C, \vec{v_g}\}$ where B is the group of SI agents, $\vec{v}$ the speed vector in the instant t, k the chromatic number of the graph and $\vec{v_g}$ the attraction to the goal.

The energy function of a configuration of the whole system is:

$$f = \left| \{\forall i, Cb_i \in C \text{ and } \nexists j \mid enemy(b_i, b_j) \wedge Cb_i = Cb_j\} \right|.$$

This energy is the count of the graph nodes which have a color assigned. The predicate *enemy* is true if a pair of SI agents have an edge between them, so they cannot have the same color. Cb_i denotes the color of the SI agent associated with the target goal. A SI agent has a color only if the SI agent is at distance below *nearenough* from a goal. Until the SI agent arrives to a goal, its color cannot be evaluated. The setting *nearenough* allows to control the speed of the algorithm: if it is small, the algorithm converges slowly but monotonically to the solution, if it is big, the algorithm is faster but it's convergence is not smooth because the system falls in local minima of the energy function and needs transitory energy increases to escape them. The dynamic of the SI agents in the world is specified by the iteration:

$$\vec{v_i}(t+1) = (1 - \frac{d}{\|d\|}) \cdot (\vec{v_i}(t) + d \cdot (\vec{v_g} \cdot (1-\lambda))) + (\lambda \vec{v_r} \cdot p_r), \qquad (11.1)$$

where d is the difference vector between the agent's position and the position of the nearest color goal, $\vec{v_g}$ represents the speed to approach to the nearest goal in Euclidean distance from current position p_i, $\vec{v_r}$ is a noise vector to avoid being stuck in spurious unstable equilibrium, and p_r is a random position. Parameter λ represents the degree of Comfort of the SI agent. When a SI agent b_i reaches to a goal in an instant t, its speed becomes 0. Every step of time that the SI agent stays in that goal without been disturbed, its Comfort increases, until reaching a maximum value *maxconfort*. When an enemy SI agent outside the goal tries to go inside that goal, the Comfort of the SI agent inside the goal decreases to 0. When Comfort is positive the parameter λ has value $\lambda = 0$. If Comfort of a SI agent is equal to 0 and enemy tries to get inside the goal then $\lambda = 1$ and the SI agent is expelled from the goal to a random point p_r with speed $\vec{v}_r$. When all the SI agents stop, $\forall i, \vec{v_i} = 0$; $f = Cb_i$ the problem is solved.

Each color goal has an attraction well spanning the entire space, therefore the gravitational analogy. The magnitude of the attraction drops proportionally with the Euclidean distance d between the goal and the SI agent, but it never disappears.

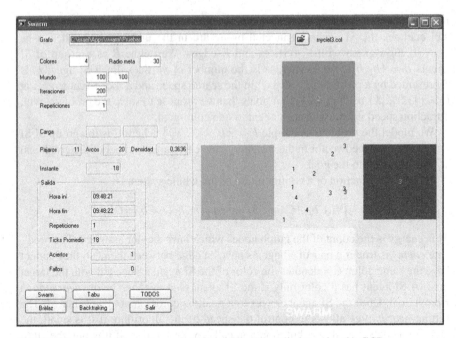

Fig. 11.1 Snapshot of the software interface of the SI applied to solve the GCP

If d is less than *nearenough* then we make $d = 0$, and the agent's speed becomes 0 stopping it. This is, again, departs from the Reynolds boid system, where boids never stop moving. Figure 11.1 shows the interface of the software implementation of this system. In the flowchart of Figure 11.2 we can see how the algorithm works for each SI agent.

11.2.1 A Supervised Coloring

The algorithm can be thought of as a supervised algorithm in the sense that we provide the number of colors when we specify the number and places of color goals. The chromatic number K is the minimum number of colors needed to color the graph. When specifying C color goals we assume that the chromatic number is equal or lower than k. If the system converges to a stationary state then we are right, if not we assume that the chromatic number is greater than k. Search for the graph's chromatic number can be done starting with a large k and decreasing down to the value until the algorithm fails to converge, or the other way, starting with low values of k increasing them until the system converges for the first time. If the chromatic number is known, then we can validate our algorithm against this knowledge.

For validation, it's a good idea to use well-known benchmarking graphs, whose chromatic number is known. The Mycielsky graphs [23] constitute a family of graphs whose chromatic number is equal to the degree of the graph plus one $K = m + 1$ where m is the Mycielsky number. There are also collections of benchmark graphs,

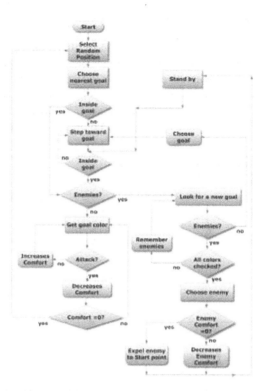

Fig. 11.2 SI Agent behavior flowchart for GCP

such as the DIMACS graphs [18, 19]. For graphs whose chromatic number is unknown the algorithm validation comes from the comparison to other graph coloring
algorithms [5, 27].

11.2.2   Unsupervised Coloring

Another way to attack the problem of determining the chromatic number is to remove the color goals and let the SI agents move according to their friend/foe relations. After a long period of time the movement of the SI agents becomes stable,
so we can say that the actual configuration as the solution of the problem. The SI
agents will never stop, because in the speed function, there is no term to approach
a specific place. The solution to the coloring problem is given by the detection of
clusters of agents moving together. The attraction to the color goals is substituted by
the attraction to friendly agents. Here again we obtain only an approximation to the
real solution, because the SI agents can generate a heterogeneous group of clusters.
We think that the unsupervised coloring can be used to find a starting upper bound
of the chromatic number of a graph, taking into account that there is no stopping
condition.

11.2.3 Coloring Waterfall

We can use a mixed strategy to find the solution for the coloring of a graph. First, apply the unsupervised coloring method to establish an upper bound on the chromatic number. Second, in an iterative way, apply the supervised coloring method to look for solutions with lower chromatic number, decreasing the number of goals, until the algorithm cannot stop and give a solution in a given computational time. The pseudo code of the Algorithm specifies this coloring waterfall.

Algorithm 11.1. Coloring waterfall

Input: G the graph to color.
$Upper\_bound = unsupervised\_coloring(G)$
while $Upper\_bound > 1$ do
let $Upper\_bound = Upper\_bound - 1$
$solution = supervised\_coloring(G, (Upper\_bound), max\_time)$
if not solution then
        $chromatic\_number = (Upper\_bound + 1)$
        exit while
end if
end while

11.3 Convergence Issues

We discuss in this section the convergence of the algorithm from an intuitive point of view. The SI agents start in a position $p_0 = \{x_0, y_0\}$ and with an initial speed $\vec{v_0}$. The direction and value of the speed vector changes with the dynamics of the system. The value of the SI agent's speed is in the range $[0, 1]$. It depends on the distance between the SI agent and its target color goal, when the distance is below a threshold the SI agent stop. Initially each SI agent tries to get to the nearest goal. The attraction of the goals is strong when the SI agent is far away, and weak when the SI agent is near the goal, becoming 0 when the SI agent is inside the goal radius. If antagonistic SI agents are already inside the goal, it tries to expel them out of the goal. An expelled SI agent tries to go to the second nearest goal and so on, until it can enter in a goal without enemies.

If all the SI agents' speed is zero, then the algorithm has converged to some fixed state where all of them are inside a goal. Thus, we have solved a part of the convergence problem, after that we have to demonstrate that the color assignment is a solution for the problem. That is, the cost function value is equal to the number of nodes when the system converges.

When a SI agent tries to enter a color goal, if there is one or more enemies inside the color goal it will try to find another color goal empty of enemies. If there is no one, then it will proceed to expel the enemies from one of the color goals. The SI agent selects a random foe and evaluates its λ parameter. If it is zero, then that enemy is expelled from the color goal to a random point p_r with a speed vector

$\overrightarrow{v_r}$ towards a random goal. All the enemies' Comfort inside that goal decreases. It doesn't matter if there are other enemies with parameter $\lambda = 0$, the SI agent can only expel one SI agent at each computational step. The SI agent doesn't stop testing the agents inside the color goal because it can be still more enemies in that goal and must wait until the next step to get inside. With this behavior, when a SI agent is inside a goal, it's sure that there are no enemies in the goal. So when our algorithm stops because all the agents have stopped, it has reached a GCP solution, because no two adjacent nodes are in the same color goal and all the agents are "inside" a color goal. If the agents never stop, the algorithm is unable to find a solution with the required number of colors

11.4 Experimental Results

We have implemented our algorithm in a graphic development platform trying to obtain an intuitive visualization of the process trading it for computational efficiency. Our algorithm is about SI agents moving around the search space, where each step of time all the SI agents move, or stand still. After each step of time, the cost function must be evaluated to see if the problem is solved or not. For this reason, the time is always referred as iteration steps. When we are evaluating the next position of a SI agent in the step t, we take into account the position of the other SI agents in the step $t - 1$.

Each time step is similar to a generation of a genetic algorithm, or a simulated annealing phase [16], because we have to move all the SI agents (all the SI agent with speed > 0), and evaluate the cost function. The real computing time of our algorithm depends on the programming language and the computer used, but the steps will be always the same.

For comparison, we have also implemented and applied a exhaustive backtraking searchalgorithm, a greedy DSATUR Brlaz algorithm, and a Tabu Search algoritm.Benchmark studies in the bibliography [14] report that the Mycielsky graph need between 4 and 416.000.000 backtrack whereas our GSI algorithm needs only between 21 and 417 stepsWorks reported in [4] could not find the minimal colorings of the "fullins" graph family, while the GSI has been able to find them.

In Table 11.1 we show the results of our algorithm on well known graphs [19, 20] whose chromatic number is known. We have launched the algorithm 20 times for each graph instance and we allow them 2000 steps to find a solution. The table indicates for each graph the number of vertices, edges, the graph's density calculated as #edges/#complete_graph_edges, the graph's chromatic number K, the percentage of success of GSI over 20 experiments and the average number of steps needed to get the solution.

In Table 11.2 we compare GSI with the results of applying the competing algorithms. DSATUR is a particular case of backtracking and the results are similar. Backtracking and DSATUR are deterministic algorithms.We let them a computational limit of 1.000.000 cycles. Tabu search is not a deterministic algorithm and the results can change between executions, therefore we have repeated 20 times with a

Table 11.1 Experimental results

Graph name	#Vertices	#Edges	Density	K	Success %	Average #Steps
anna	138	986	0.10	11	100	300
david	87	812	0.21	11	100	209
huck	74	662	0.22	11	100	84
jean	80	508	0.16	10	100	165
myciel3	11	20	0.36	4	100	21
myciel4	23	71	0.28	5	100	25
myciel5	47	236	0.21	6	100	97
myciel6	95	755	0.17	7	100	92
myciel7	191	2360	0.13	8	100	417
queen 5x5	25	160	0.53	5	100	302
1_fullins_3	30	100	0.23	4	100	37
1_fullins_4	93	593	0.14	5	100	76
1_fullins_5	282	3247	0.08	6	100	222
2_fullins_3	52	201	0.15	5	100	67
2_fullins_4	212	1621	0.07	6	100	176
miles250	128	387	0.04	8	100	317

maximun number of tabus of 2000. The table shows the tabus or steps and the %
of success. The performance of these methods worsens when the graph is medium
or big. Besides the Tab search needs a lot of memory to keep a large amount of no
valid solutions.

We can see in Table 11.2 that the GSI algorithm success in all graphs and repetitions of the process. We find that GSI is faster than competing algorithms, and is
able to solve all the graphs in a reasonable time. The deterministic algorithms are
very fast on the Myciel family of graphs, because their structure is appropriate for
them, however for some graphs they can not reach a solution before reaching the
computational limit.

Table 11.2 Backtracking, DSATUR, Tabu Search and GSI results

Graph name	k	#Backtracking	#DSATUR	#Tabu	%Tabu success	#GSI	%GSI success
Myciel3	4	1	1	11	100	21	100
Myciel4	5	1	1	34	100	25	100
Myciel5	6	1	1	107	100	97	100
Myciel6	7	1	1	290	100	92	100
Myciel7	8	1	1	597	100	417	100
queen5_5	5	5	5	442	95	302	100
anna	11	*	*	*	0	300	100
david	11	*	*	*	0	209	100
huck	11	*	*	*	0	84	100

The number of steps is not clearly related to the size or density of the graphs, because some small graphs require higher number of steps than bigger ones, and some sparse graphs require more steps than dense ones, and the contrary is true also. We need to perform an exhaustive exploration over a collection of random graphs in order to make any inference about time complexity. The fact is that the algorithm requires on average much less steps than the allowed maximum.

11.5 Conclusions

We proposed a new algorithm for the Graph Coloring Problem using Swarm Intelligence. We have modeled the problem as a collection of agents trying to reach some of a set of goals. Goals represent node colorings, agents represent graph's nodes. The color goals exert a kind of gravitational attraction over the entire virtual world space. With these assumptions, we have solved the GCP using a parallel evolution of the agents in the space. We have discussed the convergence of the system, and we have demonstrated empirically that it provides effective solutions in terms of precision and computational time. We will continue to test our algorithm on an extensive collection of graphs, comparing its results with state of the art heuristic algorithms.

For future work, we have to improve our implementation of the algorithm to make it faster. Even though our algorithm finds the global optimum of the problem, we will search for new nature inspired behavior rules to improve the algorithm.

References

[1] Akay, B., Karaboga, D.: A modified artificial bee colony algorithm for real-parameter optimization. Information Sciences, Corrected Proof (2010) (in press)
[2] Brlaz, D.: New methods to color the vertices of a graph. Commun. ACM 22, 251–256 (1979)
[3] Cases, B., Hernandez, C., Graña, M., D'anjou, A.: On the ability of swarms to compute the 3-coloring of graphs. In: Bullock, S., Noble, J., Watson, R., Bedau, M.A. (eds.) Artificial Life XI: Proceedings of the Eleventh International Conference on the Simulation and Synthesis of Living Systems, pp. 102–109. MIT Press, Cambridge (2008)
[4] Chiarandini, M., Süttzle, T.: An application of iterated local search to graph coloring problem. In: Proceedings of the Computational Symposium on Graph Coloring and its Generalizations, Fachgebiet Intellektik, Fachbereich Informatik, and Technische Universitt Darmstadt Darmstadt, pp. 112–125 (2002)
[5] Chvtal, V.: Coloring the queen graphs. Web repository (2004) (last visited July 2005)
[6] Corneil, D.G., Graham, B.: An algorithm for determining the chromatic number of a graph. SIAM J. Comput. 2(4), 311–318 (1973)
[7] Cui, G., Qin, L., Liu, S., Wang, Y., Zhang, X., Cao, X.: Modified pso algorithm for solving planar graph coloring problem. Progress in Natural Science 18(3), 353–357 (2008)
[8] Dutton, R.D., Brigham, R.C.: A new graph colouring algorithm. The Computer Journal 24(1), 85–86 (1981)

[9] Folino, G., Forestiero, A., Spezzano, G.: An adaptive flocking algorithm for performing approximate clustering. Information Sciences 179(18), 3059–3078 (2009)

[10] Galinier, P., Hertz, A.: A survey of local search methods for graph coloring. Comput. Oper. Res. 33(9), 2547–2562 (2006)

[11] Ge, F., Wei, Z., Tian, Y., Huang, Z.: Chaotic ant swarm for graph coloring. In: 2010 IEEE International Conference on Intelligent Computing and Intelligent Systems (ICIS), vol. 1, pp. 512–516 (2010)

[12] Graña, M., Cases, B., Hernandez, C., D'Anjou, A.: Further results on swarms solving graph coloring. In: Taniar, D., Gervasi, O., Murgante, B., Pardede, E., Apduhan, B.O. (eds.) ICCSA 2010. LNCS, vol. 6018, pp. 541–551. Springer, Heidelberg (2010)

[13] Handl, J., Meyer, B.: Ant-based and swarm-based clustering. Swarm Intelligence 1, 95–113 (2007)

[14] Herrmann, F., Hertz, A.: Finding the chromatic number by means of critical graphs. J. Exp. Algorithmics 7, 10 (2002)

[15] Hsu, L.-Y., Horng, S.-J., Fan, P., Khan, M.K., Wang, Y.-R., Run, R.-S., Lai, J.-L., Chen, R.-J.: Mtpso algorithm for solving planar graph coloring problem. Expert Syst. Appl. 38, 5525–5531 (2011)

[16] Graay, M., Hernndez, C., Rebollo, I.: Aplicaciones de algoritmos estocasticos de otimizacin al problema de la disposicin de objetos no-convexos. Revista Investigacin Operacional 22(2), 184–191 (2001)

[17] Johnson, D.S., Mehrotra, A., Trick, M. (eds.): Proceedings of the Computational Symposium on Graph Coloring and its Generalizations, Ithaca, New York, USA (2002)

[18] Johnson, D.S., Trick, M.A.: Cliques, Coloring, and Satisfiability: Second DIMACS Implementation Challenge, vol. 26. American Mathematical Society, Providence (1993)

[19] Johnson, D.S., Trick, M.A. (eds.): Proceedings of the 2nd DIMACS Implementation Challenge. DIMACS Series in Discrete Mathematics and Theoretical Computer Science, vol. 26. American Mathematical Society, Providence (1996)

[20] Lewandowski, G., Condon, A.: Experiments with parallel graph coloring heuristics and applications of graph coloring, pp. 309–334

[21] Mehrotra, A., Trick, M.: A column generation approach for graph coloring. INFORMS Journal on Computing 8(4), 344–354 (1996)

[22] Mizuno, K., Nishihara, S.: Toward ordered generation of exceptionally hard instances for graph 3-colorability, pp. 1–8

[23] Mycielski, J.: Sur le coloureage des graphes. Colloquium Mathematicum 3, 161–162 (1955)

[24] Reynolds, C.: Steering behaviors for autonomous characters (1999)

[25] Reynolds, C.W.: Flocks, herds, and schools: A distributed behavioral model. In: Computer Graphics, pp. 25–34 (1987)

[26] Sundar, S., Singh, A.: A swarm intelligence approach to the quadratic minimum spanning tree problem. Information Sciences 180(17), 3182–3191 (2010); Including Special Section on Virtual Agent and Organization Modeling: Theory and Applications

[27] Turner, J.S.: Almost all k-colorable graphs are easy to color. Journal of Algorithms 9(1), 63–82 (1988)

Chapter 12
Analysing the Adaptation Level of Parallel Hyperheuristics Applied to Mono-objective Optimisation Problems

Eduardo Segredo, Carlos Segura, and Coromoto León

Abstract. Evolutionary Algorithms (EAs) are one of the most popular strategies for solving optimisation problems. One of the main drawbacks of EAs is the complexity of their parameter setting. This setting is mandatory to obtain high quality solutions. In order to deal with the parameterisation of an EA, hyperheuristics can be applied. They manage the choice of which parameters should be applied at each stage of the optimisation process. In this work, an analysis of the robustness of a parallel strategy that hybridises hyperheuristics, and parallel island-based models has been performed. Specifically, the model has been applied to a large set of mono-objective scalable benchmark problems with different landscape features. In addition, a study of the adaptation level of the proposal has been carried out. Computational results have shown the suitability of the model with every tested benchmark problem.

12.1 Introduction

Many real world problems require the application of *Optimisation Strategies*. A wide variety of Approximation Algorithms has been designed to solve them. Metaheuristics [16] are a family of Approximation Techniques that have become popular to solve optimisation problems. Among them, Evolutionary Algorithms (EAs) [15] are one of the most widely used techniques. To define a configuration of an EA several components, as the survivor selection mechanism, and the genetic and parent selection operators must be specified. Thus, several flavours of EAs are seen to exist. Usually, the process of making the parameter setting of an EA takes too much user and computational effort [12].

Eduardo Segredo · Carlos Segura · Coromoto León
Dpto. Estadística, I. O. y Computación. Universidad de La Laguna
La Laguna, 38271, Santa Cruz de Tenerife, Spain
e-mail: {esegredo,csegura,cleon}@ull.es

D.A. Pelta et al. (Eds.): NICSO 2011, SCI 387, pp. 169–182, 2011.
springerlink.com   © Springer-Verlag Berlin Heidelberg 2011

In order to reduce the computational time invested by EAs, several studies have considered their parallelisation [1]. Parallel evolutionary algorithms (pEAs) can be classified [8] in three major computational paradigms: *master-worker*, *island-based*, and *diffusion*. In the *master-worker model*, objective function evaluations are distributed among the worker processors while a master processor executes evolutionary operators and other involved functions. Its search space is identical to the one obtained by the corresponding sequential EA. Island-based models conceptually divide the overall pEA population into a number of independent and separate populations, i.e., there are separate and simultaneously executing EAs (one per processor or island). Each island evolves in isolation for the majority of the pEA execution, but occasionally, some individuals can be migrated between neighbour islands. The *diffusion model* deals with one conceptual population, except each processor holds only one to a few individuals. When compared to the other parallel proposals, the island-based approach brings two benefits: it maps easily onto the parallel architectures (thanks to its distributed and coarse-grained structure), and it extends the search area (due to multiplicity of islands) preventing from sticking in local optima. Island-based models, also known as multi-deme models, have shown good performance and scalability in many areas [1].

Hyperheuristics are a promising approach to facilitate the application of EAs. A hyperheuristic can be viewed as a heuristic that iteratively chooses between a set of given low-level (meta)-heuristics in order to solve an optimisation problem [4]. They operate at a higher level of abstraction than (meta)-heuristics, because they have no knowledge about the problem domain. The underlying principle in using a hyperheuristic approach is that different (meta)-heuristics have different strengths and weaknesses and it makes sense to combine them in an intelligent manner. By this way, hyperheuristics can provide high quality solutions in a single run. At the same time, they can discover which are the most suitable low-level (meta)-heuristics for the considered problem. Depending on the hyperheuristic, some components must be defined. The adaptation level, which refers to the historical knowledge that is considered, is a typical component in the design of hyperheuristics. Based on the adaptation level, hyperheuristics can be classified as local or global approaches [21].

Although hyperheuristics have raised the level of generality of the decision support methodology, they have not usually achieved the solutions quality level of tailor-made approaches. Parallel hyperheuristics have been proposed [19] with the aim of reducing the gap between tailor-made schemes and hyperheuristic-based strategies. In [19], a dynamic-mapped island-based model was proposed to deal with multi-objective optimisation problems. It is a hybrid model that combines parallel island-based models and hyperheuristics. Such a hybridisation was modified to deal with a mono-objective real-world optimisation problem in [22]. High quality results were obtained for some instances of the considered problem. However, since a robustness analysis was not performed, it is difficult to predict the behaviour of the model with other optimisation problems.

The main contribution of this work is to analyse the robustness of the model proposed in [22]. The study has been performed using a set of well-known mono-objective scalable problems [20]. The applied hyperheuristic takes into account the

---

Algorithm 12.1. EA Pseudocode

1: Generate an initial population
2: Evaluate all individuals in the population
3: **while** (not stopping criterion) **do**
4:    Select parents to generate the offsprings
5:    Apply a crossover operator with a probability p_c
6:    Apply a mutation operator with a probability p_m
7:    Evaluate the offsprings
8:    Select individuals for the next generation
9: **end while**

---

previous results obtained by each low-level approach. Different configurations of an EA have been used as low-level strategies. As far as to our knowledge, this model had never been applied together with mono-objective EAs. Computational results have demonstrated the validity of the proposal. A study of the adaptation level has also been performed.

The rest of the paper is structured as follows. In Section 12.2, the set of applied low-level approaches are detailed. The background of hyperheuristics is given in Section 12.3. The parallel hybrid scheme is described in Section 12.4. Then, the experimental evaluation is presented in Section 12.5. Finally, the conclusions and some lines of future work are given in Section 12.6.

12.2   Low-Level Approaches

EAs are stochastic population-based algorithms inspired on the biological evolution. They have shown great promise for calculating solutions to large and difficult optimisation problems [15]. A general scheme of EAs is shown in Algorithm 12.1. It is constituted by a set of components that must be specified in order to define a particular configuration of an EA.

The low-level strategies used in this work have been different configurations of an EA. Specifically, 18 configurations have been considered by combining 3 crossover operators, 2 mutation operators, and 3 survival selection operators. The crossover operators have been the *Simulated Binary Crossover* (SBX), the *Uniform Crossover* (UX), and the *One-Point Crossover* (OPX). The tested mutation operators have been the *Polynomial Mutation* (PM), and the *Uniform Mutation* (UM). The following survival selection mechanisms have been considered:

- *Steady-State* (SS-EA): In each generation, one offspring is generated. If it is better than any of the individuals of the population, the worst of them is replaced by this new offspring.
- *Generational with Elitism* (GEN-EA): In each generation, $n - 1$ offsprings are generated, being n the population size. All parents, except the fittest one, are discarded, and they are replaced by the generated offsprings.

- *Replace Worst* (RW-EA): In each generation, n offsprings are generated. The n fittest individuals, among parents and offsprings, are selected to survive.

In order to complete the definition of the aforementioned low-level configurations, other components must also be specified. A direct encoding of the candidate solutions has been considered, i.e., they have been represented by a vector of D real numbers, being D the number of variables of the considered benchmark problem. Finally, the parent selection mechanism has been the well-known *Binary Tournament*.

12.3 Hyperheuristics

A hyperheuristic can be viewed as a heuristic that iteratively chooses between a set of given low-level (meta)-heuristics in order to solve an optimisation problem [4]. Hyperheuristics operates at a higher level of abstraction than heuristics, because they have no knowledge about the problem domain. The motivation behind the approach is that, ideally, once a hyperheuristic algorithm has been developed, several problem domains and instances could be tackled by only replacing the low-level (meta)-heuristics. Thus, the aim in using a hyperheuristic is to raise the level of generality at which most current (meta)-heuristic systems operate. Since the main motivation of hyperheuristics is to design problem-independent strategies, a hyperheuristic is not concerned with solving a given problem directly as is the case of most heuristics implementations. In fact, the search is on a (meta)-heuristic search space rather than a search space of potential problem solutions. The hyperheuristic solves the problem indirectly by recommending which solving method to apply at which stage of the solving process. Generally, the goal of raising the level of generality is achieved at the expense of reduced - but still acceptable - solution quality when compared to tailor-made (meta)-heuristic approaches. A diagram of a general hyperheuristic framework [4] is shown in Fig. 12.1. It shows a problem domain barrier between the low level (meta)-heuristics and the hyperheuristic itself. The data flow obtained by the hyperheuristic could include the quality of achieved solutions (average, improvement, best, worst), the resources (time, processors, memory) invested to achieve such solutions, etc. Based on such information, the hyperheuristic makes its decisions. The data flow coming from the hyperheuristic could include information about which heuristic must be executed, its parameters, stop criteria, etc.

Hyperheuristics can be classified in terms of the characteristics of the low-level metaheuristics into two groups [6], the ones which operate with constructive techniques and the ones which operate with improvement techniques. Constructive techniques are used to build solutions from scratch. At each step, they determine a subpart of the solution. Improvement metaheuristics are iterative approaches which take an initial solution, and modify it with the aim of improving the objective value. Some hyperheuristics have been designed to operate specifically with one kind of low-level metaheuristics, while other ones, can use both, constructive and improvement methods.

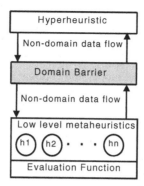

Fig. 12.1 Hyperheuristic Framework

Several ways of incorporating the ideas of hyperheuristics into an optimisation problem have been proposed. The hyperheuristics which deal with mono-objective optimisation problems are much more extensive. In [5], a hyperheuristic based on a tabu search is presented. It operates with improvement low-level heuristics. The same hyperheuristic was used inside a simulated annealing algorithm [14]. The hyperheuristic was used to combine several neighbourhood definitions. Tabu search based hyperheuristics have also been used combined with constructive methods [6]. Other metaheuristics which have inspired the creation of hyperheuristics are genetic algorithms [10] and ant colony optimisation [3, 7, 13]. In the ant-based hyperheuristics, the pheromones represent how well a (meta)-heuristic operates after the application of another (meta)-heuristic. Local search with restart [2] has also been used to implement hyperheuristics. The choice functions has been used multiple times [9, 11, 18]. In such cases, a scoring function is used to assess the performance of each low-level heuristic. The resources are granted to the heuristic which maximises such a function. In [23] a choice function is also used to score each method. However, the resources are assigned using a probability function, which is based on the assigned score.

12.4 Dynamic-Mapped Island-Based Model

The dynamic-mapped island-based model presented in [22] has been used together with the low-level configurations exposed in Section 12.2. It is a hybrid approach that combines a parallel island-based model and a hyperheuristic. A hyperheuristic based on the one presented in [23] has been used. The architecture of the dynamic-mapped model is similar to the parallel island-based model, i.e., it is constituted by a set of *worker islands* that evolve in isolation applying a certain low-level configuration to a given population. In addition, as in the island-based model, a tunable migration stage allows the exchange of individuals between neighbour islands. In order to apply the migration stage, the next components must be defined: the migration and replacement schemes, the migration frequency, the number of individuals to migrate at each step, and the migration topology. In standard island-based model there exists a static mapping among the islands and configurations, i.e., each island executes the same configuration along the complete run. In a homogeneous island-based model, there is only one configuration that is executed by every worker island. In a heterogeneous island-based model, the configurations executed by worker islands are different. However, in the considered model, a dynamic mapping among the islands and configurations is applied. Thus, configurations executed in each island along the run can vary. Such a mapping is performed using a hyperheuristic. In order to manage the dynamic mapping, i.e., to apply the hyperheuristic, a new special island, called *master island*, is introduced into the scheme.

In the standard island-based model, a global stopping criterion is defined. When this global stopping criterion is reached, every worker island sends its local solution to the master and the run ends. In the proposed model, besides the global stopping criterion, local stopping criteria are fixed for the execution of the configurations on the worker islands. When a local stopping criterion is reached, the island execution is stopped. Then, the local results are sent to the master island. At this point, the master island applies the hyperheuristic in order to decide which low-level configuration is going to be applied in the idle island.

The incorporated hyperheuristic (HH_imp) is based on using a scoring strategy and a selection strategy. First, the scoring strategy assigns a score to each configuration. This score estimates the improvement that each configuration can achieve when it breaks from the currently obtained solutions. In order to perform such an estimate, the previous fitness improvements achieved by each configuration are used. The improvement (*imp*) is defined as the difference (in objective value) between the best-achieved individual, and the best initial individual. Improvements obtained during the migration stage of the algorithm are discarded. Considering a configuration *conf*, which has been executed j times, the score ($s(conf)$) is calculated as a weighted average of the last k improvements. Depending on the value of k, the adaptation level of the hyperheuristic can be set. The weighted average assigns greater importance to the last executions. The stochastic behaviour of the involved low-level configurations may lead to variations in the results achieved by them. Therefore, it is appropriate to make some selections based on a random scheme. The hyperheuristic can be tuned by means of the parameter β, which represents the minimum selection probability that should be assigned to a configuration. Thus, being n_h the number of involved low-level configurations, a random selection following a uniform distribution is performed in $\beta * n_h$ percentage of the cases. Finally, the probability of selecting each configuration *conf* ($prob(conf)$) is given by:

$$prob(conf) = \beta + (1 - \beta * n_h) * \left[\frac{s(conf)}{\sum_{i=1}^{n_h} s(i)} \right] \tag{12.1}$$

$$s(conf) = \frac{\sum\limits_{i=1}^{max(k,j)} (k+1-i)*imp[conf][j-i]}{\sum\limits_{i=1}^{max(k,j)} i} \qquad (12.2)$$

12.5   Experimental Evaluation

In this Section, the experiments performed with the parallel model depicted in Section 12.4 are described. The model has been implemented using the METCO (*Metaheuristic-based Extensible Tool for Cooperative Optimisation*) tool [19]. A-synchronous communications for the migration scheme have been implemented. All the communications among the processes have been done using the message passing interface tool MPI. Tests have been run on a Debian GNU/Linux computer with four AMD ® Opteron ™ (model number 6164 HE) at 1.7 GHz and 64 GB RAM. The compiler that has been used is GCC 4.4.5. The MPI compiler has been OpenMPI 1.4.2. Different experiments have been applied to the F1-F11 mono-objective benchmark problems proposed in [20]. They are a set of scalable continuous optimisation problems, which combine different properties regarding the modality, the separability, and the ease of optimisation dimension by dimension. The number of variables D has been fixed to 500. The population size n in the worker islands has been fixed to 5 individuals. For all experiments, the following parameterisation has been used: $p_c = 1$, and $p_m = \frac{1}{D}$. For the parallel model, the migration stage has used the following parameterisation. An elitist migration scheme has been performed. Specifically, a subpopulation individual is migrated when it is better than any member of its previous generation. An elitist replacement scheme has also been applied. It replaces the worst fit individual. Finally, an all to all connected migration topology has been used. In every experiment, the number of worker islands (n_p) has been fixed to 4.

Since experiments have involved the use of stochastic algorithms, each execution has been repeated 30 times. Comparisons have been performed applying the following statistical analysis. First, a *Shapiro-Wilk test* is performed in order to check whether the values of the results follow a normal (Gaussian) distribution or not. If so, the *Levene test* checks for the homogeneity of the variances. If samples have equal variance, an ANOVA *test* is done. Otherwise, a *Welch test* is performed. For non-Gaussian distributions, the non-parametric *Kruskal-Wallis* test is used to compare the medians of the algorithms. A confidence level of 95% has been considered.

The first experiment has analysed the behaviour of the dynamic-mapped island-based model (DYN), in comparison with the homogeneous island-based model (HOMO). In this experiment, DYN has been executed using the 18 low-level configurations ($n_h = 18$) described in the Section 12.2. It has been compared with executions of 18 homogeneous island-based models. Each homogeneous model has used one of the aforementioned low-level configurations. In every execution, the global stopping criterion has been fixed to 10^7 evaluations. In the DYN model, the local stopping criterion has been fixed to $5 \cdot 10^4$ evaluations. The value of β has

Table 12.1 Statistical Differences among the HOMO and the DYN Model (F1-F6)

	F1		F2		F3		F4		F5		F6	
	↑	↓	↑	↓	↑	↓	↑	↓	↑	↓	↑	↓
$k=5$	16	0	14	4	11	1	17	0	17	0	17	0
$k=10$	16	0	14	4	9	1	17	0	17	0	17	0
$k=50$	16	2	14	4	9	3	17	1	16	0	16	1
$k=100$	16	2	14	4	9	2	17	1	17	0	16	1
$k=500$	16	2	14	3	11	1	17	1	16	0	16	1
$k=\infty$	16	0	14	4	10	1	17	0	17	0	17	0

Table 12.2 Statistical Differences among the HOMO and the DYN Model (F7-F11)

	F7		F8		F9		F10		F11	
	↑	↓	↑	↓	↑	↓	↑	↓	↑	↓
$k=5$	16	2	11	1	17	0	16	2	17	1
$k=10$	16	2	9	3	17	1	16	2	17	1
$k=50$	16	2	11	3	17	1	16	2	17	1
$k=100$	16	2	11	1	17	1	16	2	17	1
$k=500$	16	1	15	0	17	1	16	2	17	1
$k=\infty$	17	1	15	0	18	0	17	1	18	0

been fixed in a way that the 10% of the decisions performed by the hyperheuristic follows a uniform distribution, i.e., $\beta * n_h = 0.1$. In order to test different adaptation levels, several values for the parameter k have been tested: 5, 10, 50, 100, 500, and ∞. Thus, the hyperheuristic has been tested with both local and global ($k = \infty$) adaptation levels.

Tables 12.1 and 12.2 show for each value of k, and for each benchmark problem, the number of homogeneous models that have been significantly worse (↑) or better (↓) than the corresponding DYN. The comparison has been performed considering the fitness obtained at the end of each run. The superiority of the DYN model is clear. The number of times that the DYN model has been statistically better than the homogeneous ones is very high for every value of k. The DYN model with a global adaptation level has been the best-behaved. From a total number of 198 statistical tests (18 for each problem), its superiority has been demonstrated in 176 cases. In the homogeneous model, the applied configuration must be selected. Thus, in order to attain high-quality solutions with the homogeneous models, a preliminary study about the behaviour of the low-level configurations must be performed. However, applying the DYN model, high-quality solutions can be obtained in a single run.

Table 12.3 Speedup Factors for the Best-behaved Parallel Models (F1-F6)

	F1		F2		F3		F4		F5		F6	
	HOMO	DYN	HOMO	DYN	HOMO	DYN	HOMO	DYN	HOMO	DYN	HOMO	DYN
CONF1	2.27	2.70	2.32	1.96	2.63	2.32	2.63	4.54	2.38	2.94	2.85	3.03
CONF2	2.36	2.81	2.41	2.03	2.73	2.41	4.10	7.09	2.47	3.05	-	-
CONF4	2.36	2.81	2.51	2.11	2.94	2.60	4.31	7.45	2.57	3.17	-	-
CONF6	2.45	2.91	2.51	2.11	3.26	2.88	4.42	7.63	-	-	-	-
CONF8	2.54	3.02	3.72	3.13	3.57	3.16	-	-	-	-	-	-

Table 12.4 Speedup Factors for the Best-behaved Parallel Models (F7-F11)

	F7		F8		F9		F10		F11	
	HOMO	DYN	HOMO	DYN	HOMO	DYN	HOMO	DYN	HOMO	DYN
CONF1	2.22	2.32	2.38	2.08	2.17	3.70	2.32	1.96	2.12	3.70
CONF2	2.75	2.88	2.38	2.08	-	-	2.88	2.43	-	-
CONF4	4.08	4.27	2.47	2.16	-	-	4.09	3.45	-	-
CONF6	4.35	4.55	2.47	2.16	-	-	4.27	3.60	-	-
CONF8	4.35	4.55	2.66	2.33	-	-	4.37	3.68	-	-

The previous experiment has demonstrated the superiority of the DYN model in the majority of the cases, when it has been compared with the homogeneous models. However, it would be interesting to measure the improvement. For doing it, the second experiment has also considered the results obtained by the sequential low-level configurations. Considering the obtained results, the low-level configurations have been ordered based on the median of the fitness achieved at the end of the executions. An index based on such an order is assigned to each configuration. Thus, for each instance, the best low-level configuration is referred as *conf1*, while the worst one is referred as *conf18*. A subset of the low-level configurations has been compared with the DYN model with a global adaptation level, and with the best-behaved homogeneous model (HOMO) for each problem. The improvement of the parallel models has been quantified by using the Run-length Distributions [17]. They show the relation between success ratios and time. Success ratio is defined as the probability of achieving a certain quality level. Run-length distributions have been calculated for the considered low-level configurations, and for the parallel models. In order to establish a high enough quality level for each problem, it has been fixed as the median of the fitness obtained by *conf1* in $1.25 \cdot 10^6$ evaluations. Tables 12.3 and 12.4 show the speedup factors for the parallel models in relation with a subset of low-level configurations, for each problem. When island-based models are applied to real-world problems, the number of evaluations is n_p times higher than the evaluations carried out by the sequential approaches. Thus, the speedup factor can be calculated as:

$$speedup = \frac{SeqEval}{ParEval} \cdot n_p \tag{12.3}$$

SeqEval and *ParEval* are the number of evaluations required to obtain the desired success ratio for a sequential and a parallel approach, respectively. In both cases, the desired success ratio has been fixed to 50%. In cases where the sequential configuration has not been able to achieve the considered quality level in $2.5 \cdot 10^6$ evaluations at least in a 50% of the executions, a "-" is shown. The speedup obtained by the DYN model has been higher than the one obtained by HOMO in 7 problems. In every case, the DYN model has provided the results faster than *conf1*. In fact, in some cases, superlinear speedups have appeared. The reason is that by combining several low-level configurations, a higher area of the decision space might be explored. Given

Table 12.5 Resource Assigment breaking from Fitness Value equal to 800 (F11)

Name	Mutation	Crossover	Selection	Achieved Fitness	Resources
Seq1	PM	OPX	SS	336.53	2.57 %
Seq2	PM	SBX	SS	338.76	21.54 %
Seq3	PM	SBX	RW	339.59	9.75 %
Seq4	PM	OPX	RW	339.94	5.90 %
Seq5	PM	UX	RW	340.56	14.36 %
Seq6	PM	UX	SS	344.60	5.39 %
Seq7	PM	SBX	GEN	404.14	10.65 %
Seq8	PM	UX	GEN	439.76	6.42 %
Seq9	PM	OPX	GEN	443.69	8.58 %
Seq10	UM	SBX	GEN	580.06	2.17 %
Seq11	UM	OPX	RW	615.17	1.67 %
Seq12	UM	OPX	SS	623.82	1.41 %
Seq13	UM	SBX	RW	635.90	1.92 %
Seq14	UM	SBX	SS	638.70	2.82 %
Seq15	UM	UX	RW	643.57	1.15 %
Seq16	UM	UX	SS	647.16	1.53 %
Seq17	UM	OPX	GEN	692.88	0.76 %
Seq18	UM	UX	GEN	697.94	1.41 %

that when solving a problem, the best configuration is not known a priori, the time saving is greater than the calculated speedup.

In order to deeper analyse the benefits of the usage of the DYN model, an analysis of the resource assignment when solving the F11 problem has been performed. F11 has been selected because it is the one in which the ratio between the speedup obtained by DYN, and the speedup obtained by HOMO is the highest one. This analysis has revealed that the way in which the resources are assigned for the F11 problem, vary along the execution. Mainly, three different stages have been identified.

In the first stage, every configuration is executed one time. After it, the median of the achieved fitness value has been 800. At this stage, the hyperheuristic does not perform any decision. In the following stages, the hyperheuristic performs its decisions based on the quality of the low-level approaches. Thus, an analysis of the behaviour of the sequential low-level approaches in each of the stages have been performed.

Table 12.5 shows for each of the low-level sequential configurations the median of the fitness achieved when they break from a population with individuals whose fitness values are close to 800. Executions has been performed taking into consideration a stopping criterion of 10^5 evaluations. It also shows the resources assigned by the hyperheuristic to each one of them. It can be observed that most of the resources have been granted among the best-behaved configurations. Figure 12.2 shows the box-plots of the achieved fitness for the seven best configurations. It shows the similarities between the best six configurations. In fact, differences among them are not statistically significant. The hyperheuristic has granted the 59.5% of the resources to such set of configurations. In addition, *seq7* has been also granted with a large

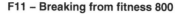

F11 – Breaking from fitness 800

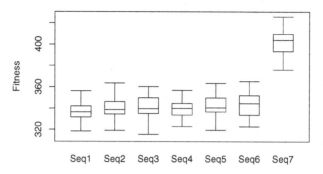

Fig. 12.2 Achieved fitness in 10^5 evaluations

Table 12.6 Resource Assigment breaking from Fitness Value equal to 46 (F11)

Name	Mutation	Crossover	Selection	Achieved Fitness	Resources
Seq1	PM	SBX	GEN	42.83	56.71 %
Seq2	PM	OPX	RW	43.76	0.52 %
Seq3	PM	OPX	SS	43.89	0 %
Seq4	UM	OPX	RW	44.11	1.55 %
Seq5	UM	SBX	GEN	44.41	1.04 %
Seq6	UM	OPX	SS	44.50	1.55 %
Seq7	UM	OPX	GEN	44.91	0 %
Seq8	PM	OPX	GEN	45.03	2.06 %
Seq9	PM	UX	RW	45.48	3.60 %
Seq10	PM	UX	SS	45.49	0 %
Seq11	PM	SBX	RW	45.51	0.51 %
Seq12	PM	SBX	SS	45.53	22.68 %
Seq13	PM	UX	GEN	45.71	0 %
Seq14	UM	SBX	SS	45.86	5.15 %
Seq15	UM	UX	SS	45.87	0 %
Seq16	UM	UX	RW	45.87	0.51 %
Seq17	UM	SBX	RW	45.87	3.09 %
Seq18	UM	UX	GEN	45.87	1.03 %

amount of resources. The reason is that such a configuration is the best in the last stage of the optimisation process.

In the last stage, after reaching a fitness value close to 46, the behaviour of the configurations is completely different. Table 12.6 shows that, when breaking from a population with individuals whose fitness values are close to 46, some of the configurations which behaved poorly in the previous stages, are now the best-behaved ones. The shown fitness values have been obtained taking into consideration a stopping criterion of 10^5 evaluations. The hyperheuristic has been able to identify such a change and has granted a large amount of resources to them. In fact, the best configuration has been granted with the 56.7 % of the resources. However, in this stage, the 22.68% of the resources has been granted to *seq12*. The reason is that such a configuration has been one of the best-behaved in the previous stage. Since

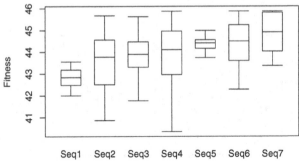

F11 – Breaking from fitness 46

Fig. 12.3 Achieved fitness in 10^5 evaluations

the hyperheuristic requires some time to identfy the configurations which better fit to the new stage, several resources might be granted to not optimal configurations. Figure 12.3 shows the box-plots of the achieved fitness for the seven best configurations. In this case, differences among the best configuration and the remaining ones are statistically significant. Therefore, the differences among the resources granted to each of these configurations are noticeable, demonstrating the suitability of the proposal.

12.6 Conclusions and Future Work

EAs are one of the most popular metaheuristics for solving optimisation problems. Their parameterisation is an arduous task. Hyperheuristics can help in the parameter setting of an EA. In the present work, an analysis of the robustness of a hybridisation among hyperheuristics and a parallel island-based model has been performed. In such a hybrid model, a master island executes the hyperheuristic, dynamically mapping different low-level configurations of an EA to a set of worker islands. The hyperheuristic scores and selects the most promising low-level configurations during the different stages of the execution. This score estimates the improvement that each configuration can achieve when it breaks from the currently obtained solutions. In order to perform such an estimate, the k previous fitness improvements achieved by each configuration are used. The adaptation level, which refers to the historical knowledge that is considered, can be tuned by using the parameter k. The experimental evaluation has been performed with a set of well-known scalable benchmark problems. It has demonstrated the validity of the proposal, in terms of robustness and performance. The adaptation level has been analysed by making a comparison with several homogeneous island-based models. Statistical tests have shown the superiority of the dynamic-mapped island-based model in the majority of the cases. The dynamic-mapped island-based model with a global adaptation level has obtained the best results. However, regardless of the adaptation level, the benefits of the approach have been shown. The comparison with the best sequential low-level configuration has measured the improvement over the performance. The speedup obtained by the dynamic-mapped island-based model with a global adaptation level has been higher than the one obtained by the best homogeneous island-based model in 7 problems. For every problem, the dynamic-mapped island-based model has provided the results faster than the best sequential approach. Since the best parameterisation is not known a priori when solving a problem, the time saving is greater than the calculated speedup. The analysis of the resources assignment for the F11 problem has demonstrated the capability of the hyperheuristic to identify the most promising approaches in the different stages that can appear along the optimisation process. It has shown that, since the most promising approach is applied in each stage, hyperheuristics can provide better results than any of the low-level approaches running isolated.

Future work targets the incorporation and the analysis of other hyperheuristics inside the dynamic-mapped island-based model. In addition, it would be interesting to perform an analysis of the robustness of the hyperheuristic in terms of the parameter β. Currently, the incorporated hyperheuristic works with a set of prefixed configurations. A great line of research would be to develop a hyperheuristic capable of performing the parameter setting without the requirement of specifying such a set of configurations.

Acknowledgements. This work has been supported by the EC (FEDER) and the Spanish Ministry of Science and Innovation as part of the 'Plan Nacional de I+D+i', with contract number TIN2008-06491-C04-02 and by Canary Government project PI2007/015. The work of Eduardo Segredo and Carlos Segura has been funded by grants FPU-AP2009-0457 and FPU-AP2008-03213.

References

[1] Alba, E.: Parallel Metaheuristics: A New Class of Algorithms. Wiley Interscience, Hoboken (2005)

[2] Araya, I., Neveu, B., Riff, M.C.: An Efficient Hyperheuristic for Strip-Packing Problems. In: Cotta, C., Sörensen, K. (eds.) Adaptive and Multilevel Metaheuristics. SCI, vol. 136, pp. 61–76. Springer, Heidelberg (2008)

[3] Burke, E., Kendall, G., Silva, J.L., O'Brien, R., Soubeiga, E.: An Ant Algorithm Hyperheuristic for the Project Presentation Scheduling Problem. In: Proceedings of the 2005 IEEE Congress on Evolutionary Computation (CEC 2005), Edinburgh, Scotland, vol. 3, pp. 2263–2270 (2005)

[4] Burke, E.K., Kendall, G., Newall, J., Hart, E., Ross, P., Schulenburg, S.: Handbook of Meta-heuristics. In: Hyper-heuristics: An Emerging Direction in Modern Search Technology. Kluwer, Dordrecht (2003a)

[5] Burke, E.K., Kendall, G., Soubeiga, E.: A Tabu-Search Hyperheuristic for Timetabling and Rostering. Journal of Heuristics 9(6), 451–470 (2003)

[6] Burke, E.K., McCollum, B., Meisels, A., Petrovic, S., Qu, R.: A graph-based hyperheuristic for educational timetabling problems. European Journal of Operational Research 176(1), 177–192 (2007)

[7] Chen, P.C., Kendall, G., Vanden Berghe, G.: An Ant Based Hyper-heuristic for the Travelling Tournament Problem. In: Proceedings of IEEE Symposium of Computational Intelligence in Scheduling (CISched 2007), Honolulu, Hawaii, pp. 19–26 (2007)

[8] Coello, C.A., Lamont, G.B., Veldhuizen, D.A.V.: Evolutionary Algorithms for Solving Multi-Objective Problems. In: Genetic and Evolutionary Computation (2007)

[9] Cowling, P., Kendall, G., Soubeiga, E.: A parameter-free hyperheuristic for scheduling a sales summit. In: Proceedings of 4th Metahuristics International Conference (MIC 2001), Porto, Portugal, pp. 127–131 (2001)

[10] Cowling, P., Kendall, G., Han, L.: An Investigation of a Hyperheuristic Genetic Algorithm Applied to a Trainer Scheduling Problem. In: Proceedings of the 2002 IEEE Congress on Evolutionary Computation (CEC 2002), pp. 1185–1190. IEEE Computer Society, Honolulu (2002)

[11] Cowling, P.I., Kendall, G., Soubeiga, E.: Hyperheuristics: A Robust Optimisation Method Applied to Nurse Scheduling. In: Guervós, J.J.M., Adamidis, P.A., Beyer, H.-G., Fernández-Villacañas, J.-L., Schwefel, H.-P. (eds.) PPSN 2002. LNCS, vol. 2439, pp. 851–860. Springer, Heidelberg (2002)

[12] De Jong, K.: Parameter Setting in EAs: a 30 Year Perspective. In: Lobo, F., Lima, C., Michalewicz, Z. (eds.) Parameter Setting in Evolutionary Algorithms, pp. 1–18. Springer, Heidelberg (2007)

[13] Dorigo, M., Maniezzo, V., Colorni, A.: The ant system: Optimization by a colony of cooperating agents. IEEE Transactions on Systems, Man, and Cybernetics-Part B 26(1), 29–41 (1996)

[14] Dowsland, K., Soubeiga, E., Burke, E.: A Simulated Annealing Hyper-heuristic for Determining Shipper Sizes. European Journal of Operational Research 179(3), 759–774 (2007)

[15] Eiben, A.E., Smith, J.E.: Introduction to Evolutionary Computing. Natural Computing Series. Springer, Heidelberg (2008)

[16] Glover, F.W., Kochenberger, G.A.: Handbook of Metaheuristics. International Series in Operations Research & Management Science. Springer, Heidelberg (2003)

[17] Hoos, H., Informatik, F., Hoos, H.H., Stutzle, T., Stutzle, T., Intellektik, F., Intellektik, F.: On the Run-time Behavior of Stochastic Local Search Algorithms for SAT. In: Proceedings AAAI 1999, pp. 661–666 (1999)

[18] Kendall, G., Cowling, P., Soubeiga, E.: Choice function and random hyperheuristics. In: Proceedings of the 4th Asia-Pacific Conference on Simulated Evolution And Learning (SEAL 2002), Singapore, pp. 667–671 (2002)

[19] León, C., Miranda, G., Segura, C.: METCO: A Parallel Plugin-Based Framework for Multi-Objective Optimization. International Journal on Artificial Intelligence Tools 18(4), 569–588 (2009)

[20] Lozano, M., Molina, D., Herrera, F.: Editorial Scalability of Evolutionary Algorithms and Other Metaheuristics for Large-scale Continuous Optimization Problems. In: Soft Computing - A Fusion of Foundations, Methodologies and Applications, pp. 1–3 (2010)

[21] Ong, Y.S., Lim, M.H., Zhu, N., Wong, K.W.: Classification of Adaptive Memetic Algorithms: A Comparative Study. IEEE Transactions on Systems, Man, and Cybernetics - Part B 36(1), 141–152 (2006)

[22] Segura, C., Miranda, G., León, C.: Parallel Hyperheuristics for the Frequency Assignment Problem. In: Memetic Computing, pp. 1–17 (2010)

[23] Vink, T., Izzo, D.: Learning the best combination of solvers in a distributed global optimization environment. In: Proceedings of Advances in Global Optimization: Methods and Applications (AGO), Mykonos, Greece, pp. 13–17 (2007)

Chapter 13
Estimation of Distribution Algorithm for the Quay Crane Scheduling Problem

Christopher Expósito Izquierdo, José Luis González Velarde,
Belén Melián Batista, and J. Marcos Moreno-Vega

Abstract. Estimation of Distribution Algorithms (EDA) are a type of optimization techniques that belong to evolutionary computation. Its operation is based on the use of a probabilistic model, which tries to reach promising regions through statistical information concerning to the individuals that belong to the population. In this work, several solution approaches based on the EDA field are presented in order to solve the Quay Crane Scheduling Problem (QCSP). QCSP consists of obtaining a schedule that minimizes the service time of a container vessel given a set of tasks (loading and unloading operations to/from) by means of the available quay cranes at a container terminal. The experimental results confirm that such algorithms are suitable for solving the QCSP and perform a wide exploration of the solution space using reduced computational times.

Christopher Expósito Izquierdo
Dpto. de Estadística, IO y Computación, ETS de Ingeniería Informática,
Universidad de La Laguna, Spain
e-mail: cexposit@ull.es

José Luis González Velarde
Centro de Manufactura y Calidad, Tecnológico de Monterrey, México
e-mail: gonzalez.velarde@itesm.mx

Belén Melián Batista
Dpto. de Estadística, IO y Computación, ETS de Ingeniería Informática,
Universidad de La Laguna, Spain
e-mail: mbmelian@ull.es

J. Marcos Moreno-Vega
Dpto. de Estadística, IO y Computación, ETS de Ingeniería Informática,
Universidad de La Laguna, Spain
e-mail: jmmoreno@ull.es

D.A. Pelta et al. (Eds.): NICSO 2011, SCI 387, pp. 183–194, 2011.
springerlink.com   © Springer-Verlag Berlin Heidelberg 2011

13.1   Introduction

The current state of globalization would have been impossible without the integration of containers in global supply chains. Occurrence of containerization has promoted the homogenization of facilities and means of transport involved within an intermodal freight transportation system. Containers allow end-to-end transport without the direct access to the freight, ensuring their integrity and speeding up their handling at intermediate nodes of the transportation system. One of the highlighted advantages of a container is its ability to transport goods of several types. There are containers with standard features for the transportation of hazardous materials, liquids, perishable goods, etc.

Transportation by ship is the predominant way of exchange of freight between different places around the world. Maritime transport is attractive due to its large cargo capacity, which makes the cost per unit smaller than in other means of transport of goods, and its suitable level of service. This hegemony of shipping has led to the development of container vessels with increasing capacities (in excess of 18.000 containers) to service the main sources of demand for goods. At the same time, it has required the creation of major container terminals in seaports capable of handling the large number of containers exchanged worldwide and provide an effective service to shipping companies.

Maritime container terminals are infrastructures to ease an effective transfer of containers within an intermodal network, usually composed by sea transport (container vessels) and hinterland transport (trucks and trains). Typically, a container terminal is divided into several sections: seaside, yard and landside, where major container flows are interrelated. The large number of processes, heterogeneity of the means of transport and the dynamic nature of the decisions to take at a container terminal make it particularly complex to manage. The main objective of container terminals is to perform the process of loading and unloading the proper containers to/from vessels; that is, unload the import containers and load the export containers.

In general terms, after the arrival of a container vessel at a port, it is necessary to establish its location within the facility; that is, providing a berth according to their technical characteristics (length, beam, draft, etc.) and its cargo. In order to carry out the container loading and unloading processes a sufficient set of the available quay cranes are assigned. Loading and unloading of containers should be performed in accordance with the intrinsic restrictions marked by the location of the containers along the vessel and the possible interference among the quay cranes. Unloaded containers are stored temporarily on the container yard until their subsequent collection by other container vessel or some hinterland transport.

Current container terminals cope with a very hard competition with nearby ports. High remote seaports can compete for the same customers because of the expansion of container trade. This fact has led to maritime terminals to bet decidedly on innovation, automation and the improvement of their infrastructures. In order to increase the productivity of a container terminal a careful analysis about the operational activities must be carried out. The improvement process leads to container terminals to be more attractive to shipping companies and allows to increase their level of competitiveness, based on the operating cost reduction and the offered quality of service enhancing. Optimization methods take a prominent role in this regard due to the fact that their use encourages better use of available resources.

Shipping companies base their business on the economies of scale, so that they pursue to increase the commercial activities in order to maximize profits. One of the main problems faced by shipping companies is that their profits are heavily influenced by the overall cargo transit time of their fleets in port. Most of the time spent by a container vessel in a container terminal is used to perform the loading and unloading tasks, so that its minimization is particularly relevant for the industry. Quay Crane Scheduling Problem pursues to achieve a proper scheduling of the allocated quay cranes for loading and unloading of containers to/from a vessel in the shortest possible service time.

The main aim of this work is to define some basic schemes of Estimation of Distribution Algorithm for the resolution of the Quay Crane Scheduling Problem. The rest of the paper is organized as follows. Section 13.2 introduces and defines the constraints considered in the Quay Crane Scheduling Problem. Section 13.3 describes the approaches proposed to solve the Quay Crane Scheduling Problem. Section 13.4 illustrates the computational experiments carried out in this paper. Finally, Section 13.5 presents the concluding remarks extracted from the work.

13.2   Quay Crane Scheduling Problem

Each one of the berthed container vessels at a container terminal must have a proper stowage plan that allows to effectively carry out the loading and unloading of containers. A stowage plan is composed by several cross-sectional layouts of the vessel bays. The stowage plan aims to establish the specific location of each container onto the vessel (these can be placed in the hold or on the deck), usually according to their weights, types or destination ports. Thus, containers with the same characteristics are included in the same container group. Usually, containers belonging to the same container group are stacked into adjacent slots (stack and tier) within a vessel bay.

Quay cranes of a container terminal have the objetive to perform the loading and unloading tasks defined by the corresponding stowage plan. In this work, a *task* refers to the loading or unloading of a set of containers belonging to the same group. In the literature this problem is referred to as Quay Crane Scheduling Problem (QCSP), whose aim is to carry out the tasks associated with a container vessel using the shortest possible service time (makespan).

There are several outstanding works that address the QCSP. In the first place, Kim and Park [2] introduced its definition and constraints. They proposed a Branch & Bound method and a GRASP technique for its resolution. Moccia et al. [4] formulated the QCSP as a vehicle routing problem with side constraints, including precedence relationships between vertices. They also proposed a Branch & Cut method to solve large instances. Samarra et al. [5] decomposed the QCSP into a routing problem solved by a Tabu Search and a scheduling problem solved by a Local Search. Finally, Bierwirth and Meisel [1] demostrated the incorrectness of the previous mathematical models and presented new interference contraints. In addition, they developed the Unidirectional Scheduling (UDS) heuristic in order to carry out an exhaustive search of unidirectional schedules. UDS employs a tree search to establish the performed set of tasks by every quay crane and then sets the order of their realization to meet the unidirectional movement. UDS is able to find optimal unidirectional schedulings for small instances using short computational times, its performance is decremented when the problem size grows.

In the QCSP, a set of tasks $\Omega = \{1, \ldots, n\}$ and a set of quay cranes $Q = \{1, \ldots, m\}$ with similar technical characteristics are considered. For convenience, dummy tasks 0 and $T = n + 1$ are incorporated to represent the starting and the finishing of the service of the container vessel. Let us define $\overline{\Omega} = \Omega \cup \{0, T\}$. Every task has a processing time, p_i, indicating the neccesary time to load or unload its containers by means of one available quay crane. The processing time of the dummy tasks are $p_0 = p_T = 0$. Let $l_i \in Z^+$, the position of the task $i \in \Omega$, expressed as a bay position of the container vessel.

Every quay crane q has a ready time, r^q, that specifies its earliest possible activity. The initial and final bay positions of the quay crane q are denoted as $l_0^q, l_T^q \in Z^+$, respectively. The required time for a quay crane to move itself between two adjacent bays is $\hat{t}$. The required time for a quay crane to move itself between the container bays i and j is $t_{ij} = \hat{t}|l_i - l_j|$ (so, $t_{0j}^q = \hat{t}|l_q^0 - l_j|$ and $t_{iT}^q = \hat{t}|l_i - l_q^T|$ is the time required by the quay crane q to move from its initial position to the bay j and from the bay i to its final position). Quay cranes can be moved along the length of the container vessel by means of a pair of rails, so that they cannot cross each other and must keep a safety distance δ (measured in container bay units).

There are tasks that need to be done before other ones because they are placed within the same container bay. For example, unloading tasks on the deck must be carried out before unloading tasks in the hold. Let Φ be the set of task pairs within the same container bay for which exists a precedence relationship. On the other hand, Ψ is the set of task pairs that cannot be processed simultaneously. That is,

$$\Phi = \{(i, j) : i \text{ has to be completed before the starting of } j\}$$
$$\Psi = \{(i, j) : i \text{ and } j \text{ cannot be done simultaneously}\}$$

Note that $\Phi \subseteq \Psi$.

The objective of the QCSP is to determine the completition times c_i of all tasks $i \in \overline{\Omega}$ so that the completition time of the last task T (c_T) is minimized; that is, minimizing the makespan.

Figure 13.1 shows an instance of the QCSP with $n = 8$ tasks, $m = 2$ quay cranes, $l_0^1 = 2, l_0^2 = 5, r^1 = r^2 = 0, \hat{t} = 1$ and $\delta = 1$. Processing times and the location of the tasks on the bays are shown in Table 13.1. Additionally,

$$\Phi = \{(1,2), (5,6), (5,7), (6,7)\}$$
$$\Psi = \{(1,2), (3,4), (5,6), (5,7), (6,7)\}$$

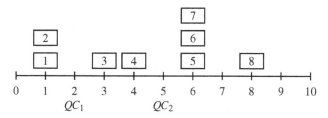

Fig. 13.1 Example of an instance of the Quay Crane Scheduling Problem

Table 13.1 Input data of the example instance

Task i	1	2	3	4	5	6	7	8
Bay position, l_i	1	1	3	4	6	6	6	8
Processing time, p_i	10	8	10	15	7	6	5	10

In this work, the search space is limited to unidirectional schedules. A scheduling is denominated *unidirectional* if the quay cranes have the same sense of moving and this is not changed after the initial positioning (see [1]). The best unidirectional scheduling may be not the optimal scheduling but, usually, it has high quality. On the other hand, without loss of generality, a lexicographic order of the tasks is assumed. That is, tasks are ordered according to their bay position along the container vessel. Figure 13.2 depicts a unidirectional scheduling for the previous instance, where tasks 1, 2, 3, 5, 6 and 7 are done by the quay crane 1 and tasks 4 and 8 are done by the quay crane 2. After the initial positioning of the quay crane 1 on the bay 1 and the quay crane 2 on the bay 4, both quay cranes move unidirectionally from left to right. Note that tasks within the container vessel are sequenced according to the lexicographic order. The lexicographic order indicates that tasks are ordered from left to right and, within each bay, in order of precedence.

The set of feasible solutions for the QCSP is composed by schedules σ that satisfy the precedence and non-simultaneity relations among tasks and safety and interference restrictions among quay cranes. In this sense, the objective is to determine the feasible schedule σ^*, which minimizes the completion time of the last task. The evaluation of every schedule is performed following the scheme based on the disjunctive graph model proposed in [1] so that $f(\sigma)$ denotes the makespan of the schedule σ.

13.3   Estimation of Distribution Algorithm

Estimation of Distribution Algorithms (EDA) [3] are a type of optimization techniques that belong to evolutionary computation. Its operation is based on the use of a probabilistic model that generates new solutions for the population. The probabilistic model is updated through statistical information concerning to the individuals that belong to the population and with the intention of reaching the most promising

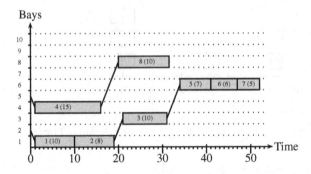

Fig. 13.2 An unidirectional scheduling

Algorithm 13.1. Template of the EDA structure

Initialize model
$p \leftarrow$ Generate initial population
Update probability model
while Stopping criteria is not meet **do**
   $p' \leftarrow$ Create new population
   Copy best individuals from p to p'
   Fill population p'
   Update probability model
   $p \leftarrow p'$
end while

---

regions of the solution space. Unlike other classical evolutionary solution schemes, EDA dismiss the use of mutation or recombination operators by the sampling of the probabilistic model.

Broadly speaking, the EDA scheme used in this work for the solution of the QCSP starts with the initialization of the probabilistic model; that is, for each pair of quay crane-task, an initial probability is assigned. From the initial probabilistic model, a population with a defined number of individuals is generated. The population size, N, is considered as fixed during the life cycle of the algorithm. For each generation of the process a new population is created. This new population consists of a defined percentage, α, of the individuals belonging to the previous population and the rest is completed with random-generated solutions from the probabilistic model. The set of solutions defined by the parameter α is denoted as *top solutions* because they are those with the best objective function value in the population. After the generation of each population the probabilistic model update is performed. These steps are repeated while a stopping criteria is not met. The whole pseudocode of the EDA is sketched in Algorithm 13.1.

Since the solution space is limited to unidirectional schedules, the EDA is applied twice, once to find schedules where the sense of movement of the quay cranes is from left to right along the length of container vessel and another in the reverse sense. The solution provided by the algorithm is the best found in both searches.

In the following several, solution proposals based on the previous scheme are analized.

13.3.1 Probabilistic Model

In order to identify promising regions in the search space, EDA keeps a probabilistic model. In this case, at each generation $g \in G$ (G is the maximum number of generations) a probability matrix $p(g)$ with $|Q|$ rows and $|\Omega|$ columns is used, where each value $p_{qt}(g)$ represents the probability of assigning the task t to the quay crane q. That is,

$$p(g) = \begin{pmatrix} p_{11}(g) & p_{12}(g) & \cdots & p_{1\Omega}(g) \\ p_{21}(g) & p_{22}(g) & \cdots & p_{2\Omega}(g) \\ \vdots & \vdots & \ddots & \vdots \\ p_{Q1}(g) & p_{Q2}(g) & \cdots & p_{Q\Omega}(g) \end{pmatrix}$$

The following constraint must be satisfied during the execution of the algorithm:

$$\sum_{t \in \Omega} p_{qt}(g) = 1, \forall q \in Q, \forall g \in G \tag{13.1}$$

The model initialization is performed setting probabilities to each possible task-quay crane assignment. In this sense, there are many options that can be considered. A first approach is to establish similar probabilities to each value, $p_{qt}(0)$. The specific characteristics of the problem suggest that the most promising solutions are those in which tasks are performed by a quay crane that is near to their bay positions along the container vessel in order to reduce its moving distance. In this case, the probabilistic model initialization aims to reflect this fact. Specifically, the initial probability of each $p_{qt}(0)$ is set to the value of the gaussian function with $\mu = l_0^q$, $\sigma^2 = b_{max} - b_{min}$ (where b_{max} and b_{min} are the right-most and the left-most bay position with at least a task, respectively). This is

$$p_{qt}(0) = \frac{1}{\sqrt{b_{max} - b_{min}}\sqrt{2\pi}} \exp \left\{ \frac{-(x - l_0^q)^2}{2(b_{max} - b_{min})} \right\}$$

with

$$x = |l_0^q - l_t|$$

Initial values of the probabilities model need to be normalized to satisfy (13.1).

13.3.2 EDA Based on Counting

One possible assumption to explore the solution space in an intensive way is that, if a task has been previously assigned to a specific quay crane in a large number

Table 13.2 Characteristics of the benchmark instances

	Categories								
	A	B	C	D	E	F	G	H	I
n	10	15	20	25	30	35	40	45	50
m	2	2	3	3	4	4	5	5	6

of high-quality solutions from the population, this assignment should take place in new generated solutions with a high probability.

At each generation $g \in G$ this scheme keeps a matrix $c(g)$ with $|Q|$ rows and $|\Omega|$ columns. Each value $c_{qt}(g)$ is the number of times that task t has been assigned to the quay crane q in one of the *top solutions* from the population. From the values of this matrix the probabilities of the model for the next generation are calculated.

13.3.3   EDA Based on Fitness

Unlike the previous scheme, the quality of the obtained solutions can take a more prominent role in the updating of the probabilistic model. In this case, changes in the probabilities associated to possible assignments are determined by the value of the objective function of the solutions where they are present.

At each generation $g \in G$ this scheme maintains a matrix $f(g)$ with $|Q|$ rows and $|\Omega|$ columns. Each value $f_{qt}(g)$ corresponds with the sum of the relative objective function values of the *top solutions* from the population in which the task t is assigned to the quay crane q. That is, for each solution σ in *top solutions* the value of the proper indexes in $f_{qt}(g)$ are incremented in an amount of $f(\sigma)^{-1}$ with the intention of encouraging the assignment of tasks to quay cranes that have led to solutions with better objective function value.

13.4   Numerical Results

The aim of this section is to compare the performance of the different schemes proposed in the work with respect to the results of the UDS presented in the previous study [1] to illustrate its adequate use. The set of benchmark problems proposed in [2] is considered. This set is composed by 9 groups with 9 instances each one. The characteristics of the instances are shown in Table 13.2. Additionally, $r^q = 0, \forall q$, $\delta = 1$ and $\hat{t} = 1$ is assumed.

All experiments presented in this section with the EDA approaches have been programmed using the Java language and the executions have been carried out on a PC equipped with an Intel Core 2 Duo E8500 3.16 GHz. On the other hand, UDS was executed on a PC P4 2.8 GHz.

For each instance, the EDA approaches have been executed considering a population where $N = 100$ and $\alpha = 20\%$. Execution has been completed when there have been 50 generations without improvement in the objective function value of the best

Table 13.3 Comparison between EDA approaches and UDS. Small instances

Set	instance	UDS obj.	UDS time	EDA Counting obj.	EDA Counting time	EDA Counting dev.	EDA Fitness obj.	EDA Fitness time	EDA Fitness dev.
A	k13	453		453	4,95E-04	0,00	453	6,48E-04	0,00
	k14	546		546	3,42E-04	0,00	546	3,18E-04	0,00
	k15	513		513	2,43E-04	0,00	513	2,58E-04	0,00
	k16	312		312	3,07E-04	0,00	312	2,93E-04	0,00
	k17	453		453	3,17E-04	0,00	453	2,53E-04	0,00
	k18	375		375	2,45E-04	0,00	375	2,20E-04	0,00
	k19	543		543	2,83E-04	0,00	543	1,88E-04	0,00
	k20	399		399	3,12E-04	0,00	399	2,33E-04	0,00
	k21	465		465	3,15E-04	0,00	465	2,47E-04	0,00
	k22	540		540	4,45E-04	0,00	540	2,63E-04	0,00
			1,12E-05		**3,30E-04**	**0,00**		**2,92E-04**	**0,00**
B	k23	576		576	6,50E-04	0,00	585	5,82E-04	1,56
	k24	666		666	7,62E-04	0,00	714	5,53E-04	7,21
	k25	738		747	8,60E-04	1,22	753	6,72E-04	2,03
	k26	639		639	8,22E-04	0,00	684	5,20E-04	7,04
	k27	657		657	6,45E-04	0,00	657	4,57E-04	0,00
	k28	531		531	1,09E-02	0,00	531	4,82E-04	0,00
	k29	807		807	7,62E-04	0,00	807	4,58E-04	0,00
	k30	891		891	4,80E-04	0,00	984	4,52E-04	10,44
	k31	570		570	1,09E-03	0,00	570	6,00E-04	0,00
	k32	591		591	2,26E-03	0,00	597	6,35E-04	1,02
			3,68E-05		**1,92E-03**	**0,12**		**5,41E-04**	**2,93**
C	k33	603		603	0,03	0,00	603	0,03	0,00
	k34	717		717	0,02	0,00	717	0,03	0,00
	k35	684		684	0,03	0,00	684	0,03	0,00
	k36	678		678	0,03	0,00	678	0,03	0,00
	k37	510		510	0,02	0,00	510	0,02	0,00
	k38	618		618	0,02	0,00	618	0,02	0,00
	k39	513		513	0,03	0,00	519	0,03	1,17
	k40	564		564	0,02	0,00	567	0,02	0,53
	k41	588		588	0,02	0,00	588	0,02	0,00
	k42	573		573	0,03	0,00	573	0,03	0,00
			6,26E-04		**0,03**	**0,00**		**0,03**	**0,17**
D	k43	876		876	0,04	0,00	876	0,06	0,00
	k44	822		822	0,04	0,00	822	0,05	0,00
	k45	834		834	0,04	0,00	840	0,04	0,72
	k46	690		690	0,05	0,00	690	0,05	0,00
	k47	792		792	0,04	0,00	792	0,04	0,00
	k48	639		639	0,04	0,00	639	0,04	0,00
	k49	894		894	0,05	0,00	900	0,05	0,67
			3,43E-03		**0,04**	**0,00**		**0,04**	**0,20**

solution found during the search or when, for a single generation, $2 \cdot N$ random-generated solutions from the probabilistic model have been previously found in the population. These values have been extracted from the experiment carried out during the study.

The comparative analysis between the proposed EDA schemes and the UDS for small and large instances is summarized in Table 13.3 and Table 13.4, respectively. The first two columns indicate the category and the instance considered. Column *UDS* shows the objective function value and the computational time obtained by means of the UDS. UDS was run with a time limit of 1 hour. In Table 13.4 the computational time is not presented in those instances for which this time was insufficient. Column *EDA Counting* represents the objective function value, the computational time and the deviation between the UDS and the EDA based on counting. Similarly, column *EDA Fitness* shows the objective function value, the

Table 13.4 Comparison between EDA approaches and UDS. Large instances

Set	instance	UDS obj.	UDS time	EDA Counting obj.	time	dev.	EDA Fitness obj.	time	dev.
D	k50	741	< 0,01	744	0,05	0,40	759	0,04	2,43
	k51	798	< 0,01	798	0,04	0,00	798	0,03	0,00
	k52	960	< 0,01	960	0,04	0,00	969	0,04	0,94
			< 0,01		**0,04**	**0,13**		**0,04**	**1,12**
E	k53	717	-	717	0,07	0,00	717	0,06	0,00
	k54	774	0,02	774	0,06	0,00	774	0,05	0,00
	k55	684	0,01	690	0,05	0,88	684	0,05	0,00
	k56	690	0,22	693	0,08	0,43	696	0,07	0,87
	k57	705	0,24	708	0,05	0,43	708	0,04	0,43
	k58	786	0,17	789	0,07	0,38	789	0,06	0,38
	k59	687	0,01	687	0,08	0,00	687	0,06	0,00
	k60	783	0,19	789	0,07	0,77	783	0,06	0,00
	k61	639	0,04	639	0,06	0,00	639	0,05	0,00
	k62	837	0,01	855	0,07	2,15	855	0,06	2,15
			0,10		**0,07**	**0,50**		**0,06**	**0,38**
F	k63	948	1,51	954	0,11	0,63	948	0,08	0,00
	k64	741	1,06	744	0,12	0,40	780	0,08	5,26
	k65	837	1,61	843	0,11	0,72	861	0,09	2,87
	k66	924	0,63	930	0,10	0,65	933	0,10	0,97
	k67	882	0,24	885	0,08	0,34	882	0,07	0,00
	k68	963	0,03	972	0,10	0,93	963	0,09	0,00
	k69	807	1,40	807	0,10	0,00	807	0,08	0,00
	k70	957	0,61	960	0,08	0,31	957	0,08	0,00
	k71	834	3,77	840	0,13	0,72	921	0,08	10,43
	k72	744	0,35	750	0,09	0,81	744	0,07	0,00
			1,08		**0,10**	**0,55**		**0,08**	**1,95**
G	k73	870	31,71	879	0,12	1,03	879	0,10	1,03
	k74	843	4,71	855	0,14	1,42	858	0,11	1,78
	k75	675	0,37	684	0,13	1,33	675	0,11	0,00
	k76	852	0,90	864	0,10	1,41	855	0,10	0,35
	k77	699	1,27	714	0,12	2,15	708	0,09	1,29
	k78	642	8,96	651	0,16	1,40	654	0,14	1,87
	k79	744	1,52	774	0,13	4,03	759	0,14	2,02
	k80	750	1,28	771	0,11	2,80	759	0,10	1,20
	k81	738	1,28	744	0,12	0,81	744	0,10	0,81
	k82	717	1,03	738	0,12	2,93	717	0,11	0,00
			2,37		**0,13**	**1,93**		**0,11**	**1,04**
H	k83	948	6,37	957	0,17	0,95	948	0,12	0,00
	k84	897	3,29	909	0,21	1,34	897	0,17	0,00
	k85	972	5,82	990	0,16	1,85	984	0,14	1,23
	k86	816	-	825	0,17	1,10	819	0,14	0,37
	k87	867	-	885	0,20	2,08	885	0,17	2,08
	k88	768	43,73	780	0,15	1,56	771	0,15	0,39
	k89	843	10,96	849	0,17	0,71	843	0,13	0,00
	k90	1053	24,95	1080	0,19	2,56	1068	0,15	1,42
	k91	837	10,74	849	0,19	1,43	849	0,12	1,43
	k92	897	34,61	918	0,15	2,34	912	0,15	1,67
			19,16		**0,17**	**1,59**		**0,14**	**0,86**
I	k93	816	-	846	0,20	3,68	831	0,20	1,84
	k94	786	-	816	0,24	3,82	834	0,22	6,11
	k95	834	-	867	0,21	3,96	843	0,19	1,08
	k96	819	-	840	0,23	2,56	837	0,21	2,20
	k97	720	-	732	0,25	1,67	738	0,22	2,50
	k98	735	23,79	765	0,22	4,08	747	0,17	1,63
	k99	852	-	867	0,23	1,76	864	0,18	1,41
	k100	900	-	906	0,25	0,67	897	0,22	-0,33
	k101	813	-	831	0,20	2,21	819	0,19	0,74
	k102	903	-	921	0,23	1,99	900	0,21	-0,33
			23,79		**0,22**	**2,64**		**0,20**	**1,68**

computational time and the deviation between the UDS and the EDA based on fitness. Computational times reported in the tables are measured in minutes while the deviation between an EDA scheme and the UDS is defined as follows

$$dev. = 100 \cdot \frac{(\text{EDA objective - UDS objective})}{\text{UDS objective}}$$

Average computational times and average deviations are reported for each category of instances (depicted in bold).

As can be drawn from an analysis of the presented results, the application of the proposed algorithms can achieve satisfactory results for the instances considered in this study. In both cases, the average deviation for each set of instances is less than 3% compared to the results achieved with UDS. Despite the difference in the characteristics of the computers used in the study, the computational times used by the EDA approaches are significantly lower than those employed by the UDS, except for very small instances. In this sense, the largest computational time spent at solving an instance has been 0.25 minutes.

Making a comparison between the two proposals presented in this work, it is possible to observe that performance of EDA based on fitness is higher than the performance of EDA based on counting as the size of instances increases. The reason can be found in the fact that the direct consideration of the quality of the solutions in the update of the probabilistic model allows a more proper distinction in the assignments of tasks to more promising quay cranes than those made by simple counting. This has improved a 0.33% the objective function value in two of the largest instances between those considered.

13.5 Concluding Remarks and Future Work

EDA are a type of techniques that allow to achieve satisfactory results when solving the Quay Crane Scheduling Problem using simple exploration components. In this study, two resolution schemes have been proposed based on the counting of tasks-to-quay cranes assignments and information about the quality of the solutions from the population during the updating process of the probabilistic model. At the same time, with the intention of achieving high quality solutions inherent characteristics of the problem are exploited in order to establish a probabilistic model that can generate a priori initial solutions with good quality.

The experiments carried out in this paper show that some encouraging preliminary experimental results have been achieved and can be used as a basis for future works. Future studies should consider including restarting systems of the probabilistic model to avoid situations in which the generation of new solutions become ineffective, intelligent stopping criteria and updates of the probabilistic model with more precise information about the characteristics of the solutions reached in previous generations.

Acknowledgements. This work has been partially funded by the European Regional Development Fund, Spanish Ministry of Science and Technology (projects TIN2009-13363 and TIN2008-06872-C04-01), Canary Government (project PI2007/019) and University of La Laguna.

References

[1] Bierwirth, C., Meisel, F.: A fast heuristic for quay crane scheduling with interference constraints. Journal of Scheduling 12(4), 345–360 (2009)
[2] Kim, K.H., Park, Y.-M.: A crane scheduling method for port container terminals. European Journal of Operational Research 156(3), 752–768 (2004)
[3] Larrñaga, P., Lozano, J.A. (eds.): Estimation of Distribution Algorithms: A New Tool for Evolutionary Computation. Kluwer, Boston (2002)
[4] Moccia, L., Cordeau, J.-F., Gaudioso, M., Laporte, G.: A branch and cut algorithm for the quay crane scheduling problem in a containera terminal. Naval Research Logistics 53(1), 45–59 (2006)
[5] Sammarra, M., Cordeau, J.-F., Laporte, G., Flavia Monaco, M.: A tabu search heuristic for the quay crane scheduling problem. Journal of Scheduling 10(4-5), 327–336 (2007)

Chapter 14
Nash Extremal Optimization and Large Cournot Games

Rodica Ioana Lung, Tudor Dan Mihoc, and D. Dumitrescu

Abstract. Equilibria detection in large games represents an important challenge in computational game theory. A solution based on generative relations defined on the strategy set and the standard Extremal Optimization algorithm is proposed. The Cournot oligopoly model involving up to 1000 players is used to test the proposed methods. Results are compared with those obtained by a Crowding Differential Evolution algorithm.

14.1 Introduction

Similarities and differences between mathematical games and multiobjective optimization problems (MOPs) have long been a source of challenges for computer scientists due to their complex features. A game is defined as the ensemble formed by a set of players, a set of actions available to each player and, most important, a set of payoff functions that each player wishes to maximize. The multi-objective (or many-objective) [4] optimization problem consist of a set of functions - most often conflicting - to be maximized (minimized) subject to some constraints. The most popular solution concept for games is the Nash equilibrium (NE) [8] while for MOPs is the Pareto set.

The Nash equilibrium concept is fundamentally different from that of Pareto optimality. The Nash equilibrium implies that no player can improve its payoff while all

Rodica Ioana Lung
Babes-Bolyai University
e-mail: rodica.lung@econ.ubbcluj.ro

Tudor Dan Mihoc
Babes-Bolyai University
e-mail: mihoct@cs.ubbcluj.ro

Dan Dumitrescu
Babes-Bolyai University
e-mail: ddumitr@cs.ubbcluj.ro

D.A. Pelta et al. (Eds.): NICSO 2011, SCI 387, pp. 195–203, 2011.
springerlink.com   © Springer-Verlag Berlin Heidelberg 2011

the others keeps their strategies unchanged. Pareto efficient solutions are such that no objective function can be improved without decreasing other ones. While the problem of detecting the Pareto frontier of a MOP has been widely addressed from an evolutionary point of view, either through the Pareto or non-Pareto approaches, the Nash equilibria of a non-cooperative game was until recently unaccessible because of the lack of a similar relation for strategy profiles.

The generative relation [5] that allows Nash equilibria detection by using a adapted version of an evolutionary algorithm has already been used for Nash equilibria detection in large Cournot games by adapting a Crowding based Differential Evolution algorithm for multimodal optimization.

A new approach based on an Extremal Optimization algorithm combined with the generative relation for NEs detection, called Nash Extremal Optimization (NEO) is proposed. Basicaly, NEO uses the generative relation instead of the usual order relation from $\mathbb{R}^n$ whenever two individuals are compared.

14.2 Generative Relations for Nash Equilibria

A generative relation for Nash equilibria is a relation between two strategy profiles that enables their comparison with respect to the Nash solution concept, i.e. it evaluates which one is 'closer' to equilibrium. In [5] such a generative relation has been introduced. Solutions that are non-dominated/ascended with respect to this relation are exactly the Nash equilibria of the game [5].

14.2.1 Nash Ascendancy Relation

A finite strategic game is defined by $\Gamma = ((N, S_i, u_i), i = 1, n)$ where:

- N represents the set of players, $N = \{1, ..., n\}$, n is the number of players;
- for each player $i \in N$, S_i represents the set of actions available to him, $S_i = \{s_{i_1}, s_{i_2}, ..., s_{i_{m_i}}\}$ where m_i represents the number of strategies available to player i and $S = S_1 \times S_2 \times ... \times S_N$ is the set of all possible situations of the game;
- for each player $i \in N$, $u_i : S \to \mathbb{R}$ represents the payoff function.

Denote by (s_{i_j}, s^*_{-i}) the strategy profile obtained from s^* by replacing the strategy of player i with s_{i_j} i.e.

$$(s_{i_j}, s^*_{-i}) = (s^*_1, s^*_2, ..., s^*_{i-1}, s_{i_j}, s^*_{i+1}, ..., s^*_1).$$

A strategy profile $s \in S$ for the game Γ represents a Nash equilibrium [7, 8] if no player has anything to gain by changing his own strategy while the others do not modify theirs.

Several methods to compute NE of a game have been developed. For a review on computing techniques for the NE see [7].

Consider two strategy profiles s and s' from S. An operator $k : S \times S \to \mathbb{N}$ that associates the cardinality of the set

$$k(s,s') = |(\{i \in \{1,...,n\}|u_i(s'_i,s_{-i}) \geq u_i(s), s'_i \neq s_i\}|$$

to the pair (s,s') is introduced.

This set is composed by the players i that would benefit if - given the strategy profile s - would change their strategy from s_i to s'_i, i.e.

$$u_i(s'_i,s_{-i}) \geq u_i(s).$$

Let $s,s' \in S$. We say the strategy profile s *Nash ascends* the strategy profile s' in and we write $s \prec s'$ if the inequality

$$k(s,s') < k(s',s)$$

holds.

Thus a strategy profile s ascends strategy profile s' if there are less players that can increase their payoffs by switching their strategy from s_i to s'_i than vice-versa. It can be said that strategy profile s is more stable (closer to equilibrium) then strategy s'.

Two strategy profiles $s,s' \in S$ may have the following relation:

1. either s dominates s', $s \prec s'$ $(k(s,s') < k(s',s))$
2. either s' dominates s, $s' \prec s$ $(k(s,s') > k(s',s))$
3. or $k(s,s') = k(s',s)$ and s and s' are considered indifferent (neither s dominates s' nor s' dominates s).

The strategy profile $s^* \in S$ is called non-ascended in Nash sense (NAS) if

$$\nexists s \in S, s \neq s^* \text{such that } s \prec s^*.$$

In [5] it is shown that all non-ascended strategies are NE and also all NE are non-ascended strategies. Thus the Nash ascendancy relation can be used to characterize the equilibria of a game and can be considered as a generative relation for NEs.

14.3   Nash Extremal Optimization

Extremal Optimization (EO) [1, 2] is a general-purpose heuristic for finding high-quality solutions for hard optimization problems. EO has been adapted to detect NEs of noncooperative games resulting in a new method called Nash Extremal Optimization (NEO).

The main feature of EO is that the value of undesirable variables in a sub-optimal solution are replaced with new, random ones. A single candidate solution is used to search the space. Depending on the problem and the representation used this solution may be formed of several components. EO assigns a fitness to each individual component of the candidate solution and rankes them accordingly. Each iteration of the EO the component having the worst fitness is randomly and unconditionally altered; if the new solution is better then the best so found so far, it will replace it.

Within NEO the candidate solution represents a strategy profile $s \in S$, $s = (s_1, s_2, ..., s_n)$ of the game to be solved. Each component $j, j = 1, ..., n$ of strategy profile s represents the strategy of player j in that situation of the game. A natural fitness for each player i is its payoff $u_i(s)$ of wich the 'worst' $u_j(s)$ is identified:

$$u_j(s) \leq u_i(s), \forall i \in \{1, ...n\}, i \neq j.$$

The only measure of quality on s is provided by the ranking of the player j, implying that all the other players are gaining more than this player for this state of the game.

Aside from this ranking there are no other parameters to adjust for selecting better solutions. The strategy of player j is randomly altered irrespective the new strategy is better or not than the previous one. Thus the near - optimal solutions are enforced by the bias against the worst solutions of the game.

Using the generative relation for Nash equilibria the new profile strategy s' is compared with the best candidate for a solution found so far s_{best}. If s' dominates s_{best}, with respect to the Nash ascendancy relation, s' becomes the new best candidate, and the search continues until a termination condition is met.

The Nash Extremal Optimization algorithm is described in Aglorithm 14.1.

Algorithm 14.1. Nash Extremal Optimization procedure

1: Initialize configuration $s = (s_1, ..., s_n)$ at will; set $s_{best} := s$;
2: **repeat**
3:    For the 'current' configuration s evaluate u_i for each player i;
4:    find j satisfying $u_j \leq u_i$ for all $i, i \neq j$, i.e., j has the "worst payoff";
5:    change s_j randomly;
6:    **if** (s _Nash ascends_ s_{best}) **then**
7:       set $s_{best} := s$;
8:    **end if**
9: **until** the desired number of iterations;
10: Return s_{best}.

However, within NEO, as well as in EOs, the worst solution may be blocked (no further improvment is possible). This could represent a problem as the step 5 always selects only the worst solution/player to be altered. If this happens the search process may stop without reaching even a local optimum.

Crowding Differential Evolution (CrDE) extends the Differential Evolution (DE) algorithm with a crowding scheme [9]. The only modification to the conventional

DE is made regarding the individual (parent) being replaced. Usually, the parent producing the offspring is substituted, whereas in CrDE the offspring replaces the most similar individual among the population if it *Nash ascends* it. A *DE/rand/1/exp* scheme is used. In the form presented here, CrDE has already been used in Nash equilibria detection for large Cournot games [6].

Within CrDE individuals from population P represent strategy profiles of the game that are randomly initialized in the first generation.

CrDE is described in Algorithm 14.3 As long as the final condition is not fulfilled (e.g. the current number of fitness evaluations performed is below the maximum number of evaluations allowed) for each individual i from the population, an offspring $O[i]$ is created using the scheme presented in 14.2, where $U(0,x)$ is a uniformly distributed number between 0 and x, pc denotes the probability of crossover, F is the scaling factor, and dim is the number of problem parameters (problem dimensionality).

Algorithm 14.2. CrDE - the *DE/rand/1/exp* scheme

procedure *create offspring $O[i]$ from parent $P[i]$*

1: $O[i] = P[i]$
2: randomly select parents $P[i_1], P[i_2], P[i_3]$, where $i_1 \neq i_2 \neq i_3 \neq i$
3: $n = U(0, dim)$
4: **for** $j = 0; j < dim \wedge U(0,1) < pc; j = j+1$ **do**
5:     $O[i][n] = P[i_1][n] + F * (P[i_2][n] - P[i_3][n])$
6:     $n = (n+1) \mod dim$
7: **end for**

Algorithm 14.3. CrDE

1: Randomly generate initial population P_0 of strategies;
2: **while** (termination condition) **do**
3:     **for** each $i = \{1,...,n\}$ **do**
4:         create offspring $O[i]$ from parent i;
5:         **if** $O[i]$ *Nash ascends* the most similar parent j **then**
6:             $O[i]$ replaces parent j;
7:         **end if**
8:     **end for**
9: **end while**

In the traditional CrDE, the offspring $O[i]$ replaces the closest parent $P[i]$ if it is fitter. Otherwise, the parent survives and is passed on to the next generation (iteration of the algorithm). Since on our approach the fitness is determined by non-domination with respect to the generative relation, an offspring replaces the parent only if it dominates it with respect to Nash ascendancy.

14.4   Numerical Experiments

Numerical experiments aim at illustrating how NEs can be computed by means
of evolutionary algorithms based on appropriate generative relations for the large
Cournot oligopoly model [3]. Results illustrate the use of the two search operators
for different number of players, up to 1000.

14.4.1   Cournot Oligopoly

Let q_i, $i = 1,...,N$ denote the quantities of an homogeneous product - produced by
N companies respectively. The market clearing price is

$$P(Q) = a - Q,$$

where Q is the aggregate quantity on the market. Hence we have

$$P(Q) = \begin{cases} a - Q, \text{ for } Q < a, \\ 0, \quad \text{ for } Q \geq a. \end{cases}$$

Let us assume that the total cost for the company i of producing quantity q_i is
$C(q_i) = cq_i$. Therefore, there are no fixed costs and the marginal cost c is constant,
$c < a$. Suppose that the companies choose their quantities simultaneously. The pay-
off for the company i is its profit, which can be expressed as:

$$u_i(q_1, q_2, ..., q_N) = q_i P(Q) - C(q_i)$$

If we consider

$$Q = \sum_{i=1}^{N} q_i,$$

then the Cournot oligopoly has one Nash equilibria that can be computed by

$$q_i = \frac{a-c}{N+1}, \forall i \in \{1,...,N\}.$$

Apart from its applications in economy the Cournot oligopoly model can be used
to test the behavior of different evolutionary approaches computing Nash equilibria
for a large number of players.

14.4.2   Parameter Settings

The operators are tested for 10, 50, 100, 250, 500, 750, and 1000 players. Because
the number of payoff functions is equal to the number of players, this setting creates
the equivalent of seven many-objective optimization problem which are known to
be difficult to solve by evolutionary algorithms.

CrDE uses a population of 50 individuals. DE parameters are $F = 0.5$ and $p_c = 0.9$ as used in [9]. NEO does not use any parameters.

Each operator was run 10 times and the average and standard deviation of the minimum distance to the NE for the 30 runs is computed. The stopping criterion is the maximum number of individual payoff functions evaluation of $2 \cdot 10^7$.

14.4.3   Numerical Results

Numerical results present the average and standard deviation of the minimum distance to the NE for both search operators.

Table 14.1 Average and St. Deviation of the distance to Nash equilibria 10, 50, and respectively 100 players

	10 players		50 players		100 players	
	Average	St. Dev.	Average	St. Dev.	Average	St. Dev.
NEO	2.71022	0.018559	2.5459	0.004414	2.53156	0.0279299
CrDE	**0.431526**	0.161009	**1.44694**	0.271021	2.5606	0.102061

For games with a relative small number of players the CrDE algorithm clearly outperforms the NEO (Table 14.1 and Figure 14.1). One of the reasons might be the limitations of the EO – premature convergence to a local optima. In oligopolies with one hundred players there is no statistical difference between the two methods. When more players are involved, the situation changes: NEO clearly outperforms CrDE for all instances tested as illustrated in Table 14.2 and also by the box-plots in Figure 14.2. In fact the results obtained by NEO are very good but this may be related to particularities of Cournot games.

Table 14.2 Average and St. Deviation of the distance to Nash equilibria 250, 500, 750 and respectively 1000 players

	250 players		500 players		750 players		750 players	
	Average	St. Dev.	Average	St. Dev.	Average	St. Dev.	Average	St. Dev.
NEO	**2.53043**	0.103513	**3.3709**	0.463604	**6.84252**	0.429245	**12.3943**	0.421896
CrDE	33.3496	0.966743	58.6151	0.475654	74.6178	0.640812	87.0525	0.430948

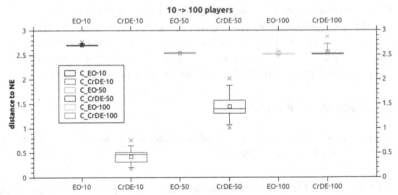

Fig. 14.1 Distance to NE for 10, 50 and 100 players. Box-plots indicate better performance for CrDE for 10 and 50 players

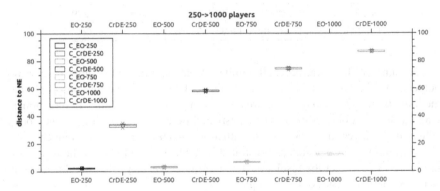

Fig. 14.2 Distance to NE for 250, 500, 750 and 1000 players. Box-plots indicate better performance for NEO in all cases

14.5 Conclusions and Further Work

The performance of the newly proposed Nash Extremal Optimization algorithm has been compared with that of the CrDE on large Cournot oligopolies.

Numerical results are most interesting as they indicate that the although the performance of NEO is not very good for small number of players (up to 100) it copes much better than CrDE with higher number of players for Cournot oligopolies. This indicates the potential of the proposed approach in surpassing the problem of solving games with large number of players. However further studies to determine if NEO performs as well for different types of games have to be carried on.

Acknowledgements. This publication was made possible through the support of a grant from the John Templeton Foundation. The opinions expressed in this publication are those of the authors and do not necessarily reflect the views of the John Templeton Foundation. This research is also supported by Grant TE 320 - Emergence, auto-organization and evolution: New computational models in the study of complex systems, funded by CNCSIS, Romania and from the SECTORAL OPERATIONAL PROGRAMME HUMAN RESOURCES DEVELOPMENT, Contract POSDRU 6/1.5/S/3 Doctoral studies: through science towards society, Babeş - Bolyai University, Cluj - Napoca, România.

References

[1] Boettcher, S., Percus, A.G.: Optimization with Extremal Dynamics. Physical Review Letters 86, 5211–5214 (2001), doi:10.1103/PhysRevLett.86.5211
[2] Boettcher, S., Percus, A.G.: Extremal optimization: an evolutionary local-search algorithm. CoRR cs.NE/0209030 (2002)
[3] Daughety, A.F.: Cournot oligopoly: characterization and applications. Cambridge University Press, Cambridge (1988)
[4] Ishibuchi, H., Tsukamoto, N., Nojima, Y.: Evolutionary many-objective optimization. In: 3rd International Workshop on Genetic and Evolving Systems, GEFS 2008, pp. 47–52 (2008), doi:10.1109/GEFS.2008.4484566
[5] Lung, R.I., Dumitrescu, D.: Computing nash equilibria by means of evolutionary computation. Int. J. of Computers, Communications & Control III(suppl. issue), 364–368 (2008)
[6] Lung, R.I., Mihoc, T.D., Dumitrescu, D.: Nash equilibria detection for multi-player games. In: IEEE Congress on Evolutionary Computation, pp. 1–5 (2010)
[7] McKelvey, R.D., McLennan, A.: Computation of equilibria in finite games. In: Amman, H.M., Kendrick, D.A., Rust, J. (eds.) Handbook of Computational Economics, vol. 1, ch. 2, pp. 87–142. Elsevier, Amsterdam (1996)
[8] Nash, J.F.: Non-cooperative games. Annals of Mathematics 54, 286–295 (1951)
[9] Thomsen, R.: Multimodal optimization using crowding-based differential evolution. In: Proceedings of the 2004 IEEE Congress on Evolutionary Computation, pp. 1382–1389. IEEE Press, Portland (2004)

Chapter 15
An Adaptive Tie Breaking and Hybridisation Hyper-Heuristic for Exam Timetabling Problems

E.K. Burke, R. Qu, and A. Soghier

Abstract. Graph colouring heuristics have long been applied successfully to the exam timetabling problem. Despite the success of a few heuristic ordering criteria developed in the literature, the approaches lack the ability to handle the situations where ties occur. In this paper, we investigate the effectiveness of applying tie breakers to orderings used in graph colouring heuristics. We propose an approach to construct solutions for our problem after defining which heuristics to combine and the amount of each heuristic to be used in the orderings. Heuristic sequences are then adapted to help guide the search to find better quality solutions. We have tested the approach on the Toronto benchmark problems and are able to obtain results which are within the range of the best reported in the literature. In addition, to test the generality of our approach we introduced an exam timetabling instance generator and a new benchmark data set which has a similar format to the Toronto benchmark. The instances generated vary in size and conflict density. The publication of this problem data to the research community is aimed to provide researchers with a data set which covers a full range of conflict densities. Furthermore, it is possible using the instance generator to create random data sets with different characteristics to test the performance of approaches which rely on problem characteristics. We present the first results for the benchmark and the results obtained show that the approach is adaptive to all the

E.K. Burke
Automated Scheduling, Optimisation and Planning (ASAP) Group School of CSIT,
University of Nottingham, Nottingham, NG8 1BB, U.K.
e-mail: ekb@cs.nott.ac.uk

R. Qu
Automated Scheduling, Optimisation and Planning (ASAP) Group School of CSIT,
University of Nottingham, Nottingham, NG8 1BB, U.K.
e-mail: rxq@cs.nott.ac.uk

A. Soghier
Automated Scheduling, Optimisation and Planning (ASAP) Group School of CSIT,
University of Nottingham, Nottingham, NG8 1BB, U.K.
e-mail: azs@cs.nott.ac.uk

D.A. Pelta et al. (Eds.): NICSO 2011, SCI 387, pp. 205–223, 2011.
springerlink.com   © Springer-Verlag Berlin Heidelberg 2011

problem instances that we address. We also encourage the use of the data set and generator to produce tailored instances and to investigate various methods on them.

15.1 Introduction

The Operational Research and Artificial Intelligence research communities have been addressing timetabling problems and research issues since the 1960s [8]. Exam timetabling has been particularly well studied. Survey papers and overviews of different approaches developed are presented in [5, 8, 17]. In [5], the importance of graph colouring methods when applied to timetabling problems is highlighted. A survey which includes the different approaches developed in exam timetabling from 1986 to 1996 is presented in [8]. Furthermore, a recent survey is presented in [17] summarising the recent approaches developed in the past decade.

An exam timetabling problem involves assigning a number of exams to a certain number of timeslots taking into account different constraints, some of which have to be satisfied (hard constraints) and some of which we would like to satisfy if possible (soft constraints). A solution without hard constraint violations is called a feasible solution. The quality of a feasible solution is determined by measuring the level of soft constraint violation.

A wide variety of search methods have been investigated for exam timetabling over the years. Examples include Tabu Search (e.g. [11, 20], Simulated Annealing (e.g. [12, 19]), Evolutionary Algorithms (e.g. [10, 15, 21]), Case Based Reasoning (e.g. [6]) and Fuzzy Methodologies (e.g. [1]). In particular, graph colouring heuristics have been widely used to solve exam timetabling problems by underpinning constructive heuristic solution methods (e.g. [5, 9]).

Many of the approaches mentioned above involve the fine-tuning of parameters and so may not work as effectively on different problems and even different instances. This observation has motivated the development of hyper-heuristics [4, 18]. One of the goals of hyper-heuristics is to increase the level of generality at which search methods can automatically solve a range of problems. A hyper-heuristic can choose the appropriate heuristics to be applied to the problem in hand. Hence, it is concerned with the exploration of a search space of heuristics rather than directly acting upon a problem to find a solution. In the literature, several hyper-heuristic techniques have been developed to solve exam timetabling problems (e.g. [7, 13]).

Graph heuristics have been hybridised in an iterative adaptive approach developed in [3]. In each iteration, the exams are ordered using different graph heuristics and the exam to be scheduled is chosen. In addition, the exams that are found to be difficult to schedule in previous iterations are adaptively moved forward. Another approach has been developed in [16] where different graph heuristics are adaptively hybridised. Different ordering strategies in graph heuristics are adaptively called at a higher level. However, a random exam is selected in the situation where a tie appears. Hence, instead of choosing these exams randomly, tie breaking is a way to help guide a search in scenarios where two or more exams have the same weight in a specific ordering.

In this paper, we investigate the effectiveness of applying tie breakers to heuristic orderings. Furthermore, we present an adaptive approach where tie breakers are applied to the Saturation Degree heuristic followed by a hybridisation with another heuristic to solve exam timetabling problems. In addition, we develop an exam timetabling instance generator to test the adaptiveness of our approach on randomly generated data.

The remainder of this paper is structured as follows. Section 15.2 gives a short description of graph heuristics and previously developed approaches for exam timetabling problems. Section 15.3 presents the Toronto benchmark and the instance generator we developed to test our approach. In Section 15.4, we present an investigation on the effectiveness of hybridisations and tie breaking in graph heuristics. We describe our approach and discuss our results in Section 15.5. In Section 15.6 we give some concluding remarks.

15.2   The Graph Based Hyper-Heuristic (GHH)

Graph heuristics are simple constructive approaches which have been widely applied in exam timetabling. Exam timetabling problems with only hard constraints can be represented as graph colouring models. Vertices in the graph represent exams in the problem, and edges represent the conflicts between exams. Colours correspond to time slots. Hence, if the exam timetabling problem is considered as a graph colouring problem, the aim is to find the minimum number of time slots which are able to accommodate all the exams without any clashes. Several heuristics (e.g. [2]) have been developed to solve graph colouring problems.

By analysing the student enrolment list, the exams that are in conflict (exams that have at least one common student) can be identified. The graph heuristics are usually used to order exams that are not yet scheduled according to different difficulty criteria. For example, by using Largest Enrolment (see Table 15.1), the exams are ordered in a descending order according to the number of students taking them and then they are scheduled one by one to construct a timetable. The objective of creating such orderings is to deal with the most difficult exams in the problem first, to construct good quality timetables.

Hyper-heuristic research is concerned with building systems which can automatically guide and adapt a search according to the structure of the problem in hand. This can be achieved through the exploitation of the structure of a problem, and the creation of new heuristics for that problem, or intelligently choosing from a set of pre-defined heuristics. In other words, hyper-heuristic research aims to automate the heuristic design process through automating the decision of which heuristics to employ to explore the solution space for a new problem. The aim here is to automate the heuristic design process and produce optimisation tools available to stakeholders who require quick and cheap solutions for their optimisation problems.

In our previous work [7], tabu search was used to search a set of heuristic sequences (rather than the solutions themselves) which consisted of the first five graph heuristics presented in Table 15.1 and a random ordering strategy. In every step of

Table 15.1 Ordering Criteria in Graph Heuristics for Exam Timetabling

Graph Heuristic	Ordering Criteria
LD (Largest Degree)	Descending order by the number of conflicts each exam has with others (i.e. conflicted exams are the ones that have students in common)
LWD (Largest Weighted Degree)	Similar to LD but is weighted by the number of students involved in the conflict
LE (Largest Enrolment)	Descending order of the number of students taking each exam
SD (Saturation Degree)	Ascending order of the number of remaining valid timeslots to assign an exam without causing conflicts
CD (Colour Degree)	Descending order of the number of conflicts each exam has with the exams already scheduled
LUD (Largest Uncoloured Degree)	Descending order of the number of conflicts each exam has with the unscheduled exams
LUWD (Largest Uncoloured Weighted Degree)	Similar to LUD but is weighted by the number of students involved in the conflict

tabu search, a sequence of these heuristics is used to construct a solution. Therefore, the search space of the tabu search consists of all the possible sequences of the low-level graph heuristics described in Table 15.1. A tabu search move in this case was to create a new sequence by changing two heuristics in the previous heuristic list. A parameter was used to decide the number of exams scheduled in every iteration of the solution construction. The list is then applied to the problem in hand and the solution obtained is evaluated. If a better solution is obtained it is saved and the heuristic list is stored in the tabu list or otherwise it is discarded. The process continues for a number of pre-defined iterations depending on the problem size. The exams are scheduled to the first timeslot that leads to the least cost. Figure 15.1 presents the pseudocode of the approach. This approach and another approach developed in [16] can adaptively search and hybridise different low-level heuristics.

From these previously developed approaches, it was observed that Saturation Degree performs the best in most cases. The saturation degree of each unscheduled exam is calculated in each iteration and the exams are ordered according to the number of remaining timeslots where an exam can be scheduled without violating any hard constraint. However, all the exams have the same saturation degree at the beginning of the solution construction as the timetable is empty. Ties could also occur anytime during the construction process when the same number of timeslots is available for two exams. In this work, we analyse the effectiveness of applying different tie breakers to the Saturation Degree heuristic (Section 15.4). In addition, we develop an intelligent adaptive approach (Section 15.5) which determines the heuristic to use in a hybridisation with Saturation Degree and the amount of hybridisation required to achieve the best solutions. First, we describe the data we used to test our approach in the next section.

```
Create a heuristic sequence  h1 = {h₁,h₂,h₃......hₖ}  //k: number of exams/exams scheduled by each heuristic
//Tabu Search on Heuristic Sequences
for i = 1 to i = number of iterations
        h =  change two heuristics from the sequence h1
        if h is not in the tabu list
                Construct a solution using h
                if a feasible solution (s_c) is obtained & s_c < s  //s: best solution so far
                        save the best solution, s = s_c
                        update the tabu list
                        h1 = h
        //end if
//end of Tabu Search
Output the best solution s
```

Fig. 15.1 The pseudo-code of the Tabu Search graph based hyper-heuristic [7]

15.3   Benchmark Exam Timetabling Problems

15.3.1   The Toronto Benchmark Exam Timetabling Problem

This data set is widely used in exam timetabling research by different state-of-the-art approaches and can therefore be considered as a benchmark [9]. It consists of 13 real-world problems from 3 Canadian high schools, 5 Canadian, 1 American, 1 UK and 1 mid-east universities. The problem has one hard constraint where conflicting exams cannot be assigned to the same timeslot. In addition, a soft constraint is present where conflicting exams should be spread throughout the timetable as far as possible from each other. The goal here is to minimise the sum of proximity costs given as follows:

$$\sum_{i=0}^{4}(w_i \times N)/S$$

where

- $w_i = 2^{4-i}$ is the cost of assigning two exams with i time slots apart. Only exams with common students and are four or less time slots apart cause violations
- N is the number of students involved in the conflict
- S is the total number of students in the problem

The characteristics of the problem instances used are presented in Table 15.2. Conflict density represents the ratio between the number of exams in conflict to the total number of exams.

15.3.2   The Exam Timetabling Instance Generator

Different exam timetabling benchmark problems were collected and widely studied over the years due to the high-level of research interest in this area [17]. The data was collected from real world problems present in several institutions over the world. The benchmarks provide a means of comparison and evaluation to the different approaches developed. In addition to the Totonto benchmark described in the previous

Table 15.2 Characteristics of the Toronto Benchmark data set [17]

Problem	Exams I/II	Students I/II	Enrolments I/II	Density	Timeslots
car91	682	16925	56877	0.13	35
car92	543	18419	55522	0.14	32
ear83 I	190	1125	8109	0.27	24
ear83 II	189	1108	8057	0.27	24
hec92 I	81	2823	10632	0.42	18
hec92 II	80	2823	10625	0.42	18
kfu93	461	5349	25113	0.06	20
lse91	381	2726	10918	0.06	18
sta83 I	139	611	5751	0.14	13
sta83 II	138	549	5417	0.19	35
tre92	261	4360	14901	0.18	23
uta92 I	622	21266	58979	0.13	35
uta92 II	638	21329	59144	0.12	35
ute92	184	2750	11793	0.08	10
yor83 I	181	941	6034	0.29	21
yor83 II	180	919	6002	0.3	21

section, the international timetabling competition (ITC2007) exam timetabling track was recently introduced [14]. It is a data set which is more complex as it contains a larger number of constraints. As the focus of this work is on tiebreaking and hybridisations, the approach developed in this paper was not applied to the ITC2007 data set which requires different heuristics to schedule exams in rooms following certain constraints.

Since the Toronto benchmark is collected from real-world problems present in several institutions, they do not provide a full coverage of all possible problem scenarios. For example, the benchmark data set does not contain instances with conflict density in the range from 19%-27% and 29%-42% (see Table 15.2).

In this paper, we introduce an exam timetabling instance generator and a benchmark exam timetabling problem data set that consists of 18 different problem instances. The 18 problems consist of 9 small and 9 large problems. To be more precise, a problem is considered small if it does not contain more than 100 exams while a large problem exceeds 500 exams. The problems are generated with conflict density values starting from 6% to 47% using 5% intervals. The number of students and their enrolments are variables according to the problem size and conflict density. The problems have similar constraints and use the same objective function as the Toronto benchmark data set stated in the previous section.

Table 15.3 presents the characteristics of the 18 problem instances in the data set. They vary in several characteristics, not only with respect to the number of exams (ranging from 80 - 567), the conflict density (ranging from 6%-47%), but also the number of enrolments (ranging from 194 - 60168) and the number of timeslots to assign the exams (ranging from 15-70). There are also different numbers of students (ranging from 66 - 15326) across different instances.

The data set is available at http://www.asap.cs.nott.ac.uk / resources/data.shtml where each problem consists of 3 text files. The file names describe the size (SP for small problem & LP for large problem) and conflict density of the problem instances. The representation is extendable to add new characteristics. At the same website, we also provide a solution evaluator to evaluate the quality of the timetables produced. In addition, the website presents the instance generator we developed to produce the data set.

We investigate the approach developed in this paper by applying it to the Toronto benchmark exam timetabling problems and the data set we generated in Table 15.3. Our approach is tested on the newly developed data set to highlight its generality on problems with a full range of conflict density. There is a large set of problem instances in the existing literature. However, none concerns the generation of a full range of problems for testing approaches which are dependent on problem characteristics.

Table 15.3 Characteristics of the Benchmark data set produced by our problem instance generator

Problem	Exams	Students	Enrolments	Density	Timeslots
SP5	80	66	194	0.07	15
SP10	100	100	359	0.11	15
SP15	80	81	314	0.17	15
SP20	80	83	344	0.19	15
SP25	80	119	503	0.26	15
SP30	80	126	577	0.32	15
SP35	100	145	811	0.36	19
SP40	81	168	798	0.42	19
SP45	80	180	901	0.47	19
LP5	526	1643	5840	0.06	20
LP10	511	2838	10659	0.12	20
LP15	508	3683	14026	0.16	24
LP20	533	5496	21148	0.21	30
LP25	542	7275	27948	0.26	35
LP30	550	8798	34502	0.31	35
LP35	524	9973	38839	0.37	50
LP40	513	10826	42201	0.41	60
LP45	567	15326	60168	0.46	70

15.4 Hybridising and Tie Breaking Graph Heuristics Using the Conflict Density of Exam Timetabling Problems

As previously stated, many ties can appear in a Saturation Degree ordering of exams during solution construction. In previous work [3, 16], an exam is randomly chosen in such situations. The random choice in the earlier stages of the search has a great effect on the quality of the solutions produced. Therefore, a means of breaking these ties is essential. We initially developed a method of breaking ties in Saturation Degree using other graph heuristics. The Saturation Degree after breaking the ties is then hybridised with another heuristic in a sequence where random heuristic sequences with different percentages of hybridisation are generated. Each heuristic in the sequence is used to schedule a single exam. This technique was applied to four instances (HEC92 I, STA83 I, YOR83 I and CAR91 I) of the benchmark exam timetabling problems described in Section 15.3.1 for off-line learning of the best tie breakers and heuristic hybridisations leading to the best solutions for problems with different conflict densities. These instances were chosen as they vary in size and cover a range of conflict densities. The effect of tie breaking and hybridising different low-level heuristics on the quality of the solutions produced is analysed. Figure 15.2 presents the pseudocode of the random graph heuristic sequence generator. The approach starts by constructing different heuristic sequences which consist of two graph heuristics (Saturation Degree and one of the other heuristics described in Table 15.1 using different pre-defined percentages of hybridisation). The sequences are then applied to the instances to construct a solution. For each percentage of hybridisation, 50 heuristic sequences are generated using different seeds to construct solutions. Only the sequences that produce feasible solutions are saved and the rest are discarded (see Figure 15.2). According to the quality of the solutions obtained for each instance, the best tie breaker and low-level heuristic for hybridisation are identified. Further analysis is performed to find the relation of conflict density with the low-level heuristics used and the amount of hybridisation required to guide the search to the best solutions.

In our previous work and work undertaken in [3, 7, 9], it was shown that using Saturation Degree on its own usually outperforms other graph heuristics and results in a better outcome. Therefore, we investigated applying Saturation Degree without breaking ties as the primary heuristic in a sequence generator in one set of experiments. In addition, we applied Saturation Degree while breaking the ties using some other heuristic orderings such as LWD, LD and LE in another set of experiments. Finally, the effect of hybridising Saturation Degree with different percentages of another graph heuristic (i.e. LWD, LD, LUWD, LUD, CD or LE) in both experiments was analysed.

```
Set h₁ as the first graph heuristic to be used (SD, SD tb LWD, SD tb LUWD, SD tb LD)
Set h₂ as the second graph heuristic to hybridise with h₁ (LWD, LD, LUWD, LUD, CD)
for i = 0 to i = 1         // i : Percentage of hybridisation
{
        for n = 1 to n = 50
        {
                Initialise the heuristic sequence h = {h₁ h₁ ......h₁}
                h = randomly change (i x size of h) heuristics in h from h₁ to h₂
                Construct a solution c using h
                if solution c is feasible
                {
                        Calculate the penalty of the solution and save h
                }
                else
                {
                        Discard the sequence h
                }
                n = n + 1
        }
        i = i + 0.05
}
```

Fig. 15.2 The pseudo-code of the random graph heuristic sequence generator using a tie breaker for Saturation Degree (tb:tie-breaker)

15.4.1 Analysis of Saturation Degree Tie Breakers

Several heuristic orderings can be used to break ties that occur in the Saturation Degree list during solution construction. An illustrative example of breaking ties in a Saturation Degree list is shown in Figure 15.3. We assume that the exams are ordered according to their LWD and Saturation Degree as shown. We also assume that the saturation degree is the same for the first 3 exams (i.e. e1, e2 and e3) in the list. In this case, LWD is used to break the ties. The tied exams are re-ordered according to their position in the LWD list. Therefore, e2 is scheduled first and removed from both lists. The Saturation Degree list is then re-ordered again in the next iteration of solution construction.

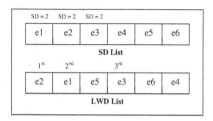

Fig. 15.3 An illustrative example of breaking Saturation Degree ties

We applied several tie breakers to four instances of the Toronto benchmark exam timetabling problems [9]. This was run for 20 times on each instance with different random seeds. Table 15.4 presents a comparison between the results obtained using Saturation Degree without breaking ties and Saturation Degree using different tie breakers.

It is observed that a Saturation Degree ordering without any tiebreakers does not produce a feasible solution when applied to HEC92 I and YOR83 I. After breaking the ties in Saturation Degree, a feasible solution could be obtained for all instances. The results also show that LWD is overall the best tie breaker for Saturation Degree as the exams are ordered according to their largest weighted degree when they have the same saturation degree.

A t-test is also carried out to give an indication if the results using the three different tie breakers are significantly different. Tables 15.5, 15.6 and 15.7 summarise the p-values of the t-tests carried out between the different tie breakers results. It can be seen that the results between the different tiebreakers are significantly different in all the cases except for hec92 I and sta83 I where LD and CD have no significant difference.

Table 15.4 Results using SD without tie breakers and with several different tie breakers. A (-) indicates that a feasible solution could not be obtained. The notation X tb Y denotes Y is used to break ties in X

	hec92 I	yor83 I	sta83 I	tre92
SD without breaking ties	-	-	178.24	9.68
SD tb LWD Best	**13.40**	**43.84**	**166.88**	**9.16**
SD tb LD Best	13.42	44.44	170.20	9.59
SD tb CD Best	13.67	46.78	168.21	9.19

Table 15.5 t-test on the results from breaking the ties using LWD and LD

	hec92 I	yor83 I	sta83 I	tre92
p-value	0.03	1.2E-04	0.01	6.71E-06
t Stat	2.14	6.25	2.59	7.05

Table 15.6 t-test on the results from breaking the ties using LWD and CD

	hec92 I	yor83 I	sta83 I	tre92
p-value	6.6E-03	6.42E-11	0.03	6.4E-06
t Stat	3.30	41.37	1.99	7.08

Table 15.7 t-test on the results from breaking the ties using LD and CD

	hec92 I	yor83 I	sta83 I	tre92
p-value	0.47	6.12E-08	0.27	0.02
t Stat	0.07	17.38	0.61	2.41

15.4.2   Hybridising Heuristic Sequences after Breaking the SD Ties

To analyse the effect of hybridisations, we decided to hybridise the SD ordering using a LWD tie breaker with different graph heuristics and to apply the sequences to the same four instances of the Toronto benchmark exam timetabling problems. Table 15.8 presents the results of applying these different hybridisations as well as a comparison against a hybridisation without using any tie breakers. As shown in Table 15.8, hybridising the tie breaking SD ordering with other graph heuristics produced better results.

For problems where a feasible solution could not be obtained using SD without breaking ties (i.e. HEC92 I and YOR83 I), breaking the ties using LWD and a hybridisation using LD produced the best results. However, for problems where a feasible solution was obtained without handling ties in SD (i.e. STA83 I and TRE92), breaking the ties using LWD and a hybridisation using CD performed better. A possible reason could be the occurrence of many ties in the SD ordering which prevents it from producing a feasible solution. Therefore, when tie breaking the SD ordering and hybridising it with LD, some exams will be ordered according to their number of conflicts with unscheduled exams. Generating a feasible solution using SD without handling ties is an indication that no ties or a very small number of ties occur. Therefore, applying a tie breaker and hybridising it with CD will order some exams according to the conflicts with the exams already scheduled, producing better results than a LD hybridisation in this case.

A t-test is also carried out to give an indication if the results of using different hybridisations are significantly different. Tables 15.9, 15.10 and 15.11 summarise the p-values of the t-tests. It can be seen that the results between the different hybridisations are significantly different in all the cases except for hec92 I and tre92 when comparing hybridising SD tb LWD with LD and CD where no significant difference can be seen.

15.4.3   Relating Conflict Density of Exam Timetabling Problems to SD Tie Breaking

By analysing the properties of the four problems and the results obtained using the different tie breakers and hybridisations, a relationship becomes apparent between the conflict density of exams in a problem and the percentage of hybridisations required (see Figure 15.4). As shown in Table 15.8, the problems with conflict density

Table 15.8 Results of hybridising SD with other graph heuristics with and without breaking ties. The notation X tb Y denotes Y is used to break ties in X

	hec92 I	yor83 I	sta83 I	tre92 I
SD + LWD Average	13.02	44.59	170.08	9.25
SD + LWD Best	12.20	42.49	164.65	9.02
best % hybridisation	54	18	7	49
SD tb LWD + LD Average	12.60	43.03	168.21	9.03
SD tb LWD + LD Best	**11.89**	**42.16**	168.21	8.94
best % hybridisation	28	24	3	18
SD tb LWD + CD Average	12.69	43.34	163.32	9.00
SD tb LWD + CD Best	12.25	42.61	**159.50**	**8.83**
best % hybridisation	15	9	78	84

Table 15.9 t-test on the results from hybridising SD with LWD and SD tb LWD with LD

	hec92 I	yor83 I	sta83 I	tre92
p-value	2.19E-08	2.45E-08	4.83E-04	2.91E-10
t Stat	7.52	7.47	3.70	9.36

Table 15.10 t-test on the results from hybridising SD with LWD and SD tb LWD with CD

	hec92 I	yor83 I	sta83 I	tre92
p-value	3.54E-04	9.53E-07	1.55E-13	5.26E-10
t Stat	3.82	6.05	13.13	9.09

Table 15.11 t-test on the results from hybridising SD tb LWD with LD and SD tb LWD with CD

	hec92 I	yor83 I	sta83 I	tre92
p-value	0.14	0.003	1.03E-16	0.1
t Stat	1.11	2.98	17.75	1.33

of less than 25% (i.e. less than half of the exams are in conflict) obtained the best results when a hybridisation of less than 50% is used (i.e. more tie breaking SD is used than CD or LD). As the conflict density exceeds 25%, the percentage of the tie breaking SD used increases and the hybridised graph heuristic appears less in the sequences generating the best results. Having a higher conflict density means a higher probability of ties occurring in the SD ordering during the solution construction. Therefore, using SD and breaking the ties proves to be more effective. In contrast, instances with a lower conflict density will have a lower probability of ties occurring in the SD ordering during construction and using a lower percentage of the tie breaking SD is more effective. The conclusions made here using four problems proved to be enough as they were verified later in the paper after running the

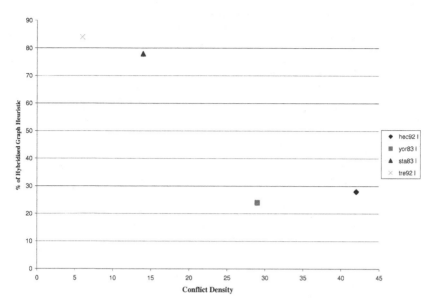

Fig. 15.4 The relation between conflict density and the percentage of hybridisations obtaining the best solutions

approach on the rest of the Toronto benchmark instances and the instances randomly generated. The results are shown in Section 15.5.

15.5   Adaptive Tie Breaking and Hybridisation for Benchmark Exam Timetabling Problems

The above observations indicate that there are two important factors to choose the most effective SD tie breaker and the percentage of hybridisation. The two factors are stated as follows:

- Whether a problem is solved using pure SD or not. If it is solved using pure SD then the tie breaking SD should be hybridised with CD. Otherwise it should be hybridised using LD.
- According to the conflict density of the problem, the percentage range of the tie breaking SD to be used could be defined.

Based on the above off-line learning we developed an approach which adapts to the problem in hand, choosing the most appropriate heuristic to be hybridised with the tie breaking SD and search for a solution within the percentage range of the tie breaking SD in a sequence. Figure 15.5 presents the pseudocode of the Adaptive Tie Breaking (ATB) approach we developed.

```
Schedule all exams using SD only
Set h₁ to SD and set the tiebreaker as LWD      // h₁ : 1ˢᵗ heuristic in hybridisation
if(feasible solution could be obtained using SD only)
{
        Set h₂ to CD      //h₂: 2ⁿᵈ heuristic in hybridisation
}
else
{
        Set h₂ to LD
}
if(Conflict density > 25 %)
{
        i₁ = 0.5;
        i₂ = 1;
}
else
{
        i₁ = 0;
        i₂ = 0.5;
}
for i₁ to i₂        // i₁ to i₂: range of percentages of hybridisation
{
        i = i₁;                // i: percentage of hybridisation
        for  n = 1 to n = number of exams x 10
        {
                Initialise heuristic sequence h = {h₁ h₁ ......h₁}
                h = randomly change (i x size of h) heuristics in h to h₂
                Construct a solution c using h
                if a feasible solution (s_c) is obtained & s_c < s
                {
                        save the best solution, s = s_c
                        save h
                }
                else
                {
                        Discard sequence h
                }
        }
        i = i + 0.05;
        Save the smallest s and the corresponding h
}
```

Fig. 15.5 The psuedocode for the Adaptive Tie Breaking (ATB) approach

According to the conflict density of the problem, the range of percentages of hybridisations is defined. For problems with a conflict density greater than 25%, a percentage of 50% or greater of tie breaking SD is used. For problems with conflict density less than 25%, a percentage of 50% or less of tie breaking SD is used.

We tested this approach on the Toronto benchmark exam timetabling problems and present the results in Table 15.12. Furthermore, we tested the approach on the data set we developed to test the generality of the approach and present the results in Table 15.13. The average computational time across the instances is also presented for 5 runs on a Pentium IV machine with a 1 GB memory. Please note that this run time is acceptable in university timetabling problems because the timetables are usually produced months before the actual schedule is required [16].

To verify the results obtained, we reversed the decisions taken by our approach in choosing a low-level heuristic and a percentage range of hybridisation, and applied

Table 15.12 Results from the the adaptive tie breaking (ATB) approach on the Toronto Benchmark data set, Percentage of tie breaking SD (% of tb SD).Computational time is presented in seconds.

	ATB Average	% of tb SD Average	ATB best	% of tb SD best	Time (s)
hec92 I	12.61	75	11.89	85	274
yor83 I	43.03	71	42.16	81	1683
ear83 I	38.35	90	38.16	77	1475
sta83 I	163.33	42	159.50	56	357
car92	4.5	49	4.44	67	35812
car91	5.35	57	5.23	62	100304
uta92 I	3.64	52	3.50	53	54893
ute92	29.72	43	28.81	48	1469
lse91	11.89	47	11.73	39	2053
tre92	9.01	43	8.83	45	4293
kfu93	16.39	35	15.38	35	2697
hec92 II	12.52	69	11.84	73	346
yor83 II	51.24	91	50.40	84	1144
ear83 II	38.35	90	38.16	82	1457
sta83 II	36.10	47	35.00	42	1415
uta92 II	3.56	46	3.48	48	66339

Table 15.13 Results from the the adaptive tie breaking (ATB) approach on the random data set, Percentage of tie breaking SD (% of tb SD).Computational time is presented in seconds.

	ATB Average	% of tb SD Average	ATB best	% of tb SD best	Time (s)
SP5	3.71	34	3.55	34	10
SP10	11.91	48	10.54	45	30
SP15	17.15	37	15.56	35	29
SP20	20.33	44	18.69	41	33
SP25	25.18	54	23.22	52	43
SP30	33.89	52	31.56	60	78
SP35	47.42	64	45.19	74	141
SP40	28.58	53	27.28	61	109
SP45	32.76	78	31.08	81	100
LP5	9.27	42	9.12	39	4610
LP10	15.15	45	14.96	40	1803
LP15	13.31	29	13.05	24	2540
LP20	11.39	12	11.30	10	5708
LP25	9.93	89	9.90	86	16396
LP30	7.45	72	7.43	72	29367
LP35	6.81	90	6.76	91	40160
LP40	5.23	69	5.22	73	42306
LP45	4.85	72	4.77	81	59045

Table 15.14 Best results obtained by the adaptive tie breaking (ATB) approach compared to other Hyper-Heuristic approaches and the best reported in the literature

Problems	ATB Best	σ	GHH Best (Burke et al., 2007)	Tabu search HH (Kendall, 2005)	Best Reported (Qu et al., 2009a)
hec92 I	11.89	1.90	12.72	11.86	**9.2**
sta83 I	159.50	1.56	158.19	157.38	**157.3**
yor83 I	42.16	4.21	40.13	-	**36.20**
ute92 I	28.81	3.11	31.65	27.60	**24.4**
ear83 I	38.16	6.26	38.19	40.18	**29.3**
tre92	8.83	0.65	8.85	8.39	**7.9**
lse91	11.73	1.50	13.15	-	**9.6**
kfu93	15.38	1.68	15.76	15.84	**13.0**
car92 I	4.44	0.36	4.84	4.67	**3.93**
uta92 I	3.50	0.25	3.88	-	**3.14**
car91 I	5.23	0.52	5.41	5.37	**4.5**

the reversed approach to some of the problems with different characteristics and present a comparison in Table 15.15.

The results obtained indicate the generality of our adaptive approach to all these exam timetabling instances regardless of the problem size. We compare our approach with other approaches where a random exam is chosen in the situation where a tie occurs. The results are presented in Table 15.14. We also highlight the best results reported in the literature. In addition, the standard deviation from the best reported results obtained is shown (σ in Table 15.14). Recall that the aim of the paper is to illustrate the effect of breaking ties in heuristic orderings and automatically hybridising and adapting heuristics. We do not expect to outperform other heuristic and meta-heuristic approaches which are tailored specially for specific instances of this exam timetabling benchmark. However, we demonstrate that we can achieve results that are competitive with the best in the literature.

In comparison with the tabu search based hyper-heuristic in [13], our tie breaking approach performs better in 8 out of 11 cases reported. Furthermore, it outperforms in all the cases in comparison with the pure GHH investigated in [7]. Only the problems presented in Table 15.14 were compared to other results since the others were not reported in the literature. Computational time was also not compared for the same reason.

To validate our results we present the average, best and standard deviation of the results obtained when using different combinations of the low-level heuristic and the percentage range used in the hybridisations in Table 15.15. A paired t-test obtained a stat of 2.1, which is close to 2.11 ($p = 0.05$), demonstrating the adaptiveness of our approach during solution construction, indicating that hybridising the tie breaking SD with CD improves the quality of the solutions for problems solved using only SD. Otherwise a LD hybridisation is better. In addition, the comparison

Table 15.15 A comparison of the results obtained by the adaptive tie breaking and the reverse of the approach

Problems	ATB Average	Reverse ATB Average	ATB Best	Reverse ATB Best	ATB σ	Reverse ATB σ
ute92 I	29.72	30.27	28.81	29.94	0.531	0.120
ear83 I	38.35	38.41	38.16	38.27	0.116	0.091
lse91	11.89	13.04	11.73	12.17	0.099	0.511
kfu93	16.39	17.13	15.38	16.90	0.589	0.142
car92 I	4.5	4.64	4.44	4.49	0.042	0.096
uta92 I	3.56	3.82	3.50	3.52	0.042	0.182
car91 I	5.35	5.51	5.23	5.24	0.076	0.165
SP5	3.71	4.24	3.55	3.82	0.099	0.252
SP10	11.91	11.98	10.54	10.78	0.797	0.702
SP15	17.15	17.71	15.56	15.98	0.924	1.008
SP20	20.33	20.52	18.69	18.89	0.953	0.950
SP25	25.18	25.21	24.17	25.18	0.589	0.030
SP30	33.89	34.22	31.56	32.02	1.351	1.279
SP35	47.42	47.85	45.19	46.00	1.293	1.077
SP40	28.58	29.57	27.28	27.61	0.756	1.140
SP45	32.76	32.86	31.08	31.18	0.976	0.979
LP5	9.27	9.35	9.12	9.29	0.093	0.046
LP25	9.93	9.99	9.90	9.96	0.025	0.03
LP45	4.85	5.03	4.77	4.78	0.053	0.154

indicates that for problems with conflict density greater than 25% more tie breaking SD should be used than any low-level heuristic used in the hybridisation.

15.6   Conclusions

This paper presents an adaptive approach where a Saturation Degree heuristic, using Largest Weighted Degree to break ties, is dynamically hybridised with another low-level heuristic for exam timetabling problems. The hybridisation is performed according to the conflict density of the problem and the ability of the problem to be solved using a pure Saturation Degree (SD) heuristic. Largest Colour Degree First (CD) and Largest Degree First (LD) are used in the hybridisation process. CD is used for hybridisation if the problem is solved using a pure SD heuristic or LD is used otherwise. The amount of heuristic hybridisation is determined according to the conflict density of the problem. If the conflict density of a problem is greater than 25%, using more SD in a hybridisation with LD or CD and breaking the ties was more effective. On the other hand, for problems with conflict density less than 25% using more LD or CD than SD in a hybridisation was more effective. Our approach is simple and performs the same regardless of the problem instance size although large instances take more time to solve. It performs better than a pure graph based hyper-heuristic and obtains reasonably good results when compared to a tabu search hyper-heuristic.

To test the generality of our approach and verify our results, we developed an exam timetabling instance generator and introduced a set of benchmark exam timetabling problems and reported the first results on this data. The generator takes the problem size and conflict density as inputs and generates a random problem with the required characteristics. To encourage scientific comparisons we made the generator and data set available at `http://www.asap.cs.nott.ac.uk/resources/data.shtml`. The aim is to have a larger variety of exam timetabling problem instances with different characteristics.

Future research directions include observing the effect of more problem characteristics such as the number of students and enrolments on the performance of the approaches developed. Using hybridisations of more than two heuristics and different tie breakers during solution construction could also be investigated.

Finally, the solutions obtained in this paper could be further improved by applying a meta-heuristic approach. Therefore, it would be interesting to study the effect of different improvement meta-heuristics on exam timetabling problems. The objective would be choosing the meta-heuristic that would produce the best improvement and automating this process.

References

[1] Asmuni, H., Burke, E.K., Garibaldi, J.M., McCollum, B.: Fuzzy multiple heuristic orderings for examination timetabling. In: Burke, E.K., Trick, M.A. (eds.) PATAT 2004. LNCS, vol. 3616, pp. 334–353. Springer, Heidelberg (2005)
[2] Brelaz, D.: New methods to colour the vertices of a graph. Communications of the ACM 22, 251–256 (1979)
[3] Burke, E., Newall, J.: Solving examination timetabling problems through adaptation of heuristic orderings. Annals of Operations Research 129, 107–134 (2004)
[4] Burke, E., Kendall, G., Newall, J., Hart, E., Ross, P., Schulenburg, S.: Hyper-heuristics: An emerging direction in modern search technology. In: Glover, F., Kochenberger, G. (eds.) Handbook of Meta-Heuristics, pp. 457–474. Kluwer, Dordrecht (2003)
[5] Burke, E., de Werra, D., Kingston, J.: Applications to timetabling. In: Gross, J., Yellen, J. (eds.) Handbook of Graph Theory, ch. 5.6, pp. 445–474. Chapman Hall/CRC Press (2004)
[6] Burke, E., Petrovic, S., Qu, R.: Case based heuristic selection for timetabling problems. Journal of Scheduling 9(2), 115–132 (2006)
[7] Burke, E., McCollum, B., Meisels, A., Petrovic, S., Qu, R.: A graph-based hyper-heuristic for educational timetabling problems. European Journal of Operational Research 176, 177–192 (2007)
[8] Carter, M., Laporte, G.: Recent developments in practical exam timetabling. In: Burke, E.K., Ross, P. (eds.) PATAT 1995. LNCS, vol. 1153, pp. 3–21. Springer, Heidelberg (1996)
[9] Carter, M., Laporte, G., Lee, S.: Examination timetabling: Algorithmic strategies and applications. Journal of Operational Research Society 74, 373–383 (1996)

[10] Cote, P., Wong, T., Sabouri, R.: Application of a hybrid multi-objective evolutionary algorithm to the uncapicitated exam proximity problem. In: Burke, E.K., Trick, M.A. (eds.) PATAT 2004. LNCS, vol. 3616, pp. 151–168. Springer, Heidelberg (2005)

[11] Di Gaspero, L., Schaerf, A.: Tabu search techniques for examination timetabling. In: Burke, E., Erben, W. (eds.) PATAT 2000. LNCS, vol. 2079, pp. 104–117. Springer, Heidelberg (2001)

[12] Duong, T., Lam, K.: Combining constraint programming and simulated annealing on university exam timetabling. In: Proceedings of the 2nd International Conference in Computer Sciences, Research, Innovation & Vision for the Future (RIVF 2004), pp. 205–210 (2004)

[13] Kendall, G., Hussin, N.: An investigation of a tabu search based hyper-heuristic for examination timetabling. In: Kendall, G., Burke, E., Petrovic, S., Gendreau, M. (eds.) Selected Papers from MISTA 2003, pp. 309–328. Springer, Heidelberg (2005)

[14] McCollum, B., Schaerf, A., Paechter, B., McMullan, P., Lewis, R., Parkes, L., Di Gaspero, A. J.: Setting the research agenda in automated timetabling: The second international timetabling competition. INFORMS Journal on Computing 22(1) (2010)

[15] Mumford, C.: An order based evolutionary approach to dual objective examination timetabling. In: Proceedings of the 2007 IEEE Symposium on Computational Intelligence in Scheduling, CI-Sched 2007 (2007)

[16] Qu, R., Burke, E., McCollum, B.: Adaptive automated construction of hybrid heuristics for exam timetabling and graph colouring problems. European Journal of Operational Research 198, 392–404 (2009)

[17] Qu, R., Burke, E., McCollum, B., Merlot, L., Lee, S.: A survey of search methodologies and automated approaches for examination timetabling. Journal of Scheduling 12, 55–89 (2009b)

[18] Ross, P.: Hyper-heuristics. In: Burke, E., Kendall, G. (eds.) Search Methodologies: Introductory Tutorials in Optimisation, Decision Support Techniques, ch. 17, pp. 529–556. Springer, Heidelberg (2005)

[19] Thompson, J., Dowsland, K.: A robust simulated annealing based examination timetabling system. Computer & Operations Research 25, 637–648 (1998)

[20] White, G.M., Xie, B.S.: Examination timetables and tabu search with longer-term memory. In: Burke, E., Erben, W. (eds.) PATAT 2000. LNCS, vol. 2079, pp. 85–103. Springer, Heidelberg (2001)

[21] Wong, T., Cote, P., Gely, P.: Final exam timetabling: a practical approach. In: IEEE Canadian Conference on Electrical and Computer Engineering(CCECE 2002), vol. 2, pp. 726–731 (2002)

Chapter 16
Cooperation and Self-organized Criticality for Studying Economic Crises

Andrei Sîrghi and D. Dumitrescu

Abstract. The economics appears to have not paid the required attention and re-search effort for understanding the *destructive phenomena* in economic systems. Relying on various widely accepted models and theories that practically avoid the occurrence of *failures, depressions, crises* and *crashes* the economic science is incomplete and clearly needs a major reorientation and a change of focus from the deterministic to the holistic perspective.

The causes of unpredictable behaviour in economic systems are investigated and the solutions that can prevent these phenomena are searched for. A new economic model inspired by the *paradigm of complex systems* based on a connective structure is introduced. Interesting complex behaviour emerges from the simple interactions between the agents. A new, unconventional, cause for economic crises is identified and described.

16.1 Introduction

State-of-the art economical models do not pay enough attention to understanding the *destructive phenomena* in economic systems. Relying on various widely accepted models and theories – that practically avoid the occurrence of *failures, depressions, crises*, and *crashes* – the economic science is incomplete and needs a major reorientation and a change of focus from the deterministic to the holistic perspective.

When the world economy gives birth to some surprising situations – the occurrence of which could not have been be predicted even by the most influential economists – some revelatory moments that raise doubts about the objectivity and

Andrei Sîrghi
Centre for the Study of Complexity, Babes-Bolyai University, Cluj-Napoca, Romania
e-mail: andreisirghi@yahoo.com

D. Dumitrescu
Centre for the Study of Complexity, Babes-Bolyai University, Cluj-Napoca, Romania
e-mail: ddumitr@cs.ubbcluj.ro

D.A. Pelta et al. (Eds.): NICSO 2011, SCI 387, pp. 225–237, 2011.
springerlink.com   © Springer-Verlag Berlin Heidelberg 2011

actuality of economic science stand out. The world economic crisis which began in 2008 was such a situation.

The modern economic science, inheriting much of *Neoclassical Synthesis*, is basically build on constraint optimization mathematical models the applicability of which in economy has become more and more sophisticated. Mainstream macroeconomic theory is dominated by *dynamic optimization models*, while *strategic optimization* has become a commonplace in microeconomics. Having a deterministic character, this mathematical base created the illusion that economics is a *good child of science* having an easy to influence behaviour and being fully controllable.

But this idealistic mask of economic science is "betrayed" by its inability to demystify and prevent the appearance of *destructive phenomena* in economic systems, such as: *fluctuations*, *market failures*, *depressions*, *disequilibria*, and *crises*, in attempting to meet the theoretical expectations in the real world economic systems.

This theoretical base for modern economy is a direct influence of: (i) *the principle of reductionism* – used by the Cartesian-Newtonian paradigm which dominated the study of exact sciences until the beginning of twentieth century; (ii) *the neglecting of animal spirits* [1] as an important cause of many disturbances of economic systems; (iii) the desire of many economists to view the economy as a *science governed by exact mathematical models*.

In the last years a new paradigm for studying and thinking about systems has come to prominence, mainly in natural sciences. This paradigm states that many systems cannot be studied using standard approaches because they have a *complex behaviour*. While the goal of conventional economy was to understand *equilibrium* and to create models for finding and maintaining it, the complex systems paradigm moves the priority to understanding the *process of system activity* and its dynamics rather than the state of equilibrium [2]. Moreover, the complex systems paradigm has a different concept about the equilibrium. The conventional economic equilibrium is considered as a *fixed high state of order* described by balanced economic forces, while new paradigm treats the equilibrium as a *dynamical meta-state* of the system which can maintain system viability and follows some principles of acceptability imposed by the normal functioning of the system. Compared to complex systems, conventional systems cannot evolve since their structure cannot change due to their *restrictive* and *static* character of equilibrium.

In this paper we propose a new economic model, defined in terms of a *connective structure* composed of *elements* and *connections* called *Econnectomic Model*. The model is based on the theory of complex systems and introduces a *connective approach* and a *historical dimension* for studying economic systems. Using a revolutionary view the model is appropriate to study all aspect of various economic systems, including the phenomena which tend to disturb them.

The paper is focused primarily on the *phenomena which disturb economic activity* by creating disequilibria, thus raising collapses and crises. We demonstrate that unpredictable behaviour in economic systems can result from small independent shocks generated by the elements inside the system. No exogenous cataclysmic force is needed to create large catastrophic events. Using the proposed model we will analyze the dynamics of these phenomena in a simplified distribution network of the form: producers-distributors-consumers. Despite the simplicity of the analyzed system some complex systems phenomena may be observed: *adaptability, self-organization, fluctuations, domino effects* and *crashes*.

16.2   Related Work

The acknowledgement that almost all interesting economic systems can be classified as *complex* led to a big flux of interdisciplinary theories and models which attempt to understand the old economic problems with new approaches or to discover new aspects of the economic systems. Almost all of these new approaches can be classified using one or a combination of the following categories:

1. **Evolutionary computation.** This approach combines the evolutionary techniques from Artificial Intelligence with economic models in order to find better solutions or to introduce the historical dimension in these models. One of the most fruitful examples in this category is the combination between evolutionary techniques and game theory models. However the resulting models do not change the approach to economic systems in general. They just propose new techniques for solving old economic problems.
2. **Agent models.** This approach represents economic actors as a society of intelligent agents. Usually these models are defined as constraint optimization problems, and are used for finding optimal solutions in conditions of concurrence, or to analyze the behaviour of economic agents in various situations. When the agent societies are large this models became too complex, being very hard to find optimal solutions even with an adequate amount of computational resources.
3. **Network models.** These models use the same principles and structures as agent models but they move the focus from the agents' behaviour to the structure which link them, and study how this structure influences the activity of economic systems. These models are primarily concerned with the occurrence of blockages and tails in the systems' structures – which usually are networks – and with the influence of individual nodes on the network according to their level of connectivity.

As result, there is an emerging literature in economic theory which tries to approach it with instruments from complex systems theory. However, this literature is primarily concerned with *efficiency*, *optimality*, and *equilibrium* tradeoffs. Our model instead is a change of vision for the economic systems *dynamics*, *stability* and *critical behaviour*.

16.3   The Econnectomic Model

In each economic process some *general patterns* may be identified, along with a set of *individual characteristics*. The economic science is oriented mainly on studying general patterns, but we think that individual characteristics are equally important for discovering the real causes and particularities of the analyzed processes.

An economic process can be completely defined by the set of agents involved in it and by their ability to create connections within the economic activity. To denote these things with a single word, we will use the term *"econnectome"*, obtained from the term *"connectome"* – widely used in Neuroscience. In an econnectome each agent is viewed from two perspectives: (i) as an individual entity having its own goals, interests, and activities, and (ii) as a connected entity which is part of a complex community, is dependent on this community for achieving its own goals, and is implied in a global process of the community.

The *econnectome* – being suitable for modelling the structure of various economic systems – represents the structure of our new economic model, called Econnectomic Model (ECM).

The model structure is defined by $ECM = (A, C, \lambda)$, where:

- A is a fixed set of agents, $A = \{a_1, a_2, ..., a_n\}$, where n represents the number of agents. Agents can have different roles, e.g. producer-consumer;
- C is a dynamical set which contains the connections between agents, $C = \{c_1, c_2, ..., c_i, ...\}$. Each connection links two different agents from the set A. Connections show relations who depict the exchange of information, or material between connected agents. When agents participate in economic activity, this set is changing as result of their tendency to optimize connections, by keeping profitable connections and destroying those that generate losses;
- and λ is an incidence function that associates to each connection $c_i \in C$ an ordered pair of agents $\{u, v\}, u, v \in A, u \neq v$, thus: $\lambda(c_i) = \{u, v\}$ - is a function that connects agent u with agent v by connection c_i.

While participating in an economic process each agent in the model tracks the *profitability* (fitness) of their connections. They assign an individual fitness value to each connection. This fitness value is updated after every interaction between agents supported by the corresponding connection. The function that is used to evaluate connection fitness is dependent on the modelled problem and in most cases is an evolution function. The agents can have a particular function $fc(a_i) : C' \to R, C' \subset C, C' = \bigcup c_i, \lambda(c_i) = \{a_i, v\}, \forall a_i, v \in A$ for evaluating the fitness for each of their connections, but further we will consider that all agents use the same function fc. The fitness function governs the evolution and optimization of the ECM structure. Moreover it is a simple but effective mechanism for the implementation of *interactional historical dimension* in our model, which is a key factor in understanding economy as a process.

Additionally, each economic agent has a fitness value which shows if the agent is prepared to take new connections. These fitness values depend on the agents' activities and form a *law of preferential attachment* in the econnectome which govern the process of new connection forming. The agents with a higher fitness are preferred by other agents when trying to connect.

Although the econnectome is a general concept, it can be also represented as a network of economic agents.

16.4   Producers-Distributors-Consumers Network

Using the ECM we will analyze the activity of a producers-distributors-consumers (PDC) distribution network. In the PDC network are involved three types of agents:

- **producers** – the agents who produce economic goods, and generate the supply of economic goods on the market;
- **distributors** – the agents who connect consumers with producers, distributing economic goods from producers to consumers;
- **consumers** – the agents who generate the demand on the market by requesting economic goods.

The connections between consumers, distributors and producers form the distribution chains of the economic goods from producers to consumers. We presume that in PDC network (i) a consumer cannot be directly connected to a producer, so it can connect only to distributors, and (ii) a distributor can be connected to other distributors or to producers. Distributors do not generate demand or supply for economic goods, they just transfer the consumers demand to producers, and economic goods from producers to consumers.

The law of preferential attachment that governs the formation of new connections is based on the nodes fitness. In the PDC network the producers' fitness value is the difference between the quantity of produced economic goods and the quantity of distributed economic goods in current iteration. In other words, the producers' fitness is equal to their quantity of undistributed goods. Thus, if all goods produced by a producer are distributed to consumers, then its fitness is zero. Having the fitness values of producers, the fitness of distributors are recursively calculated as the sum of the reports between the fitness of distributors' sources (agents which provides

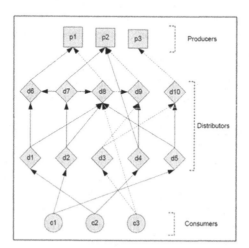

Fig. 16.1 A simplified producers-distributors-consumers network. *The connections with dotted line indicate the complete distribution chain for the consumer c3 in the network.*

distributors with goods) and the number of agents supplied by these sources, minus a penalty for the average distance of distributor from producers. Consumers does not have a node fitness value, they have just outgoing connections.

In Fig. 16.1, we show a small PDC network. The demand for economic goods is transferred from consumers to distributors, and finally to the producers. The economic goods are transferred inversely through the same chains.

16.5 Model Activity

The activity of ECM is organized in iterations. In the PDC network, each iteration producers generate a quantity of goods conforming to their production capacities. At the same time all consumers, or a part of them, generate the demand for goods. The goal of ECM in this case is to cover the demand generated by the clients with goods produced by producers, thus finding and maintaining an effective set of distribution chains in the network.

The activity of ECM is governed by a simple law:*"in the econnectome, each connection should be profitable"*. Therefore, having a limited amount of connections, each agent in the econnectome should *maintain profitable connections* and *destroy unprofitable ones*. The connections profitability is evaluated using connections' fitness function. When the PDC network is generated, a constant fitness value k is assigned to each connection. During model activity, the fitness of each connection in PDC network evolves:

$$fc(c_i, i) = fc(c_i, i-1) + (\frac{q_p}{q_r} - 1),$$

where i represents iteration number, q_p represents the quantity of goods provided by this connection, and q_r is the quantity of goods this connection should provide (i.e. if consumer $c2$ requires 2 units of economic goods, the connections $(c2, d1)$ and $(c2, d4)$ should provide one unite each, when the number of required units cannot be divided equally between all connections, the remained units are divided randomly between some selected connections).

At the end of iteration the agents remove connections with a zero or negative fitness and create new connections to replace the removed ones. The repeated application of the simple law has a major effect. Initially distribution network is generated randomly, and obviously forms an ineffective set of distribution chains. But letting model to evolve some iterations, the distribution chains are gradually improved to a level where consumer demand is *"nearly fully"* covered and no further improvement is possible. Such configuration of the econnectome is called an *optimized meta-state*.

Giving agents control over their connections – which is based on their goals and interests – leads to long term cooperation between them. As a consequence, these relations of cooperation involve network in a process of *adaptability* which gradually improves the distribution chains in the network therefore entering an optimized meta-state.

We say that the demand is *nearly fully* covered, because it is fully covered in some specific cases but in most cases the demand coverage is close to optimum. Thus, the econnectome evolves close to an optimum configuration but reaches it in very specific cases. This situation is not new to economics, there are many economic processes that meet this situation. For instance, when the labour market is in equilibrium, there is still a small unemployment rate, called *natural unemployment rate*. So, the labour market is very close to optimum, but it does not reach it.

We call this level of network evolution *optimized meta-state*, because it is not a stationary configuration. It is rather a continuous chain of states or configurations which changes network structure but at the same time preserves its ability to nearly fully cover client demand. Each configuration in this chain is close to optimum and is called *optimized configuration*.

Analyzing network evolution, an important property of the model stands out: the interactions between the agents engage the system in a complex process of learning and adaptation. Moreover, the agents interactions generates a tendency of the model to self organize (see [3–5] for an introduction to self-organization in complex systems) the network globally and to maintain optimized meta-state at a close to optimum level.

Despite the *"prima facie"* similarity between the *optimized meta-state* of the ECM and an *equilibrium state* from neoclassical economic models, there are some crucial differences between these two concepts. First difference is in the behaviour of these states. Neoclassical equilibrium is a strong static state which never changes – the agents have no incentives to change their strategies. While the optimized meta-state from our model is a chain of states, which is more similar to a process than to a configuration of agents' preferences. The second difference between these concepts stands in their stability. The neoclassical equilibrium acts as a strong static attractor where are excluded the situations to move or destabilize it, while the optimized meta-state in our model is in continuous exploration of space, being possible jumps from optimization to unpredictable behaviour.

The apparition of optimized meta-states in our model is very similar to the *emergence of complex* behaviour observed in the models which analyze the sandpile formation, apparition of earthquakes, organization of traffic flow, etc. Our model alongside with all the enumerated models evolves in a state of *self-organized criticality* [6], which in ECM is not a stationary state but is an *optimized meta-state*. These *critical* states behave as an attractor of the system dynamics, and can be viewed as a *complement of the chaos concept*. However, reaching certain levels of critical configuration and interconnectivity, the interdependences between system elements also makes the system very susceptible to small variations or shocks. At the same time, the system cannot be too sensitive since the current state is the result of a long chain of optimization and evolution. The presence of this balanced configuration is considered an important part of the "critical" systems. As result, small variations in the system environment may destabilize these *critical* states which in turn will give rise to unpredictable behaviour in the whole system behaving as a source of avalanches and catastrophes which easily can be distributed by the connective structure creating domino effects. Further we shall see that this qualitative concept of criticality can be put on a firm numerical basis.

16.6   Critical Behaviour in the Model

Instabilities and crises in economic systems are well known surprising situations. While many neoclassical economists associated these phenomena with exceptions of economic systems, there are many economists that claim that these phenomena are part of real economic processes and should be intensively studied and included in actual economic science [8, 9].

As a conventional cause of these phenomena are viewed external shocks that affect the entire economy in a similar way by creating business cycles. As instances of this cause serves changes in government policy that affect directly financial markets or budgets of many people and through them the entire economic system.

An alternative scenario is that economic systems possess some intrinsic deterministic dynamics, which – even in the absence of external shocks – involve persistent fluctuations, having a cyclic behaviour, or even chaotic behaviour related to strange attractors. But the analysis of economic time series has not found evidence of this.

A different approach that causes these phenomena is that, being in an optimized meta-state, the effects of small independent shocks from different agents in the economic system (econnectome) cannot be cancelled out by the whole structure, and these small shocks may take control of the whole system generating unpredictable behaviour and crises. These shocks can be some variations in agents' interests and preferences (in PDC network these shock can be variations of the consumers' need).

Although the most important properties of our model is its adaptability and self-improvement of the econnectome, entering an optimized meta-state, which is also critical, the system becomes more vulnerable to small internal shocks. Possessing a dynamical structure, and an optimized highly interconnected society of agents, these states train the model in a continuous exploration of space, thus, maintaining normal functionality of the model even in many critical conditions. But on the other hand these optimized critical meta-states and the highly interconnected structures can also be the source of unpredicted behaviour of the model which can raise destructive avalanches in the system.

When the model evolves to an optimized meta-state the connectivity of each agent is configured to satisfy some specific criteria. Therefore, when these criteria change, the configuration of the optimized meta-state may fail to satisfy them using the same resources as before. So, the econnectome will imply other resources (agents and connections) to solve this problem, thus splitting and spreading it in space and time. Hence it is very possible for these optimized meta-states to devolve in a catastrophic behaviour of the econnectome, preceded by multiple avalanches that unbalance normal operation of an important part of the agents. In the PDC network this event can occur in the situation when a small set of the clients increase their need. This situation affect the operation of a limited number of agents, but the network structure – using cascade effects – can extend its effect on a big part of the network thus creating a global crisis in the model. Moreover, accepting that a big part of real economic networks follow power low distribution of connections, when the normal function of a hub or of a set of hubs in the econnectome is disturbed, all the nodes connected with the hub are affected [7].

Taking into account that small variations inside the economic systems and in their environments are a natural occurrence, for preventing economic crises the model should somehow prepare for dealing with these shocks, by absorbing them avoiding crashes in the system. There are two modalities to anticipate the variations: (i) *to maintain a surplus of material and information* which can be used to cover additional requirements; (ii) to let the model evolve simultaneously with the sources of shocks (e.g. environment), including a *mechanism of innovation in the model*.

The first modality is widely used in the real economy, through a set of institutions such as: central banks, World Bank, International Monetary Fund, etc. The goal of these institutions is to regulate the activity of the economy, and when some destabilizations occur, to provide the resources needed for avoiding them. In the PDC network, this modality can be implemented by maintaining an artificial product capacity bigger than the consumers need. This method works in many cases but the maintenance of an excess of resources and information requires important costs.

The second modality for avoiding crisis is more natural, but it is also more complex. It consists in integration of an innovation mechanism in economic model. This idea belongs to Austrian School of Economics and mainly to Joseph Schumpeter [8, 10–12]. Austrian School of Economics considers the economic crises as a regular occurrence which is used to clean and evolve the economic systems. In our case, by introducing an innovation mechanism in the economic model, the model will evolve simultaneously with its environment and a continuous evolution process will make the model more robust and powerful facing the changes which can occur.

The most effective way to foresee the destabilizations in economic models is to apply both these mechanisms. First mechanism is more effective for short term solutions because it implies important costs. Moreover, using it in long term economic processes can result in some disastrous consequences generated by the increase and accumulation of causes which lead to destabilizations. The second mechanism is effective in long term solutions because it cleans and maintains a clear evolution of the economic systems thus being always prepared to face variations inside or outside the systems.

16.7 Numerical Experiments

We will analyze the behaviour of the ECM modelling PDC network in different situations.

CASE 1. In the network are involved 100 producers, 100 consumers and 300 distributors. The network is randomly generated with a uniform distribution of connections between nodes (all producers have same production capacity). The producers' production capacity is 4 units. The consumers' demand is 4 units. Initial connection fitness (k) is 3. In Fig. 16.2 is presented the dynamics of ECM for Case 1 in two different situations with different numbers of outgoing connections. If the number of outgoing connections is increased, the econnectome became more stable to shocks.

Situation a. The network enters an optimized meta-state at iteration 192, where the rate of demand coverage is 1.0. Beginning with iteration 201, 5 percent of randomly selected consumers increase their demand up to 7 units, thus generating small shocks in the econnectome. As result, the network demand coverage collapses to ≈ 0.91 after 8 iterations from optimized meta-state, at iteration 208, creating a small crash in the PDC network. Further the demand coverage collapses even more, having a decreasing character.

Situation b. The network enters an optimized meta-state at iteration 193, where the rate of demand coverage is 0.93 ± 0.04. Beginning with iteration 201, 5 percent of randomly selected consumers increase their demand up to 7 units, thus generating small shocks in the econnectome. As result, the network demand coverage collapses to ≈ 0.84 after 8 iterations from optimized meta-state, creating a crash in the PDC network. Further big fluctuations can be observed in the system behaviour, and the average demand coverage is distanced from the optimized meta-state's coverage.

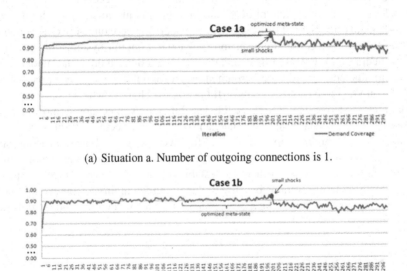

(a) Situation a. Number of outgoing connections is 1.

(b) Situation b. Number of outgoing connections is 1–2.

Fig. 16.2 Dynamics of ECM for in Case 1.

CASE 2. In the network are involved 100 producers, 300 distributors and a variable number of consumers, depending on total production capacity. The network is generated with a power low distribution of connections between nodes, depending on producer's production capacities. The producers' production capacities are approximately: 56% – 2 units, 20% – 4 units, 11% – 6 units, 7% – 8 units and 5% – 10 units. The consumers' demand is 2 units. Initial connection fitness is 3. In Fig. 16.3

is presented the dynamics of ECM for Case 2 in two different situations with different numbers of outgoing connections. If the number of outgoing connections is increased, the network became more stable to shocks.

Situation a. The network enters an optimized meta-state at iteration 109, where the rate of demand coverage 1.0. Beginning with iteration 116, 5 percent of randomly selected consumers increase their demand up to 5 units, thus generating small shocks in the econnectome. As result, the network demand coverage collapses to ≈ 0.77 at iteration 123, creating a considerable disequilibrium in the PDC network which can be considered a big crash of economic system. Further the system behaviour is on average foreseeable – fitting within some visible bounds, but having big local fluctuations.

Situation b. The network enters an optimized meta-state at iteration 98, where the rate of demand coverage is 0.92 ± 0.04. Beginning with iteration 201, 5 percent of randomly selected consumers increase their demand up to 5 units, thus generating small shocks in the econnectome. As result, the network demand coverage collapses to ≈ 0.76 after 4 iterations from optimized meta-state, creating a considerable disequilibrium in the PDC network. Further big fluctuations are observed in system dynamics.

(a) Situation a. Number of outgoing connections is 1. Number of consumers is 173.

(b) Situation b. Number of outgoing connections is 1–2. Number of consumers is 205.

Fig. 16.3 Dynamics of ECM for in Case 2.

16.8   Conclusions

The economic science should include a dedicated chapter for the study of economic crises and their overcoming. This complex phenomenon is analyzed based on a holistic approach, within the theoretical framework of complex systems. A connective structure called econnectome is introduced. Using a simplistic 'producers-distributors-consumers' distribution network, complex aspects of economic processes are addressed. Among these are identified adaptability, emergence, and self-organization.

The connectivist approach based on long term cooperation between agents provides a reliable framework for understanding the propagation of destabilization and occurrence of the domino effect in the economic systems. The theory and computational approaches of complex systems offers a new perspective on understanding and modelling the economic processes. Treating economic equilibrium as a continuous dynamical process opens new ways for understanding its stability and the potential destructive factors.

Numerical experiments indicate that small variations inside or outside an economic system may generate unpredictable behaviour. These can further lead to a global crisis covering the entire system. There is no need for big external shocks to generate catastrophic behaviour. Two methods for preventing instabilities and crises are proposed, based on the idea that small shocks may generate big crashes.

The observed dynamics of the simplistic model is very interesting and it offers a new vision for understanding the dynamics of economic systems. New explanations may also be supplied for other real world occurrences that are not completely understood.

Establishing a strong correlation between economic science and the theory of complex systems, we consider that in the new science of economy, economic crises will be usual phenomena just as market equilibrium is in neoclassical economy.

Acknowledgements. The authors acknowledge the support of a grant from the John Templeton Foundation. The opinions expressed in this publication are those of the authors and do not necessarily reflect the views of the John Templeton Foundation.

References

[1] Akerlof, G.A., Shiller, R.J.: Animal Spirits, How Human Psychology Drives the Economy and Why it Matters for Global Capitalism. Princeton, New Jersey (2009)

[2] Delorme, R.: Theorising Complexity. In: International Workshop on the Evolution and Development of Evolutionary Economics, University of Queensland, Brisbane (1999)

[3] Miller, J.H., Scott, E.: Complex Adaptive Systems: An Introduction to Computational Models of Social Life. Princeton U. Press, Princeton (2007)

[4] Holland, J.: Hidden Order: How Adaptation Builds Complexity. Addison-Wesley, Massachusetts (1995)

[5] Ashby, W.R.: Principles of the self-organizing system. In: Principles of Self-Organization: Transactions of the University of Illinois Symposium, pp. 255–278. Pergamon Press, London (1962)

[6] Bak, P.: How Nature Works: The Science of Self-Organised Criticality. Copernicus Press, New York (1996)
 [7] Barabási, A.L.: Linked - The New Science of Networks. Perseus Publishing, Cambridge (2002)
 [8] Roubini, N., Mihm, S.: Crisis economics: a crash course in the future of finance. Penguin Press (2010)
 [9] Krugman, P.R.: The return of depression economics and the crisis of 2008. W.W. Norton, New York (2009)
[10] Schumpeter, J.A.: Business Cycles: A theoretical, Historical and Statistical Analysis of the Capitalist Process. McGraw-Hill, New York (1939)
[11] Dopfer, K.: The Origins of Meso Economics, Schumpeter's Legacy, Evolutionary Economics Group, Jena (2006)
[12] Foster, J.: From Simplistic to Complex Systems in Economics. Cambridge Journal of Economics 29(6), 873–892 (2005)

Chapter 17
Correlations Involved in a Bio-inspired Classification Technique

Camelia-M. Pintea and Sorin V. Sabau

Abstract. An improved unsupervised bio-inspired clustering model is introduced. The main goal is to involve a correlation between properties of objects and some bio-inspired factors. The statistical classification biological model is based on the chemical recognition system of ants. Ants are able to create groups discriminating between nest-mates and intruders based on similar odor. Comparative analysis are performed on real data sets.

17.1 Introduction

Nowadays, biologically inspired algorithms are well known techniques. They are used with success for solving complex problem. These methods reproduce biological behaviors to solve problems. For example, in ant based models, pheromone trails are used to perform optimization [2]. Ant-based clustering algorithms model real ants abilities to sort their brood, [6, 11]. Artificial ants may carry one or more objects and may drop them according to given probabilities. Agents influence themselves through the configuration of objects on the floor. After a while they are able to build groups of similar objects. This problem is known as data clustering.

The clustering algorithm *AntClust* [7] was developed on simple rules inspired by real ants behaviors. In order to correlate the properties of given data and bio-inspired parameters, is introduced a new ant-clustering technique, called the *Correlation-Ant Clust (C-AntClust)* algorithm.

Camelia-M. Pintea
G.Cosbuc N.College, A.Iancu 70-72, Cluj-Napoca, Romania
e-mail: cmpintea@yahoo.com

Sorin V. Sabau
Tokai University, Minamisawa 5-1-1-1, Sapporo, Japan
e-mail: sorin@tspirit.tokai-u.jp

D.A. Pelta et al. (Eds.): NICSO 2011, SCI 387, pp. 239–246, 2011.
springerlink.com   © Springer-Verlag Berlin Heidelberg 2011

Different real data sets, including medical and technical data, are used to test the new ant-based clustering technique. Comparative test and statistical analysis are also performed.

17.2 Statistical Classification with an Ant-Based Technique

An optimal partitioning of a given data set is the result of a clustering program. Ant-based clustering programs are using artifical ants in order to group data sets. Ants, like many other social insects, have developed a mechanism of colonial closure that allows them to discriminate between nest-mates and intruders from the same species, [8]. The phenomenon is known as *phenotype matching*, [4]. This recognition relies on the comparison of an odor emitted by each ant, called the *label*, and a reference template, called *template*. An ant i is characterized by the following entities: $Label_i$, $Template$, M_i, M_i^+ (estimators) and A_i (age).

Each object of a dataset is associated with an artificial ant-its genome. At each iteration of the algorithm, ants are randomly selected and meet eachother. Each ant modify an individual label according to the similarity between their genomes, a learned acceptance threshold and associated behavioral rules. When each ant is well integrated in its nest, the algorithm ends.

Definition 17.1. *Label* is the individual chemical odor of an ant, spread over its cuticle, partially constructed by each ant. The label evolves according to the meetings performed by the ant.

Definition 17.2. *Template* is a model of reference indicating the type of odor nets-mates should have on their cuticle. The template is learned and then continuously updated.

A *similarity measure* takes as input a pair of objects (i,j) and outputs a value $Sim(i,j)$ between 0 and 1. Two objects are totally different if $Sim(i,j)=0$. Two objects are *identical* if $Sim(i,j)=1$.

The *Template* is defined half by $Genetic_i$ and half by an acceptance threshold $Temp_i$.

- $Genetic_i$ the genetic odor of ant_i, is initialized with i^{th} objects of the data set.
- $Temp_i$ a function of all the similarities observed during an initialization phase evaluates the genetic odor similarities; the $Temp_i$ formula is following.

$$Temp_i = \frac{\overline{Sim}(i,\cdot) + Max(Sim(i,\cdot))}{2}, \tag{17.1}$$

where $Max(Sim(i,\cdot))$ is the maximal similarity and $\overline{Sim}(i,\cdot)$ the mean similarity observed during meeting with other ants.

Increasing a variable x means:

$$x \leftarrow (1-\alpha) \times x + \alpha; \tag{17.2}$$

Decreasing a variable x means:

$$x \leftarrow (1 - \alpha) \times x, \tag{17.3}$$

where $\alpha \in (0,1)$.

M_i, M_i^+ estimators and A_i age entities of an ant are following. These parameters, including $Label_i$ are initialized with 0.

- M_i reflects if ant_i is successful during its meetings with encountered ants or not.

  - M_i is increased when ant_i meets an ant with the same $Label$.
  - M_i is decreased when $Labels$ are different.

- M_i^+ measures how well accepted is ant_i in a nest.

  - M_i^+ is increased when ant_i meets an ant with the same $Label$ and when both ants accept each other.
  - M_i^+ is decreased when there is no acceptance between ants.

- A_i is an age used when updating acceptance threshold.

The meeting process between two ants is the main principle of the *AntClust* method, [7]. It allows ants to exchange their labels if they accept each other. The *acceptance rule* is following:

$$Accept(i,j) \leftrightarrow (Sim(i,j) > Temp_i) \wedge (Sim(i,j) > Temp_j). \tag{17.4}$$

The rules based on the ants behavior [7], associated with meetings are following.

1. **New nest creation: is generating a new $Label$ for a new nest.**
   If $Label_i = Label_j = 0$ and $Accept(i,j)$ then *Create* a new label, $Label_{NEW}$ and $Label_i \leftarrow Label_{NEW}$, $Label_j \leftarrow Label_{NEW}$.

   - The rule gathers similar ants in the first clusters; will be used as *seeds* to generate the final clusters.
   - A cluster contains at least two objects.

2. **Adding** an ant with no $Label$ to an existing nest.
   If $Label_i = 0 \wedge Label_j \neq 0$ and $Accept(i,j)$ then $Label_i \leftarrow Label_j$.
   If $Label_j = 0 \wedge Label_i \neq 0$ and $Accept(i,j)$ then $Label_j \leftarrow Label_i$.

3. **Positive** meeting between two nest-mates
   If $Label_i = Label_j \wedge Label_i \neq 0 \wedge Label_j \neq 0$ and $Accept(i,j)$ then increase M_i, M_j, M_i^+ and M_j^+.
   In case of acceptance between two ants M and M^+ are increased.

4. **Negative** meeting between two nest-mates
   If $Label_i = Label_j \wedge Label_i \neq 0 \wedge Label_j \neq 0$ and $Accept(i,j) = False$ then increase M_i, M_j and decrease M_i^+ and M_j^+.
   Ant_x, with the worst integration in the nest, $M_x^+ = Min_{k \in [i,j]} M_k$, losses its label and now has no more nest: $Label_x \leftarrow 0$, $M_x \leftarrow 0$ and $M_x^+ \leftarrow 0$.

- Remove ants that were accepted when the nest profile was not clearly defined, because there were not enough ants in the nest.
- The worst integrated ants in a nest can be rejected and then reset.

5. **Meeting** between two ants of different nests
   If $Label_i \neq Label_j$ and $Accept(i,j)$ then decrease M_i and M_j.
   The ant_x with the lowest M_x, the ant belonging to the smallest nest, changes its nest and belong now to the nest of the encountered ant. This rule decreases the initially large number of clusters by gathering small sub-clusters into one bigger.
6. **Default rule**: If no other rule applies then nothing happens.

17.3 The New Statistical Classification Model: *C-AntClust*

The *Correlation-Ant Clust* (*C-AntClust*) is an improved version of the *AntClust* algorithm, [7]. The first approach of a similar algorithm is in [12].

Two main modifications of *AntClust* are made allowing the *C-AntClust* to be adapted to the data sets, optimizing the *AntClust* technique. The first one concerns the number of iterations in order to ensure a good convergence of the algorithm.

The number of iterations N_{iter} is computed using the maximal, *Max* and the minimal, *Min*, values between the number of objects and the number of attributes; n is the number of objects within a given data set.

$$N_{iter} = \frac{n \cdot Max}{Min}. \tag{17.5}$$

In the first version of the *AntClust*, the number of iterations was specified as a fixed value.

The second modification concerns the third step in the *C-AntClust* procedure the method used to delete clusters-nests, that are not enough representative to be kept in the final partition.

In *AntClust* all the nests whose sizes do not exceed a threshold were deleted. The threshold was equal to a percentage of the number of data. This approach was limited, because the algorithm could not find more than a given number of clusters of the same size.

In the *C-AntClust* are deleted the clusters whose M^+ average, the mean integration of the ants on the nest, is less than the same average parameter values from all clusters.

```
Procedure C-AntClust
```

- *Initialize* the ants parameters
- *Simulate* N_{iter} iterations during which two ants randomly chosen, meet
- *Delete* nests that are not enough representative to be kept in the final partition.
- *Re-assign* each ant having no more nest, to the nest of the most similar ant found that have nest

The algorithm does not required an initial number of classes and is able to find from small to large number of clusters on real datasets.

17.4   Numerical Experiments and Statistic Analysis

In this section, we compare *C-AntClust* with the *K-Means* algorithm [5] and the *AntClust* algorithm [7]. *K-Means* algorithm will be initialized with 10 clusters randomly generated, because the data sets used for the tests do not have more than 10 expected clusters.

All evaluations are produced over 50 runs for each data set and each method, during each run every artificial ants encounters an other ant randomly selected.

There are used real data sets from [1] with attributes-based representations in order to evaluate the clustering abilities of the two methods.

In Table 17.1 and Table 17.3 introduced fields are: for each data the number of objects (*Objects.*) and their associated number of attributes (*Attributes*) and the number of clusters expected to found in the data (*Clusters*).

Table 17.2 shows the results of the introduced ant-based algorithm. Standard Deviation table for medical and technical data is also computed.

Table 17.1 Technical and medical data sets.

		Technical dataset			Medical dataset			
	Austr.	SatImg.	Segment	Vehicle	Diabetes	Heart	Hepatitis	Thyroid
Objects	690	6435	2310	846	768	270	80	215
Attributes	14	36	19	18	8	13	19	5
Clusters	2.00	7.00	7.00	4.00	2.00	2.00	2.00	3.00

Table 17.2 Standard deviation values for technical and medical data sets.

		Technical dataset			Medical dataset			
	Austr.	SatImg.	Segment	Vehicle	Diabetes	Heart	Hepatitis	Thyroid
No.Clust.	2.00	7.00	7.00	4.00	2.00	2.00	2.00	3.00
C-AntClust.	2.00	7.28	7.04	2.78	2.60	3.00	3.62	3.30
StdDev	0.00	0.1979	0.0283	0.8627	0.42426	0.7071	1.1455	0.2121

From Table 17.2 *C-AntClust* has good solutions for the technical data (e.g. *SatImage*). For medical data (e.g. *Diabetes, Thyroid*) are also encouraging results, the clustering values being close to the real clustering values.

Other real data series use *Iris, Glass, Pima* and *Soybean*, see [10].

Table 17.3 Real data characteristics.

Real data	Iris	Glass	Pima	Soybean
Objects	150	214	768	47
Atributes	4	9	8	35
Clusters	3	7	2	4

Follows the analysis of Iris, Glass, Pima and Soybean.

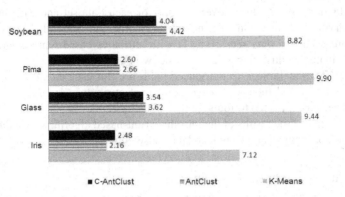

Fig. 17.1 Average results of number of clusters found after 50 runs of each method, applied over each of the first series data.

The *AntClust* algorithms tends to perform better than the *K-Means* method for $K = 10$. *AntClust* algorithms have, a better appreciation of the number of clusters in the data. Between objects are little differences, that is way *K-Means*, starting from 10 does not manage to reduce significantly the number of clusters.

K-Means has better results than *AntClust* algorithms for *Glass* because the number of clusters expected is near 10 and *AntClust* and *C-AntClust* not manage to reach this number.

AntClust and *C-AntClust* can treat from small to large sets of data with success. An example is in [9] where are some good results on web session clustering.

A statistical analysis is performed in the following. The Expected Utility Approach [3] technique has been employed to determine the most accurate heuristic. Let x be the percentage deviation of the heuristic solution and the best known solution of a particular heuristic on a given problem:

$$x = \frac{heuristic solution - best known solution}{best known solution} \times 100. \tag{17.6}$$

The expected utility function is $\gamma - \beta(1 - \bar{b}t)^{-\bar{c}}$, where $\gamma = 500$, $\beta = 100$ and $t = 0.05$. $\bar{b}$ and $\bar{c}$ are the estimated parameters of the Gamma function.

Table 17.4 Statistical analysis. Calculations for the expected utility function for the compared heuristics.

Heuristic	$\bar{x}$	s^2	$\bar{b}$	$\bar{c}$	$\gamma - \beta(1-\bar{b}t)^{-\bar{c}}$	Rank
$K-means$	1.7192	1.8098	1.0527	1.6332	390.7677	3
$AntClust$	-0.0819	3.3454	-40.8158	0.0020	400.2231	2
$C-AntClust$	-0.0894	3.3543	-37.5178	0.0024	400.2514	1

The following notations are used for Table 17.4

$$\bar{x} = \frac{1}{4}\sum_{j=1}^{4} x_j, \quad s^2 = \frac{1}{4}\sum_{j=1}^{4}(x_j - \bar{x})^2, \quad \bar{b} = \frac{s^2}{\bar{x}}, \quad \bar{c} = (\frac{\bar{x}}{s})^2. \tag{17.7}$$

The last column provides the rank 1 to 3 of the entries. As indicated in Table 17.4, *C-AntClust* has Rank 1 being the most accurate algorithm within the compared algorithms. Table 17.4 has statistical data for Iris, Glass, Pima and Soybean.

The compared results from Table 17.4 indicated that the newly introduced *C-AntClust* algorithm outperforms the other heuristics considered.

Ant-based algorithms had to be improved because they do not manage to find the right number of clusters when there are overlapping groups of objects or when data sets are made of uniform noise.

The new model has to be improved in terms of parameter values or/and an efficient hybridization with other algorithms.

17.5   Summary and Conclusions

The paper introduces an improved ant-based clustering technique. The principle of the new algorithm is based on *AntClust* [7]. The goal is to generate meetings between all ants. After a while, agents with the same type of objects, are gathered in the same groups.

The newly algorithm is called, *C-AntClust* and it is tested with good results on some real dataset. This approach does not make any assumption about the nature of the data to be clustered.

The technique had potential because does not required an initial number of classes and involves properties of objects in setting some parameters value.

References

[1] Blake, C.L., Merz, C.J.: Machine learning repository (1998)
[2] Colorni, A., Dorigo, M., Maniezzo, V.: Distributed Optimization by Ant Colonies. In: Proceedings of the First European Conference on Artificial Life, pp. 134–142. Elsevier Publishing, Amsterdam (1991)

[3] Golden, B.L., Assad, A.A.: A decision-theoretic framework for comparing heuristics. European J. of Oper. Res. 18, 167–171 (1984)

[4] Holldobler, B., Wilson, E.O.: The Ants. Springer, Heidelberg (1990)

[5] Jain, A.K., Dubes, R.C.: Square-Error Clustering method. In: Algorithms for clustering data. Prentice Hall Advanced References series, pp. 96–101 (1988)

[6] Kuntz, P., Snyers, D.: Emergent Colonization and Graph Partitioning. In: Proceedings of the Third Int. Conf. on Simulation of Adaptive Behaviour, pp. 494–500. MIT Press, Cambridge (1994)

[7] Labroche, N., Monmarché, N., Venturini, G.: A new clustering algorithm based on the chemical recognition system of ants. In: Proceedings of the European Conference of Artificial Intelligence, pp. 345–349. IOS Press, Amsterdam (2002)

[8] Labroche, N., Monmarché, N., Lenoir, A., Venturini, G.: Modélisation du système de reconnaissance chimique des fourmis. Rapport interne, Lab. d'Informatique de l'Universite du Tours (2002)

[9] Labroche, N., Monmarché, N., Venturini, G.: Web session clustering with artificial ants colonies. In: Proceedings of Conference WWW 2003, Budapest, Hungary (2003)

[10] Monmarché, N.: Algorithmes de fourmis artificielles: applications à la classification et à l'optimisation. Thèse de doctorat, Lab. d'Informatique de l'Universite de Tours (2000)

[11] Monmarché, N., Slimane, M., Venturini, G.: On Improving Clustering in Numerical Databases with Artificial Ants. LNCS (LNAI), pp. 626–635 (1999)

[12] Pintea, C.-M.: A clustering model based on ant-system. In: Proceeding of Int. Conf. of Applied Mathematics (ICAM5), p. 38 (2006)

Chapter 18
Degeneracy Reduction or Duplicate Elimination? An Analysis on the Performance of Attributed Grammatical Evolution with Lookahead to Solve the Multiple Knapsack Problem

Muhammad Rezaul Karim and Conor Ryan

Abstract. This paper analyzes the impact of having degenerate code and duplicate elimination in an attribute grammar with lookahead (AG+LA) approach, a recently proposed mapping process for Grammatical Evolution (GE) using attribute grammar (AG) with a lookahead feature to solve heavily constrained multiple knapsack problems (MKP). Degenerate code, as used in DNA, is code in which different codons can represent the same thing. Many developmental systems, such as (GE), use a degenerate encoding to help promote neutral mutations, that is, minor genetic changes that do not result in a phenotypic change. Early work on GE suggested that at least some level of degeneracy has a significant impact on the quality of search when compared to the system with none. Duplicate elimination techniques, as opposed to degenerate encoding, are employed in decoder-based Evolutionary Algorithms (EAs) to ensure that the newly generated solutions are not already contained in the current population. The results and analysis show that it is crucial to incorporate duplicate elimination to improve the performance of AG+LA. Reducing level of degeneracy is also important to improve search performance, specially for the large instances of the MKP.

18.1 Introduction

Grammatical Evolution (GE) is an evolutionary automatic programming system (see Sect. 18.3) that uses a grammar (typically a context-free grammar (CFG)) to map individuals from binary strings to higher level structures. Although problems with constraints are of general interest to the EC community, because they occur

Muhammad Rezaul Karim
University of Limerick, Ireland
e-mail: rezaul.karim@ul.ie

Conor Ryan
University of Limerick, Ireland
e-mail: conor.ryan@ul.ie

D.A. Pelta et al. (Eds.): NICSO 2011, SCI 387, pp. 247–266, 2011.
springerlink.com   © Springer-Verlag Berlin Heidelberg 2011

frequently in real-life situations, there has been little work in applying GE to constrained optimization problems. This is mainly because CFGs are not well suited for passing/codifying information about constraints. An Attribute grammar (AG), first proposed by Knuth [7], augments a CFG to provide context-sensitivity using a set of attributes and thereby makes it suitable for constrained problems.

Attribute grammar with lookahead (AG+LA) approach is a recently introduced AG based mapping process for GE [4] to solve multiple knapsack problem (MKP). In AG+LA, a lookahead mechanism is incorporated with the mapping process to avoid constraint violations. Unlike other attribute grammar based systems [1], AG+LA eliminates introns and repeated remapping of non-terminals by avoiding constraint violations. It was reported that this approach enhances the performance of existing attribute grammar based approach to solve the MKP without generating a single intron [4].

Genetic code degeneracy in GE give rise to silent mutations that have no effect on the phenotype. Even though the high redundancy introduced by degenerate code feature in standard GE has been shown to be effective for a large range of problems [10], the effect of redundancy on AG+LA is not well understood. This is especially true when AG+LA is employed as a decoder to solve difficult combinatorial optimization problems.

Decoders are employed in Evolutionary Algorithms (EAs) to ensure feasibility of solutions. Earlier work reported that decoder-based EAs are frequently susceptible to premature convergence due to their encoding redundancy while solving multidimensional knapsack problem [11], a constrained combinatorial optimization problem. It was shown that duplicate elimination is expected to be helpful for maintaining population diversity and, therefore, preventing premature convergence. Duplicate elimination means that newly generated solutions are only accepted if they are not already contained in the current population. This paper shows that duplicate elimination alone is not sufficient in AG+LA to get the desired results, specially for the large instances of the MKP. Due to special degenerate code employed by AG+LA, level of degeneracy should also be reduced to improve search.

For the first time this study presents an examination of the degeneracy and duplicate elimination in AG+LA while solving a difficult constrained combinatorial optimization problem. A number of experiments are performed on several instances of the MKP problem by varying the level of degeneracy, with and without incorporating duplicate elimination in the search process.

The remainder of the paper is organized as follows: Sect. 18.2 introduces MKP followed by a brief introduction to GE and AG in Sect. 18.3. An overview of the existing EA approaches for the MKP is given in Sect. 18.4; next, Sect. 18.5 introduces the attribute grammar with the lookahead extension which is used to solve the MKP. Degeneracy and duplicate elimination concept for the AG+LA is given in Sect. 18.6 followed by a description of the experimental setup in Sect. 18.7. Then, in Sect. 18.8, we present the results obtained and analyze the results. Finally, in Sect. 18.9 we conclude the paper with a scope of potential future work.

18.2   MKP

Like any multimodal function, combinatorial problems also have multiple peaks. In contrast to function optimization, combinatorial optimization problems are usually highly constrained. Thus the solution space is greatly discontinuous and both local and global optima are most likely in the boundaries of feasible regions. The MKP has these common characteristics of constrained combinatorial optimization problems and is widely used as a testbed for this kind of problems.

The MKP is a NP-complete problem of both theoretical and practical interest. It is a generalization of the standard 0-1 knapsack problem and has multiple knapsacks. There are many variants of the MKP available in the literature. The variant of the MKP that we want to solve can be formulated as:

$$\text{maximize} \quad \sum_{i=1}^{n} p_i x_i$$

$$\text{subject to} \quad \sum_{i=1}^{n} w_{ij} x_i \leq c_j \quad j = 1, \ldots, m \tag{18.1}$$

$$x_i \in \{0,1\} \quad i = 1, \ldots, n$$

$$\text{with} \quad p_i > 0, \quad w_{ij} \geq 0, \quad c_j \geq 0$$

Each item $i\{i = 1, \ldots, n\}$ is assigned a profit p_i and weight w_{ij} with respect to knapsack $j\{j = 1, \ldots, m\}$. The objective of the MKP is to determine a set of items with maximum profit to fill a series of m knapsacks each of capacity c_j, which does not exceed the knapsack capacities. Each selected item is either placed in all knapsacks or in none at all.

18.3   GE and AG

GE is an evolutionary automatic programming system that uses a combination of a variable-length binary string genome and a BNF (Backus Naur Form) grammar to evolve structures. A genotype consists of groups of eight bits (denoted as a *codon*). GE employs a genotype-phenotype mapping process where each codon dictates which production rule from the BNF grammar to apply at each state of the derivation sequence, when one or more non-terminals symbols exist in a sentence.

The mapping process creates a clear distinction between the search and solution space, so the genetic operators such as crossover and mutation are applied to the linear genotype in a typical genetic algorithm (GA) manner. GE is fully described in [14].

BNF is a convenient way of describing CFGs. A CFG can be represented with a tuple $G=< N,T,P,S >$ where N is the set of non-terminal symbols, T is the set of terminal symbols, P is the set of productions (syntactic rules), and S is the start symbol where $(S \in N)$.

We can define a CFG for the MKP defined in Eq. 18.1:

$$S \rightarrow K$$
$$K \rightarrow I \quad | \quad IK$$
$$I \rightarrow i_1 \quad | \quad i_2 \quad | \quad ...... \quad | \quad i_n$$

Here I is the terminal-generating or item-generating non-terminal symbol and each $i_1, ..., i_n$ represents a physical item for the MKP.

18.3.1 AGs

An attribute grammar is a context-free grammar augmented with attributes, semantic rules, and conditions. Similar to CFGs, AGs can be represented with a tuple $G=< N,T,P,S >$. The key difference is the form of the productions rules. Each production, $p \in P$, is written in the form

$p = X_0 \rightarrow X_1 X_2 ... X_n$ where n $\geq$ 1, $X_0 \in N$ and $X_k \in N \cup T$ for $1 \leq k \leq n$. X_0 is the head and $X_1 X_2 ... X_n$ is the body of the production p. Each symbol $X \in N \cup T$ of the grammar G has two finite disjoint sets I(X) and S(X) of inherited and synthesized attributes. Let $A(X) = I(X) \cup S(X)$ be the set of attributes of X. The notation $X.a$ is used to denote that a is an element in A(X) or a is an attribute of symbol X.

Each attribute takes a value from some semantic domain (such as the integers, group of characters, or complex data structures of some type). These values are defined by semantic functions or semantic rules associated with the productions in P. Semantic functions or semantic rules define the value of the attribute occurrence in terms of the values of other attribute occurrences that appear in the same production.

The value of a synthesized attribute $X.s$ is computed using the attributes at descendant nodes of node X of the corresponding derivation tree. The value of an inherited attribute $X.i$, on the other hand, is dependent on value of the attributes at the parent of X, X itself or the sibling nodes of node X of the corresponding derivation tree. Synthesized attributes are used to pass information up the tree, whereas inherited attributes pass information down the tree.

In an AG, there is no specific order in which attributes are evaluated. An attribute can be evaluated only when the value for the attributes on which it depends are available.

18.4 EA Approaches for the MKP

Due to the constrained nature of the MKP, various constraint handling techniques have been adopted with various EA approaches. Most of the constraint handling techniques can be classified as either direct and indirect search, depending on the space they operate in. Apart from direct and indirect search techniques, there is a hybrid GA approach based on problem-space exploration.

In direct search approaches for solving the MKP, the search algorithm operates in either the original complete search space or in the original feasible space. Indirect search approaches, on the other hand, conducts search in a transformed search space and rely on a decoder to generate feasible solutions as well as to map into original search space.

The direct approaches dealing with complete search space penalize infeasible individuals using penalty function [5, 6]. Among the penalty-based approaches, Khuri et al. [5] use a simple fitness function that uses a graded penalty term to penalize infeasible solutions. Kimbrough et al. [6] use a two-market GA where two phases are used to obtain feasible solutions and suggest two different penalty terms, violation product and sum of violations.

Repair algorithms and other specialized operators are used by some of the direct search techniques operating in the feasible space. Repair functions convert illegal solutions into legal ones restricting the search to the feasible region of the search space [12]. In [12], a GA with a repair function was suggested that is loosely analogous to RNA editing and reports good results for a set of test problems.

Decoder based techniques have been studied mainly in [1, 4]. GE was extended with AG by Cleary [1] to build a decoder to solve the MKP. This was the first decoder-based approach using GE to solve any constrained optimization problem. In [1] *AG(Full)*, a mapping approach was proposed where if a codon violates any knapsack weight constraint, the codon is skipped and considered an intron. The offending non-terminal is then tried to remap with the next codon in the genome. This remapping process stops only when a new terminal (item) is found that is not selected earlier and that does not violate any weight constraint.

The results reported with AG(Full) were moderate compared to other EA approaches applied to the same set of test problems. One of the major problems with AG(Full) is the large computational effort spent in the implementation of repeated remapping. Due to its constraint handling technique, many introns are also generated by AG(Full), even for lightly constrained instances of the MKP. Few computationally expensive operations were later integrated with AG(Full) to get further improvement like splicing, an intron removal technique (AG(Full)+Splice) and phenotype duplicate elimination [1].

Cotta et al. [2] propose a hybrid genetic algorithm based in local search. In this approach local optimization is achieved with the help of a greedy construction heuristic that chooses items of instance specific high value-weight ratios. The individuals in the population represent perturbations of the problem instance by manipulation of the items profits. Thus the search does not take place in the solution space but in a solution meta space (the problem space).

18.5   AG with Lookahead for the MKP

A constraint satisfaction problem (CSP) is typically solved by a backtracking algorithm that works by incrementally extending a partial solution to a complete solution [8]. At each step of the algorithm, the current variable is tried to assign a value; if all the values in the domain of the current variable have been tried and fail to give a consistent assignment with already assigned variables, the algorithm backtracks to the preceding variable and try alternative values in its domain.

While solving a CSP with backtracking, the search space can be easily pruned and these unnecessary backtracking can be avoided by employing domain reduction operations like forward checking [8]. The idea of forward checking is to reduce the domains of the uninstantiated variables (also called future variables) by filtering out those values which are inconsistent with the current instantiation. The algorithm backtracks only when the reduction leads to an empty domain of any variable.

AG+LA [4] is a genotype-phenotype mapping technique for GE using attribute grammar to incorporate context-sensitive information. It is inspired by forward checking. To incorporate lookahead concept like forward checking, some special attributes are included in the attribute set of the grammar to eliminate the drawbacks of previous AG based approaches (described in Sect. 18.4). During the genotype to phenotype mapping phase of GE, the set of attributes in the attribute grammar works together to make the algorithm work as a decoder for the MKP.

18.5.1   AG for the AG+LA

AG+LA generates variable-length solutions using an attribute grammar. A phenotype generated by AG+LA can be represented by a vector $\mathbf{x} = \{x_1, x_2, ..., x_k\}$ where $1 \leq k \leq n$. Each element x_i in the phenotype represents a distinct item from the set of n available items.

In AG+LA, the grammar can generate any item from the set $i_1, ..., i_n$ as well as a special purpose item ε. The special purpose item ε is a end-of-derivation marker which signals when the derivation must stop.

The attribute grammar used by AG+LA is as follows:

$$
\begin{aligned}
S \rightarrow K \quad & S.alloweditems = initialAllowedItems() \\
& S.limit = initialWeightConstraints() \\
& S.items = K.items \\
& S.weight = K.weight \\
& K.alloweditems = S.alloweditems \\
& K.limit = S.limit \\
K \rightarrow I \quad & I.alloweditems = K.alloweditems \\
& K.items = I.item \\
& K.weight = I.weight
\end{aligned}
$$

$$K_1 \rightarrow IK_2 \quad K_1.selecteditem = I.item$$
$$K_1.selecteditemweight = I.weight$$
$$K_1.usage = K_1.usage + K_1.selecteditemweight$$
$$K_1.useditems = K_1.useditems + K_1.selecteditem$$
$$K_1.items = K_2.items + I.item$$
$$K_1.weight = K_2.weight + I.weight$$
$$I.alloweditems = K_1.alloweditems$$
$$K_2.usage = K_1.usage$$
$$K_2.useditems = K_1.useditems$$
$$K_2.limit = K_1.limit$$
$$K_2.alloweditems = genAllowedItems($$
$$K_2.useditems, K_2.usage, K_2.limit)$$

$$I \rightarrow i_1 \quad I.item = ``i_1''$$
$$I.weight = getweight(i_1)$$
....
$$I \rightarrow i_n \quad I.item = ``i_n''$$
$$I.weight = getweight(i_n)$$

$$I \rightarrow \varepsilon \quad I.item = ``\varepsilon''$$
$$I.weight = 0$$

In the above grammar, when there are more than one occurrence of the same symbol in a production rule, subscript notation is used in this grammar to avoid any confusion in identifying the attributes of different occurrences of the same symbol. Mapping rule remains same for all occurrences of the same symbol. The attributes in the above grammar are written in such a way that there is dependency among the attributes from left to right in the symbols of each production rule.

18.5.2   Attributes for the AG+LA

The attribute grammar used by AG+LA has following attributes:

limit: This is an inherited attribute for non-terminal symbol K and a special inherited symbol for S to hold the individual capacity of each of the knapsacks. S does not inherit this value from any symbol, rather the value is initialized by the semantic function *initialWeightConstraints* at the start of the derivation.

selecteditem and *selecteditemweight:* Both are synthesized attributes for non-terminal symbol K. The attribute *selecteditem* is used to record the item derived by the immediate descendant I to this K, whereas *selecteditemweight* holds the weight of the corresponding derived item.

usage and *useditems:* These are inherited attributes for non-terminal symbol K. For each knapsack, *useditems* and *usage* records the list of all derived items and the total weight of all items derived so far, respectively.

items and *weight:* Both these attributes are synthesized. The attribute *items* is relevant for for non-terminal symbol S and K, whereas the attribute *weight* is applicable for symbols S, K and I. The purposes of the attributes *items* and *weight* are to hold all items that is being derived by all the descendant nodes of the current node and the corresponding total weight for each knapsack, respectively. *getweight* is a semantic function used for returning the knapsack specific physical weight of the item so that the value can be assigned to the attribute *weight* of I. For consistency, the special terminal ε is assigned zero weight for each of the knapsacks.

item: a synthesized attribute for non-terminal symbol I, which stores the physical item being derived by this non-terminal.

alloweditems: This attribute is an inherited attribute for non-terminal symbol K and I. For start symbol S, it can be considered as a special inherited attribute, as the semantic function *initialAllowedItems* generates the value of this attribute without using any other attribute.

During the derivation when this attribute comes into context, determines which items are available for selection by the next item-generating non-terminal I, without exceeding weight constraint of any knapsack.

While computing the value for the attribute *alloweditems*, if the current usage and knapsack capacity constraints leads to null value for the domain of this attribute for any symbol, only the special item ε is added to the allowed item list and the derivation stops.

18.5.3   Mapping and Evaluation of Attributes in the AG+LA

In GE, when mapping of an individual is represented as a derivation tree, all nodes corresponding to non-terminal symbols in a derivation tree are expanded in depth-first order. Since attribute evaluation is also required in AG+LA, to facilitate attribute evaluation alongside derivation using a single left-to-right depth-first generation of the derivation tree, the grammar needs to be well-designed. Each semantic function responsible for evaluating each attribute of a symbol should depend on the attributes of the symbols to its left, including the head of the production. As a result, when shown as a derivation tree, each attribute defined at a node will rely only on attributes of any sibling node(s) defined to its left or any attributes of its parent node.

In this section, we describe an example to describe the genotype-phenotype mapping process in the AG+LA. In Fig. 18.1, we show how the grammar in Sect. 18.5.1 is used to derive knapsack items for the MKP without details of attribute evaluation. For simplicity, attribute evaluation mechanism alongside derivation is described step by step in subsequent figures using derivation trees. In Fig. 18.1, like standard GE mapper, beginning from the start symbol of the grammar, integer codons are mapped to non-terminals or terminals through the iterative application of the mod rule to the leftmost non-terminal in the derivation sequence. The mod rule can be described as:

Rule No. = (codon integer value) % (number of production rules)

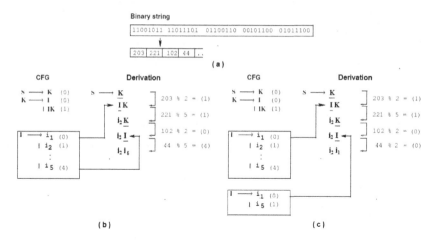

Fig. 18.1 Derivation in GE and AG+LA for a problem instance with five items and four knapsacks. Each underlined non-terminal in the derivation sequence represents the leftmost symbol which will be expanded by the next codon in the genome. (a) Genome (b) GE mapping process (c) AG+LA mapping process. Unlike GE, the number of production rules for non-terminal *I* changes based on knapsack constraints

The mod rule selects a rule from the number of available production rules for the leftmost non-terminal and the body of selected production rule is used to expand that non-terminal. In Fig. 1, each underlined non-terminal in the derivation sequence represents the leftmost symbol.

Unlike standard GE where the number of production rules for all non-terminal is fixed, in AG+LA the number of production rules for the item-generating non-terminal *I* changes at different stages of derivation and is dictated by the attribute *alloweditems* of *I*, which *I* inherits from its parent *K* during derivation. For non-terminal *I*, a production rule is created for each of the items in the allowed item list before this non-terminal can be expanded by any terminal (item). Thus the number of production rules for non-terminal *I* is always equal to the number of items in the allowed item list.

Figure 18.2, 18.3, 18.4 and 18.5 show how attributes are evaluated along side derivation for the same MKP example. In all these figures, each attribute of a symbol is shown using a rectangular box on the right side of the node representing that symbol. An outgoing arc from an attribute of a symbol points to an attribute which depends on this attribute for its own value. An outgoing solid arc points to a synthesized attribute of other symbols, whereas an outgoing dotted arc points to an inherited attribute of this symbol or other symbols. The number shown on the left of any node represents the order in which nodes are expanded during depth-first derivation and the rectangular box on the left shows how the mod rule is used to choose the production rule to expand that non-terminal.

A gray node in any derivation tree represents the state when a node corresponding to a non-terminal symbol is visited for the first time and a black node, on the

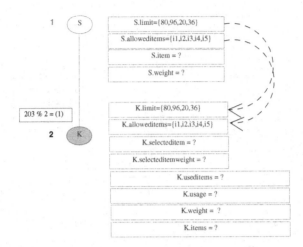

Fig. 18.2 The state of the derivation tree after start symbol S is expanded to K and is visited for the first time. Dotted arc points to some inherited attributes of K which can be evaluated at that time. The question mark shown on the right of any attribute indicates that the concerned semantic function cannot be used at this point to compute its value, as the required attributes on the right side of the semantic function is not yet evaluated.

other hand, represents the state when it is visited for the last time. If a question mark is shown on the right side of any attribute, it means that the attribute cannot be evaluated at that point of derivation due to dependency on other attributes. For the attributes dealing with knapsack capacity or usage, each element in the array represents a value for a single knapsack. To avoid clutter, arcs pointing to attributes or rectangular boxes for mod rule shown in one figure are dimmed in the subsequent figures.

When the derivation starts and the start symbol S is visited for the first time, all the special inherited attributes of S are initialized with the relevant semantic functions. The attribute *limit* is initialized to { 80, 96, 20, 36 }, with the knapsack capacity of all knapsacks. The attribute *alloweditems*, on the other hand, is set to items { i_1, i_2, i_3, i_4, i_5 }, with the list of items that will not violate constraints. When the start symbol S is expanded to the symbol K and the symbol K is visited for the first time, the semantic functions are executed at that node to inherit these attribute values through its own inherited attributes *limit* and *alloweditems* (see Fig. 18.2). After inherited attributes of K are evaluated, the next unused codon 203 is read from the genome and AG+LA mapper mod the codon value by the number of production rules available for the non-terminal K (2 in this case) to expand it by the production rule IK.

In this situation there are more than one non-terminal symbols (IK) in the current derivation sequence which needs to be transformed. Like the standard mapper in GE, AG+LA mapper always selects the leftmost non-terminal, which means in this case I. As soon as the leftmost child node corresponding to symbol I is visited for the first time (see Fig. 18.3), its inherited attribute *alloweditems* is set to

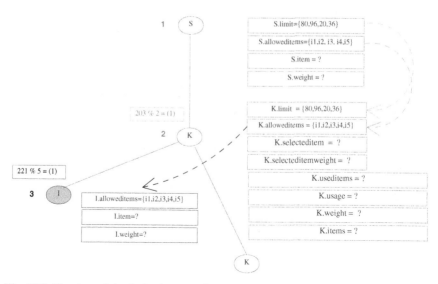

Fig. 18.3 The state of the derivation tree after non-terminal K is expanded by the codon 203 and the symbol I is visited for the first time. The arcs pointing to the attributes evaluated in the previous steps are dimmed. The information regarding the allowed items that is used to select the next item for the knapsacks is propagated through the inherited attribute *alloweditems* of I

$\{\ i_1,\ i_2,\ i_3,\ i_4,\ i_5\ \}$ by the attribute *alloweditems* of its parent node K, using the semantic rule defined at this node for this attribute. Thus the list of items that will not violate constraints if selected, is computed at the K node and is passed down to the item-generating non-terminal I through the inherited attribute *alloweditems*. At this point, the attribute *alloweditems* dictates that the number of production rules for symbol I is 5. Later the non-terminal I is transformed to item i_2, the second item from the allowed item list, using the next codon 221 (221%5=1) from the genome.

The synthesized attributes *item* and *weight* of I are then evaluated using the respective semantic functions and set to item i_2 and its weight (see Fig. 18.4). When symbol I returns control to its parent K, these two synthesized attributes are used to pass the selected item i_2 information up the derivation tree to evaluate the value of each synthesized attribute, *selecteditem* and *selecteditemweight*, at its parent node. These computed values, in turn, facilitate the immediate evaluation of other inherited attributes *usage* and *useditems* of K (see Fig. 18.4). In this way, selected item information passes bottom-up and left-to-right to facilitate computation of already selected items and capacity usage of each knapsack. These two attribute values along with each knapsack capacity are required to calculate the next allowed item list in the second child (corresponding to K) node of this parent K node.

Due to lack of space, the rest of the derivation and attribute evaluation process is shown in Fig. 18.5. When the AG+LA mapper visits the second child K, this symbol inherits usage information for each knapsack as well as previously selected items list from its parent, using its own attributes *usage* and *useditems*, respectively. Based on

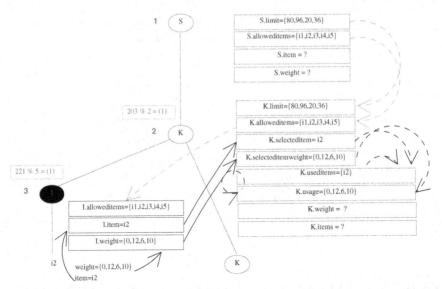

Fig. 18.4 The derivation tree after non-terminal *I* is expanded by the codon 221 to select an item i_2. A solid arc points to a synthesized attribute, whereas a dotted arc points to an inherited attribute. After item i_2 is selected and the node corresponding to non-terminal *I* is visited for the last time, the synthesized attributes *items* and *weight* of non-terminal *I* is computed based on the item i_2 information. When node *I* returns control to its parent node *K*, the synthesized attributes *selecteditem* and *selecteditemweight* of non-terminal *K* are computed, as required attributes have been evaluated. Processing of these two attributes leads to the attribute evaluation of two other inherited attributes *usage* and *useditems* of *K*, as attribute dependencies are resolved

the remaining capacity for each knapsack as well as previously selected items information, the semantic function *genAllowedItems* evaluate the attribute *alloweditems* of this symbol and set the value to { i_1, i_5 }. It excludes the already selected item i_2 to avoid duplicate items. The item i_3 is also excluded, which if selected, can exceed knapsack weight constraints. Thus using the lookahead technique, allowed item list are generated beforehand to transmit the list to the next item-generating non-terminal *I* so that constraints violation and repeated remapping problem can be avoided.

After evaluating all inherited attributes at the current node *K*, the AG+LA mapper apply the mod rule to the next codon 102 (102%2=0) to expand this node by the production rule *I*. As the inherited value for the the attribute *alloweditems* of non-terminal *I* is 2 this time, the number of production rules for *I* to be used in the mod rule is 2. Using the next codon 44 and the number of production rules, the mod rule maps the resulting non-terminal *I* to item i_1 (44%2=0), the first item from the allowed item list. Next, in the derivation process all synthesized attributes of *I* are evaluated.

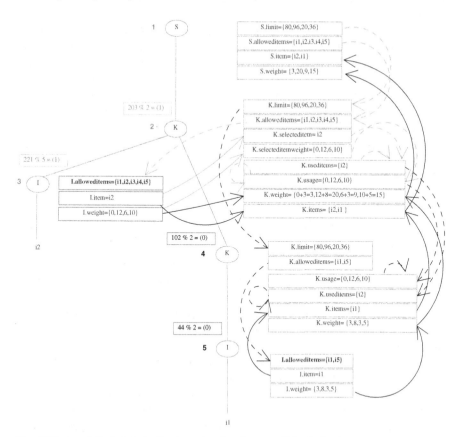

Fig. 18.5 An attributed derivation tree showing the rest of the derivation process

When the symbol *I* returns control to its parent *K*, synthesized attributes *items* and *weight* of its parent are computed. Similarly, when exploration of all child nodes of any node representing the symbol *K* or *S* are completed, synthesized attributes *items* and *weight* of that node are evaluated based on the item and weight information of its child nodes. Finally the result is collected from the attributes of the root node *S*.

18.6   Degenerate Code and Duplicate Elimination in AG+LA

Degenerate code means that codons of different values can represent the same production rule of the BNF definition. Standard GE uses 8 bit to represent each codon. That means each codon can represent 256 possible integer values. Due to the mod operator, different integer values can map to the same production for the non-terminal that is to be mapped. Taking the production rules for nonterminal *I* in Fig. 18.1(b) showing the GE Mapping process, if the current codon value is mutated and becomes 104 instead of 44, 104%5=4 would select the same rule to select item i_5. The same rule would be chosen if the codon value was 4, 14, 19 etc. Degeneracy can be controlled to a great extent by changing the codon size which is measured in bits.

Degenerate code has implications in case of GE. Due to degenerate code, subtle changes in the search space (genotype) may have no effect on the solution space (phenotype), which could ensure genotype diversity [10]. Different genotypes can also map to the same phenotype in GE due to degeneracy. Nicolau et al. later reported that introduction of degeneracy reduces the functional dependency between position specifications and their placement on the real specifications string for the GAuGE system which also employs the mod rule for the mapping process [9].

Like GE, AG+LA also uses the mod rule for selecting production rule for a non-terminal. In AG+LA, the number of production rules for all non-terminals except I, the item-generating non-terminal remains same at different stages of the genotype-to-phenotype mapping phase. So for nonterminal I, same codon value might select different rule at different point of mapping, but for other non-terminals, same codon value always select the the same rule. During mapping, as more and more items are filtered out to avoid already selected items as well as items that might violate constraints, the number of production rule choices for the nonterminal I decreases with the progress in the mapping process. Thus at the early stages of the mapping process for an individual, when many terminals (items) are available for nonterminal I, the number of different codon values representing the same terminal (item) is small in number, but this number increases as mapping progresses.

Degeneracy has been shown as a key feature of GE which helps to achieve good results for a large range of problems [10]. But the effect of degeneracy on the performance of decoders designed with AG+LA has not been investigated so far. Earlier work reported that the redundancy in decoder-based EAs frequently leads to premature convergence while solving multidimensional knapsack problem [11]. Raidl et al. investigated the effects of genotypic or phenotypic duplicate elimination in steady-state decoder based EAs as a method for preventing premature convergence [11]. Phenotypic duplicate elimination means that a child is rejected if its phenotype is already represented in the current population, while genotypic duplicate elimination rejects an offspring if the same genotype is already contained in the current population.

They showed that without phenotypic duplicate elimination, crossover operator cannot reliably produce new solutions and the considered EAs get trapped at bad local optima very early [11]. They also reported that genotypic duplicate elimination cannot avoid or reduce this premature convergence, only phenotypic duplicate elimination can reliably achieve this. A reason for the failure of genotype duplicate elimination is that considered decoder-based EAs use large genotype search spaces which are mapped to significantly smaller parts of the actual phenotype space. This high encoding redundancy results in bad performance when applied to combinatorial optimization problems.

18.7   Experiments

Standard MKP test data publicly available from SAC-94 library for multiple knapsack problems [3] are used to practically examine the AG+LA and compare the results with other relevant evolutionary approaches from the literature. Where required data is available, we determine the statistical significance of the differences in best scores using t-test requiring $p < 0.05$.

Rather than implementing the other approaches to generate unavailable data, we rely on the information available in the literature to compare performances. For comparison purposes, we use twelve problem instances from SAC-94 library. We conduct a total of 30 runs for each problem instance. The basic characteristics of these test problems and problem specific parameter settings are given in Table 18.1.

The tests have been carried out using standard experimental parameters for GE except population size and codon size. We adopt 1500 as population size (unless explicitly stated below). We use a variable length one-point crossover probability of 0.9, bit mutation probability of 0.01, roulette wheel selection and steady state GA. The steady state GA we use inserts the generated child if its fitness is better than the worst individual in the population. During fitness calculation, we convert the raw fitness f_r of each individual to scaled fitness f_s in the following way:

$$f_s = 1 - \frac{1}{1 + f_r} \qquad (18.2)$$

These scaled fitnesses are used for calculating the selection probability in the roulette wheel selection.

For initializing the population, we use Sensible Initialization (a GE specific version of the commonly used GP ramped half-and-half technique) [13] with a grow probability of 0.5. The minimum depth parameter for Sensible Initialization is 8, whereas the maximum depth is problem instance specific. The maximum depth is set to three quarters of the number of items in the problem instance to introduce diversity in the initial population. We do not allow any wrapping operation.

Instead of using standard 8 bit codon, we conduct experiments with different sizes of codon {6,7,8 and 9 bits}. For each problem instance having more than 60

Table 18.1 Properties of the test problem instances with parameter settings

Problem	Items	Knapsacks	Optimal-Value	Individual-Processed	Max-Depth
hp2	35	4	3186	100000	26
pet3	15	10	4015	20000	11
pet4	20	10	6120	20000	15
pet5	28	10	12400	50000	21
pet6	39	5	10618	100000	29
pet7	50	5	16537	150000	37
sento1	60	30	7772	150000	45
sento2	60	30	8722	150000	45
weing7	105	2	1095445	200000	78
weing8	105	2	624319	200000	78
weish12	50	5	6339	150000	37
weish17	60	5	8633	150000	45

Table 18.2 Comparison of the performance of AG+LA for various codon sizes (in bits) with and without duplicate elimination. 'DE' indicates that phenotypic duplicate elimination is incorporated, while 'No-DE' means duplicate elimination is not considered. Comparison is based on average fitness and standard deviation (shown in brackets) of the best individual in the final generation. The best average score is highlighted in bold

Problem	6 bit		7 bit		8 bit	
	No-DE	DE	No-DE	DE	No-DE	DE
hp2	3186(0)	3186(0)	3186(0)	3186(0)	3186(0)	3186(0)
pet3	4015(0)	4015(0)	4015(0)	4015(0)	4015(0)	4015(0)
pet4	6120(0)	6120(0)	6120(0)	6120(0)	6120(0)	6120(0)
pet5	12400(0)	12400(0)	12400(0)	12400(0)	12400(0)	12400(0)
pet6	10606.97(11.58)	10608.9(7.65)	10603.13(11.01)	**10611.03(8.14)**	10603.57(11.09)	10610.8(6.86)
pet7	16487.53(32.96)	16522.97(10.85)	16492.4(28.29)	**16527.47(7.92)**	16487.1(23.24)	16524.07(10.13)
sento1	7561.13(118.03)	7751.36(20.22)	7557.56 (128.19)	**7754.66 (17.55)**	7543.06(130.31)	7747.56(26.15)
sento2	8581.76(47.12)	**8684.8(24.89)**	8596.81(56.33)	8684.8(25.53)	8583.8(47.09)	8675.76(35.31)
weing7	-	-	1094149(955.32)	**1095220(333.42)**	1093481(1272.45)	1094909.16(497.68)
weing8	-	-	596567.6(16473.9)	**623764.3(1539.56)**	581393.2(18074.47)	622661.1(2584.06)
weish12	6214.1(89.13)	**6338.86(0.34)**	6221.3(91.03)	6338.3(3.46)	6204.63(71.08)	6338.9(0.30)
weish17	8580.33(30.18)	8615.1(21.10)	8557.33(43.39)	**8615.63(17.88)**	8560.733(40.61)	8612.33(16.91)

items, the minimum value of the codon size is set to the lowest possible value so that the number of integers represented by a single codon is enough to represent the number of items in the problem instance. For other small instances, we choose the minimum codon size as 6.

18.8 Results and Discussion

In this section, we describe experiments to investigate the effect of having degeneracy and incorporating phenotypic duplicate elimination on the performance of AG+LA while solving the MKP. To compare the performance of AG+LA for various parameter settings as well as with other relevant approaches, we use average fitness and standard deviation of the best individual in the final generation.

18.8.1 Effect of Phenotypic Duplicate Elimination

Table 18.2 shows the results for various codon sizes with and without duplicate elimination. We first conduct experiments to investigate the performance of the AG+LA with various codon sizes without incorporating any form of duplicate elimination in the search process. However, during population initialization using the Sensible Initialization procedure [13], we eliminate all duplicates. For this set up, AG+LA generates moderate result for each tested problem instance. The bad quality of solutions is caused by premature convergence at the phenotypic level of the population.

Due to high redundancy introduced by the many-to-one mapping process in AG+LA, exploration by the one point crossover operator gets affected negatively since significant number of phenotypic duplicates are generated. To solve instances like *weing8* or *sento1* where feasible solutions in the search space is sparse (around 30% of items can be chosen in the optimal solution without violating constraints),

Table 18.3 Average fitness and standard deviation (shown in brackets) of the best individual in the final generation for various population and codon sizes for the instance *weing8*. Phenotypic duplicate elimination is incorporated in the search process, while keeping other experimental parameters same. As we move from left to right, for any population size, the performance of AG+LA drops with increase in codon size

Population Size	7 bit	8 bit	9 bit
1500	623764.3(1539.56)	622661.1(2584.06)	616807.4667(5986.06)
2000	624052.13(697.41)	622482.27(1950.34)	617930.56(3567.21)
2500	623662.53(984.75)	619500.3(4121.74)	613523.8667(5399.43)

individuals with genotypes containing codon values that might select significant items need to be present in the population. Increased presence of these codon values maximize the inclusion probability of those items. Duplicates leads to loss of individuals containing those significant items and thereby few or insignificant items might dominate in all the phenotypes in the population.

In our next experiment, we incorporate phenotypic duplicate elimination with standard 8 bit codon. We consider a newly generated individual to be a phenotypic duplicate if any existing individual in the population derives the same set of items in any order. From Table 18.2 we find that, for each problem instance, AG+LA with phenotypic duplicate elimination and 8 bit codon size achieves better results than AG+LA without duplicate elimination incorporated, regardless of the codon size used. Our t-test results show that improvements for larger instances are statistically very significant.

18.8.2 Effect of Degeneracy

To understand the impact of degeneracy on the performance of AG+LA, with duplicate elimination incorporated in the search process, we either increase or decrease the codon size from the standard 8 bit and observe the performance for each problem instance. For challenging and large problem instances like *weing7* or *weing8*, increase in codon size from 8 bit to 9 bit results in significant performance drop, while in case of reduction in size, AG+LA shows large and statistically significant improvement.

The improvement or drop in performance can be attributed to the increased or decreased phenotypic diversity. When codon size is reduced, the number of different codon values representing the same item decreases and the probability of introducing new or significant items in the phenotypes increases. For other problem instances with smaller number of items, the difference in scores are small and statistically not significant. Our further experiments with different population sizes for the problem instance *weing8* (see Table 18.3) shows that increased codon size results in performance drop despite increase in population size, within same maximum individual evaluations limit.

Thus phenotypic duplicate elimination alone is not sufficient to achieve good result for AG+LA to tackle large instances of the MKP, proper setting of codon size is also required. Rather than using the the same codon size for each problem which

Table 18.4 Comparison is based on the average fitness of the best individual in the final generation. 'Opt.' indicates the percentage of runs yielding optimal solution. A positive number indicates by how much AG+LA performed better than the approach named on top of the column, whereas a negative number indicates an inferior performance of AG+LA. Data is highlighted in bold where the difference is large. 't*' indicates that enough data is available for this approach to perform statistical test , '*' indicates that the difference is considered to be statistically significant ($p < 0.05$)

Problem	AG+LA		AG(Full)		HGA		SGA		2-MGA		ExGA II	
	Avg.	%Opt.	Avg.	Diff.	Avg.	Diff.	Avg.	Diff.	Avg.	Diff.(t*)	Avg.	Diff.(t*)
hp2	3186	100	-	-	-	-	-	-	3186		3186	±0
pet3	4015	100	4011.33	+3.67	4015	± 0	4012.7	+2.3	-	-	4015	± 0
pet4	6120	100	6111.0	+9	6120	± 0	6102.3	+17.7	-	-	6120	± 0
pet5	12400	100	12377.66	+22.34	12400	±0	12374.7	+25.3	-	-	12400	± 0
pet6	10611.03	53	10457.63	**+153.4**	10609.8	+1.23	10536.9	**+74.13**	-	-	10618	-6.97(*)
pet7	16527.47	33.33	16266.13	**+261.34**	16512.0	+15.47	16378.0	**+149.47**	16486.6	+40.87(*)	16537	-9.53(*)
sento1	7754.66	36	7693.2	**+61.46**	7767.9	-13.24	7626	**+128.66**	7769.8	-15.14	7772	-17.34(*)
sento2	8684.8	3	8670.06	+14.74	8716.5	-31.7	8685	-0.2	8720.4	-35.6(*)	8722	-37.2(*)
weing7	1095220	6	1075950.83	**+19270**	1095386	-166	1093897	**+1323**	1094727	**+493(*)**	1095435.1	-215.1
weing8	623764.3	83	605993.63	**+17770.67**	622048.1	**+1716.2**	613383	**+10381.3**	623627.8	+136.5	624158	-393.7
weish12	6338.86	90	-	-	-	-	-	-	6339	-0.14	6339	-0.14
weish17	8615.63	40	-	-	-	-	-	-	8633	-17.37(*)	8633	-17.37(*)
Instances worse			0		3		1		4		8	
Instances equal			0		3		0		1		4	
Instances better			9		3		8		3		0	

is conventional in GE based systems, our results emphasize the need for setting problem instance specific codon size while solving any difficult combinatorial optimization problem like the MKP. We recommend that the codon size should be set to a minimum possible value so that the number of integers represented by a single codon is enough to represent the number of items in the problem instance.

18.8.3 Comparison with Other Approaches

Table 18.4 shows the average fitness of the best individual in the final generation for the AG+LA alongside other approaches from the literature. For comparison purposes, we consider the AG+LA best output with duplicate elimination from Table 18.2 as the output for AG+LA. We compare our results with AG(Full), one attribute grammar based approach [1] and two penalty-based techniques e.g. SGA, a canonical GA with graded penalty term [5] and two-market genetic algorithm (2-MGA) [6]. We also consider one hybrid GA technique (HGA) based on local search [2] and EXGA II, a repair based algorithm [12].

From our results we find that AG+LA outperforms AG(Full) [1] across all instances. It is worth noting that AG(Full) [1] uses 200000 individual evaluations for each problem instance, while AG+LA uses fewer individual evaluations for problem instances with smaller numbers of items (see Table 18.1). We do not compare our approach with any other attribute grammar based techniques reported in [1] as mean fitness is not reported for those techniques.

Among the penalty based techniques, SGA produces worse results than AG+LA for almost all problem instances, while 2-MGA performs worse for instances *pet7*, *weing7* and *weing8*. From the results we can say that AG+LA performs better than any penalty based technique and the difference in the best scores is high when the problem instance is large e.g. *weing7* or *weing8*.

The performance of AG+LA against HGA, the only approach using problem specific knowledge among all the compared, is very encouraging. In terms of the best score, AG+LA either produces same score or better score for all relevant tested problem instances except *sento1*, *sento2* and *weing7*. The amount of improvement shown in the result for one of the large instances *weing8* is significant. The last approach that we compare is the most successful one so far, EXGA II. Even though AG+LA achieves the same best scores for four instances, it produces relatively worse results than EXGA II. But it is worth noting that the difference between the best scores for each instance is very low and we believe that the difference can be reduced further with adaptive tuning of the level of degeneracy in AG+LA.

We conclude from the results and analysis that phenotypic duplicate elimination significantly improves the performance of AG+LA while solving any instance of MKP. Reduction in codon size is also crucial if the problem instance is large. The results presented in this paper also show that overall improvement shown after applying both these techniques makes AG+LA a more capable approach in either reaching the optimal score or approximating the solution, without employing any problem specific knowledge.

18.9   Conclusions

In this paper, we have shown that phenotypic duplicate elimination significantly improves the performance of attribute grammar with lookahead approach (AG+LA) for all the tested MKP problem instances. In addition to phenotypic duplicate elimination, reducing the codon size is also important for the large instances of the MKP. The significant improvement in performance as shown for *weing7* and *weing8*, the two large and difficult problem instances among all the tested, validates the need for applying both techniques and also makes AG+LA a very promising approach for approximating large problem instances of the MKP.

Rather than using a fixed value for a codon size, dynamic level of degeneracy using adaptive codon size has the potential to make AG+LA a more powerful approach. Our future work will look into that. The performance of AG+LA can be improved further if items' value-weight ratio information is used in the item selection process to reduce the number of items in the allowed item list.

References

[1] Cleary, R.: Extending grammatical evolution with attribute grammars: An application to knapsack problems. Master of science thesis in computer science, University of Limerick, Ireland (2005)
[2] Cotta, C., Troya, J.M.: A hybrid genetic algorithm for the 0-1 multiple knapsack problem. In: Artificial Neural Nets and Genetic Algorithms, vol. 3, pp. 250–254. Springer, New York (1998)

[3] Heitktter, J.: Sac-94 suite of 0/1-multiple-knapsack problem instances (2001),
    http://elib.zib.de/pub/Packages/mp-testdata/
    ip/sac94-suite

[4] Karim, M.R., Ryan, C.: A new approach to solving 0-1 multiconstraint knapsack problems using attribute grammar with lookahead. In: Silva, S., Foster, J.A., Nicolau, M., Giacobini, M., Machado, P. (eds.) EuroGP 2011. LNCS, vol. 6621, pp. 250–261. Springer, Heidelberg (2011)

[5] Khuri, S., Back, T., Heitkotter, J.: The zero/one multiple knapsack problem and genetic algorithms. In: Proceedings of the 1994 ACM Symposium on Applied Computing, pp. 188–193. ACM Press, New York (1994)

[6] Kimbrough, S.O., Lu, M., Wood, D.H., Wu, D.J.: Exploring a two-market genetic algorithm. In: Proceedings of the Genetic and Evolutionary Computation Conference, pp. 415–421. Morgan Kaufmann, New York (2002)

[7] Knuth, D.E.: Semantics of context-free languages. Theory of Computing Systems 2(2), 127–145 (1968)

[8] Kumar, V.: Algorithms for constraint satisfaction problems: A survey. AI Magazine 13(1), 32–44 (1992)

[9] Nicolau, M., Auger, A., Ryan, C.: Functional dependency and degeneracy: Detailed analysis of the gAuGE system. In: Liardet, P., Collet, P., Fonlupt, C., Lutton, E., Schoenauer, M. (eds.) EA 2003. LNCS, vol. 2936, pp. 15–26. Springer, Heidelberg (2004)

[10] O'Neill, M., Ryan, C.: Genetic code degeneracy: Implications for grammatical evolution and beyond. In: Floreano, D., Mondada, F. (eds.) ECAL 1999. LNCS, vol. 1674, pp. 149–153. Springer, Heidelberg (1999)

[11] Raidl, G.R., Gottlieb, J.: On the importance of phenotypic duplicate elimination in decoder-based evolutionary algorithms. In: Brave, S., Wu, A.S. (eds.) Late Breaking Papers at the 1999 Genetic and Evolutionary Computation Conference, Orlando, Florida, USA, pp. 204–211 (1999)

[12] Rohlfshagen, P., Bullinaria, J.A.: Exga ii: an improved exonic genetic algorithm for the multiple knapsack problem. In: Proceedings of the 9th Annual Conference on Genetic and Evolutionary Computation, pp. 1357–1364. ACM, New York (2007)

[13] Ryan, C., Azad, R.M.A.: Sensible initialisation in grammatical evolution. In: Proceedings of the Bird of a Feather Workshops, Genetic and Evolutionary Computation Conference, Chigaco, pp. 142–145 (2003)

[14] Ryan, C., Collins, J., O'Neill, M.: Grammatical evolution: Evolving programs for an arbitrary language. In: Proceedings of the First European Workshop on Genetic Programming, pp. 83–95. Springer, Heidelberg (1998)

Chapter 19
Enhancing the Computational Mechanics of Cellular Automata

David Iclănzan, Anca Gog, and Camelia Chira

Abstract. Cellular Automata are discrete dynamical systems having the ability to generate highly complex behavior starting from a simple initial configuration and set of update rules. However, the discovery of rules exhibiting a high degree of global self-organization for certain tasks is not easily achieved. In this paper, a fast compression based technique is proposed, capable of detecting promising emergent space-time patterns of cellular automata. This information can be used to automatically guide the evolutionary search toward more complex, better performing rules. Results are presented for the most widely studied cellular automata computation problem, the Density Classification Task, where incorporation of the proposed method almost always pushes the search beyond the simple block-expanding rules.

19.1 Introduction

Representing useful and important tools in the study of complex systems and interactions, Cellular Automata (CA) are decentralized structures of simple and locally interacting elements that evolve following a set of rules [22]. CA dynamics can be used to analyze and better understand the emergent behavior and computational

David Iclănzan
Department of Electrical Engineering, Sapientia Hungarian University of Transylvania,
540485, Tg-Mureş, CP 4, OP 9, Romania
e-mail: david.iclanzan@gmail.com

Anca Gog
Department of Computer Science,Babeş-Bolyai University, Kogălniceanu no. 1,
Cluj-Napoca, 400084, Romania
e-mail: anca@cs.ubbcluj.ro

Camelia Chira
Department of Computer Science, Babeş-Bolyai University, Kogălniceanu no. 1,
Cluj-Napoca, 400084, Romania
e-mail: cchira@cs.ubbcluj.ro

D.A. Pelta et al. (Eds.): NICSO 2011, SCI 387, pp. 267–283, 2011.
springerlink.com © Springer-Verlag Berlin Heidelberg 2011

complexity of a system [2]. Programming CA is not an easy task particularly when the desired computation requires global coordination. CA provide an idealized environment for studying how simulated evolution can develop systems characterized by "emergent computation" where a global, coordinated behavior results from the local interaction of simple components [14].

One of the most widely studied CA problems is the density classification task (DCT) [13]. This is a prototypical distributed computational task with the aim to find the density most present in the initial cellular state. DCT is not a trivial task since finding the density of the initial configuration is a global task while CA relies only on local interactions between cells with limited information and communication.

Evolutionary algorithms [7, 14, 15, 17, 18], genetic programming [1, 10], coevolutionary learning [8, 9] and gene expression programming [5] have been engaged to approach DCT with better results than any known human-written rule for DCT. Packard [18] made the first attempts to use genetic algorithms for DCT using the human designed Gacs-Kurdyumov-Levin rule as a starting point. Later, genetic algorithms for computational emergence in the density classification task have been extensively investigated by Mitchell et al [3, 14, 15, 19]. The three main strategies discovered are default, block-expanding and particle revealing increasing computational complexity [19]. These rules are able to facilitate global synchronization opening the prospect of using evolutionary models to automatically evolve computation for more complex systems. The potential of evolutionary models to efficiently approach the problem of detecting CA rules to facilitate a certain global behavior is confirmed by various current research results [4, 7, 12, 15, 17, 23]. To the best of our knowledge, the best reported DCT rule was evolved by a two-tier evolutionary algorithm with a performance of 0.89 [17]. The success of the method can be attributed to the smart prefiltering of the search space by employing various efficient heuristics.

In this paper, we focus on detecting at runtime the formation and propagation of "signals", the behavior that characterizes (or can lead to) particle based computing and subsequently to the discovery of high performance rules.

19.2   Related Work

A key aspect in CA research refers to evolving rules for CA able to perform computational tasks which require global coordination highlighting an interesting emergent behaviour. The one-dimensional binary-state CA capable of performing computational tasks has been extensively studied in the literature [2, 8, 12, 15–17, 20, 23]. A one-dimensional lattice of N two-state cells is used for representing the CA. The state of each cell changes according to a function depending on the current states in the neighborhood. The neighborhood of a cell is given by the cell itself and its r neighbors on both sides of the cell, where r represents the radius of the CA. The initial configuration of cell states (0s and 1s) for the lattice evolves in discrete time steps updating cells simultaneously according to the CA rule.

The density classification task (DCT) aims to find a binary one-dimensional CA able to classify the density of 1s (denoted by ρ_0) in the initial configuration. If $\rho_0 > 0.5$ (1 is dominant in the initial configuration) then the CA must reach a fixed-point configuration of 1s otherwise it must reach a fixed-point configuration of 0s within a certain number of time steps. Most studies consider the case $N = 149$ (which means that the majority is always defined) and neighborhood size of 7 (which means the radius of CA is $r = 3$).

The CA lattice starts with a given binary string called the initial configuration (IC). After a maximum number of iterations (usually set as twice the size of CA), the CA will reach a certain configuration. If this is formed of homogeneous states of all 1s or 0s, it means that the IC has been classified as density class 1, respectively 0. Otherwise, the CA makes by definition a mis-classification [19]. It has been shown that there is no rule that can correctly classify all possible ICs [11].

The performance of a rule measures the classification accuracy of a CA based on the fraction of correct classifications over 10^4 ICs selected from an unbiased distribution (ρ_0 is centered around 0.5), while the computationally more cheap fitness evaluation take into account 100 ICs, with uniformly distributed densities. The length of a CA is usually an odd number to avoid the undecidable case when $\rho_0 = 0.5$

DCT is a challenging problem extensively studied due to its simple description and potential to generate a variety of complex behaviors. In a 1978 study of reliable computation in CA, Gacs, Kurdyumov and Levin [6] proposed the well known GKL rule which approximately computes whether $\rho_0 > 0.5$. The GKL rule has a good performance (around 0.81) when tested for random 149-bits lattices but gives significant classification errors when the density of 1s in the IC is close to 0.5 [14]. The GKL rule represents the starting point for studying DCT and inspired Packard [18] to make the first attempts to use genetic algorithms for finding CA rules.

Genetic programming [1, 10] has been succesfully engaged for DCT being able to produce a new rule with a better performance (0.82) compared to previous results. Ferreira [5] used gene expression programming for the same task and discovered two new rules both with a performance around 0.8255.

The potential of genetic algorithms for computational emergence in the DCT has been extensively investigated by Mitchell et al [3, 14, 15, 19]. The three main strategies discovered are default, block-expanding (a refinement of the default strategy) and particle [19]. The performance of default strategies is around 0.5. Block-expanding strategies explore the presence of a large block of 0s/1s (correlated with overall high/low density) by expanding it to cover the whole string. The performance value for these strategies is typically between 0.55 and 0.72 [19]. Particle rules reveal increasing computational complexity having a performance value of 0.72 or higher. Typically, the evolutionary search process discovers many default strategies followed by block-expanding strategies and few particle rules on some runs [14, 19]. The best particle rule found by Mitchell et al [14] has a performance of 0.769. A similar result in terms of rule performance is reported in [7] using a circular evolutionary algorithm with asynchronous search. The efficacy of the circular evolutionary search is superior in terms of number of runs producing high-performance rules.

Coevolutionary learning [8, 9] has been shown to be able to produce really good results for DCT (the best rule has a performance of 0.86). The idea is that a population of learners coevolves with a population of problems to facilitate continuous progress. However, coevolutionary models can lead to non-successful search usually associated with the occurrence of the Red Queen effect. This means that individuals tend to specialize with respect to the current training problems as they are evaluated in a changing environment. In a comparative study of evolutionary and coevolutionary search for DCT, Pagie and Mitchell [19] indicate that an evolutionary model produces the particle strategy in only 2 out of 100 runs compared to the very high 82% efficacy obtained by coevolutionary models. The effectiveness of these models is however significantly diminished when the ability of the coevolving population to exploit the CA is removed leading to the evolution of CA with lower performance value [19].

Oliveira et al [4] present a multiobjective evolutionary approach to DCT based on the nondominated sorting genetic algorithm (NSGA). The algorithm is implemented in a distributed cooperative computational environment and was able to discover new rules with a slightly better performance of 0.8616.

In 2008, Wolz and Oliveira [21] proposed a two-tier evolutionary framework able to detect rules with a performance of 0.889 clearly superior to the previously best known discovered in 1998 using coevolutionary algorithms [8]. The approach integrates a basic evolutionary algorithm in a parallel multi-population search algorithm. To the best of our knowledge, the DCT rules presented in [21] are the ones with best performance discovered to date [17].

An important constraint included in the latter two approaches [4, 21] presented above refers to the number of state transitions of a rule following black-white, left-right (BWLR) transformations. Experiments have shown that this BWLR symmetry should be maximal to achieve high-performant DCT rules [17]. Oliveira et al [17] have shown that the bits related to the complementary-plus-reflection transformation (BWLR-symmetry) are totally symmetrical in the best rules published in [4]. The incorporation of this heuristic information in the evolutionary search has been shown to have a significant impact on the quality of results [4, 17, 21].

After all this extensive research concerning CA computation in general and the DCT in particular, it has become clear, that well performing one-dimensional CA are characterized by a higher degree of self-organization. These rules, exhibit a more complex spatiotemporal behavior, where coherent configurations emerge and interact.

These local information propagating, self-organization enabling configurations have been studied under various labels such as solitary waves, gliders, information structures, particle-like structures and domain boundaries or defects. When analyzing the spatiotemporal behavior of a CA rule, a particle-like structure generally represents a dislocation in the background, exhibiting an out of phase pattern. In large enough systems, out of phase patterns can collide, their interaction forming a compound particle-like structure. Interactions between these structures can be contrived to achieve universal computation [24]. Particle-like structure in the interaction phase can be regarded as competing sub-attractors, with the final survivors persisting in the attractor cycle [25].

The question that naturally arise is how to characterize and group rules according to the emergent space-time patterns and their attractor basin topology. Such characterization enables the development of numerical and analytical tools that discover, analyze, and filter patterns that emerge in CA.

The dynamics of CA rules exhibiting particle-like structures corresponds to Wolfram's fourth complexity class [26], at a phase transition between order and chaos. CA are grouped by Wolfram in automata whose evolution:

- from any initial condition ends in a fixed pattern.
- falls in simple repetitive patterns.
- leads to chaotic patterns.
- leads to localized structures with complex behavior.

CAs can be classified in one of these classes automatically by using entropy-density signatures [25].

Computational mechanics [27, 28] embody methods able to characterize patterns and structure occurring in natural processes by means of formal models of computation. One of the most studied connection between CA and computation theory has centered around the problem of designing CAs exhibiting particles and computing with them [29, 30].

In this work we focus on developing fast computational methods which detect rules exhibiting particle-like structures. These methods then are used to augment the evolutionary search to favor the development of particle-based computation in the best individuals.

19.3 Towards the Evolution of Particle-Based CA

Mitchell et al. [14] identifies three types of evolved strategies for the DCT. The simplest strategy classifies all bit strings to one and the same density class, thus it has a performance values of approximately 0.5. Block-expanding strategies are refinements of the default strategy where most bit strings are classified to one default density class. However, if the IC contains sufficiently large blocks of values opposing the default density classification, then these blocks are expanded to cover the whole string. Block-expanding strategies typically have performance values between 0.55 and 0.72 on bit strings of length 149. The third class, where the most complex, the so called particle strategies use interactions among larger-scale patterns in order to classify ICs. Particle rules typically have performance values of 0.72 and higher on bit strings of length 149.

When an evolutionary search is applied for rule detection, usually the first generations are dominated by default strategies, which are gradually replaced by block-expanding strategies, and finally (on some runs) particle strategies.

Particles are the main mechanisms for carrying information over long space-time distances. The information might indicate the result of some local processing which has occurred at an early time. In the case of DCT an expanding black block indicates that the local density is definitively leaning towards a majority of ones, a white block is a vote for a majority of zeros, while a propagating checkerboard like pattern indicates that a decision can not made locally, the decision must be postponed to a higher level.

In this way particles can act as signals. Higher level decision can be made by performing logical operations on the information of particles that collide-interact. For example, when a checkerboard pattern collide with a black domain wall, a higher level decision is made in the favor of a ones majority and the checkerboard like pattern carrying signal is absorbed, it disappears.

Without particles and particle interactions, the CA's computation capabilities are very limited. Thus, it is important to develop methods that can produce CA exhibiting particle-based computing. In contrast with previous approaches [4] that prune the search space apriori using various heuristics, here we focus on techniques that are able to efficiently detect and favor the formation of signals and particle interactions at runtime.

For the DCT, complex rules exhibit transient phases during which a spatial and temporal transfer of information about the density in local regions takes place. Figure 19.1 visually depicts the difference between the space-time diagram of a) a block-expanding rule dominated by large areas of homogeneous blocks of 1s and 0s, and b) a particle based rule, where the signal areas exhibit fractal like patterns, maintaining a local symmetry and balance between the density of 1s and 0s.

19.3.1 Detecting Complex Space-Time Signatures by Compression

Our automatic signal detection proposals are based on the following observations:

- In order to propagate in time, signals must maintain locally a roughly equal density of 1s and 0s. Therefore they do not compress well under run-length encoding (RLE), which replaces a long sequence of the same symbol as a single data value plus its count. Space-time diagrams of block-expanding rules compress very well with RLE.
- Along with density p close to $1/2$, signals are also characterized by repetition of patterns of 1s and 0s which are bilateral symmetric (mirror-like symmetry). This symmetry enables the propagation of the signal, by recursively transforming pattern of 1s in pattern of 0s and vice-versa.

Space-time diagrams of block expanding rules, with a lot of lengthy constant states through time, are well reduced by encoding such runs with two symbols. The first symbol represents a number corresponding to the number of repeating states in the run and is called the run count. The second symbol is the value of the state, which in the case of binary CA space-time diagram is a bit of either 0 or 1. On the other

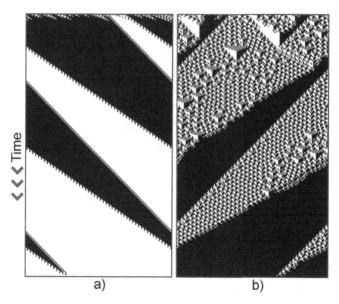

Fig. 19.1 Space-time plot of: a) a simple block-expanding rule (performance: 0.64%) b) best known rule for DCT (performance: 0.89% reported in [17]).

hand, the cells in a chaotic or complex CA often change their states through time, resulting in a highly inefficient RLE compression of the space-time diagram.

As we are not interested in the exact RLE compression rate, we use a computationally more efficient measure based on similar principles, which averages how often cells in the CA changed their states during a run:

$$C(ST) = \frac{\sum_{j=1}^{n} \sum_{i=s}^{T} (ST(i,j) \oplus ST(i-1,j))}{n*(T-s)+1}; (s < T) \qquad (19.1)$$

The input of eq. 19.1 is a space-time diagram denoted by ST, which is obtained by running a CA upon an IC state until convergence or a maximum time step T_{max}. In practice, for the DCT, ST is a binary matrix with n columns – number of cells in the CA and T rows – the actual time step values. When quantifying the changing dynamics of a CA, it is recommended to start the computation after a prefixed number of time steps, in order to eliminate the first, almost always noisy states from the space-time diagram. This offset is expressed by the parameter s which has the non-optimized, arbitrary chosen value $s = 20$ in the experiments conducted in this paper. If there are less time steps T then the prefixed generation threshold s, the value of $C(ST)$ is not computed and is assigned the default value of 0.

Using the metric given in eq. 19.1 one can discriminate between block-expanding CA and CA with high state changing dynamics by computing an average for different space-time diagrams, obtained from random initial states with density close

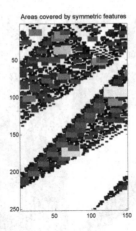

Fig. 19.2 The result of a mirror-like symmetric pattern detection algorithm, applied to the space-time diagram presented in Figure 19.1 b)

to 0.5. To differentiate between chaotic and complex rules, more sophisticated methods are needed, which can approximate the amount of information in the state alternating regions of the space-time diagram. While on the case of chaotic rules, state changes are expected to be random, in the case of complex rules these signal regions follow a pattern that are mirror-like symmetric.

Dividing the total area of mirror-like symmetric regions with the size of space-time diagram one can get a good approximate of the amount of information - patterns exhibited by the CA. For more exact measures, the counted areas of mirror-like symmetric regions should have a minimal predefined size, thus ensuring that their formation is improbable due chance. Further, bigger mirror-like symmetric should provide a bonus, as bigger sizes imply more structure in the signal. Finally, mirror-like symmetry should be searched only for blocks that contain both 0's and 1's, ensuring that the rewarded patterns differ from the trivial all 0's or all 1's.

Figure 19.2 depicts the result of a mirror-like symmetric pattern detection algorithm, applied to the space-time diagram presented in Figure 19.1 b). Lighter colors correspond to bigger detected areas.

Unfortunately, the mirror-like symmetric pattern matching search is computationally expensive in its current form, thus it can not be used to guide the evolutionary search for well performing CA.

However, augmenting the raw performance measure of a CA with a reward value which depends on the cell changing dynamics measured with eq. 19.1 can yield improved results:

- a high fitness value (measured over 100 ICs, with uniformly distributed densities) implies that the rule is not chaotic;
- a high cell changing dynamics suggest that the rule is not block-expanding.

Thus, a CA that maximizes both measures is highly likely to belong to the complex class of CAs, exhibiting particles.

19.3.2   Encouraging the Evolution of CA with State Changing Dynamics

In order to encourage the development of particle exhibiting CAs, we modify the fitness measure to include a bonus which depends on the cell changing dynamics:

$$F100_m(IC) = F100(IC) + max\left(0, 2*\left(1 - \frac{1 - F(100)}{0.55}\right)*C(ST(IC))\right) \quad (19.2)$$

$F100(IC)$ is the number of correct classification over the 100 random ICs. $C(ST(IC))$ is the averaged quantified value for cell state change dynamics. As it can be observed from the formula, a bonus is given only to individuals with an original fitness value greater than 0.55. Furthermore, the amount of bonus is designed to depend and increase along with the original fitness value. In this way we wish to avoid the forming of a local-optima with a large basin of attraction for CA with original mediocre fitness value close to 0.5 but exhibiting a higher amount random cell change dynamics.

If the non-compressibility arrises from random behavior and not useful signals, the classification performance of a rule will still be weak. Thus, the bonus rewards only good enough individuals, in proportion with their quality.

19.4   Experiments

The space-time diagrams from 1000 unbiasedly generated ICs for DCT, of 50 block-expanding rules (randomly extracted from runs of the GA reported in [14]) with performances between 0.56% and 0.64% and highly fit particle rules reported in the literature (presented in Table 19.1) were measured by using the relation defined in eq. 19.1 and an average was computed.

The block-expanding rules exhibited a high compressibility, with a maximal average value of 0.174%. The particle rules exhibited a compression rate between 0.4196% and 0.6706% as presented in the 3rd column, Table 19.1, empirically confirming the introduced measure is a good discriminant between poorer block-expanding and higher performing rules for the DCT.

In a second step, we modified the GA reported in [14] by employing a population of 200 and using our proposed fitness function defined in eq. 19.2. 20 runs of the modified algorithm on the DCT of size 51 showed, that except one case, the dominant block-expanding rules are always replaced with relatively high fitness rules, that exhibit enlarged transient portions in their space-time diagrams.

The average rule performance evolved (computed by removing two outliers, the minimum and maximum from the dataset) with the modified fitness function is 66.28% with a standard deviation of 2.97%. As a rough comparison, the average performance of block-expanding rules for DCT of size 149 (versus 51 in our case) is 65.52%, a value empirically obtained by Crutchfield and Mitchell [31]. The

Table 19.1 Various high performance rules for the DCT, in hexadecimal coding

Code	Rule (Hex)	Perf.%	$C(ST)$
GKL	050005FF050005FF05FF05FF05FF05FF05FF	81.6	0.6241
K96	00550055005500555F55FF5F5F55FF5F	82.3	0.5385
JP1	156043701700D25F15630F7714F3D77F	85.1	0.4846
JP2	050C350F05007717058CF5FFC5F375D7	85	0.6706
O08	0203330F01DF7B17028CFF0FC11F79D7	88.9	0.4196

Table 19.2 The evolved rules in hexadecimal coding, their performance and measured cell state change dynamics

Code	Rule (Hex)	Perf.%	$C(ST)$
01	1406334D3760270936058A0F5763F93F	63.62	0.4479
02	05490875118323017D43376DBB93B24F	66.26	0.4898
03	075941870BB1FDE14120C3032975CBF7	63.79	0.4354
04	160C448539E57FEB78FA81357D87B1F7	59.16	0.4054
05	05663C3E71D61B136EF2DD17EE25971F	48.34	0.7222
06	054B35292015003977F950B5A79939DF	67.09	0.3281
07	11293340496D2B653F2D37C51BB7C35F	67.4	0.5702
08	010717847B3421517335BA1F6B4D7DD7	61.75	0.6805
09	003F34897B05C51B4D350F4D075DD19F	69.47	0.4583
10	04012E611603435956218E132BF78F7F	72.28	0.6975
11	11347E014F5530015E220C2DD7C55FFF	66.21	0.6581
12	0430754E765409D1731735CB3BF7371F	67.68	0.3198
13	03163B6230633663567B3B63776F33FF	71.54	0.3942
14	021401500B11595175F9775DF4F9FF5F	67.85	0.6772
15	101473462C33205771237BBF37B9F357	68.03	0.4305
16	0006250A7F636D774B614AF16F3D6DB7	67.23	0.4459
17	021025D863379D7911FD73E7E93D6FFF	69.45	0.1528
18	10426070145DB909534719FF54CB9D7F	66.4	0.3398
19	044758095C3B6E9140F18809FDCD535F	65.09	0.4528
20	01035FB940032D116FD367C9FBE5FB57	65.02	0.7165

highest ever measured performance they observed for block-expanding rules was 68.5%. Some runs from our experiment produced a higher level of computational sophistication, above that of the block-expanding rules, with performance values around of 70%.

The current formulation of the fitness function clearly encourages cell state change dynamics: the average of this measure is 0.4911 with a standard deviation of 0.1581. This value is much higher than the ones typically exhibited by block-expanding rules and is in the range with the dynamics measured on high performance particle rules reported in literature. The evolved rules, their performance and cell state change dynamics are presented in Table 19.2.

In the following we take a closer look to the space-time diagrams of some of the evolved rules.

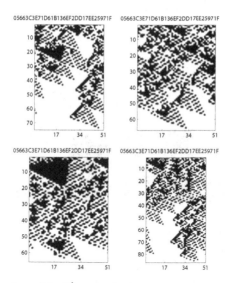

Fig. 19.3 Space-time plots of the 5th evolved rule
(05663C3E71D61B136EF2DD17EE25971F).

The rule with the lowest performance of 48.34% was evolved in the 5th run.
The selection of a rule with a performance below 50% is explained by the high
cell dynamics exhibited by this rule. Some of the space-time diagrams of this rule
are presented in Fig. 19.4. The rule exhibits some interestingly spreading patterns,
however regardless of the initial configuration it always converges to a state of all
zeros. This poor classification strategy explains the low performance.

The highest obtained performances were 72.28% and 71.54% obtained in runs
10 and 13. From the figures of their space-time diagrams (Fig. 19.4 and Fig. 19.5)
one can see that different they work with different strategies. Rule 10 propagates an
alternate signal when expanding blocks of zeros, however block of ones are straight-
forwardly expanded.

Fig. 19.6 and 19.7 depicts space-time diagrams of rules with a low cell dynamics
obtained in runs 6 and 17. The lowest dynamics is 0.1528 of rule 17. The rule is
characterized by a quick convergence and simple block-expanding strategy. The rule
obtained in run 6 exhibits a slightly richer dynamics, with cells rarely alternating
states.

Space-time dynamics of other interesting rules are showed in Fig. 19.8, 19.9,
19.10 obtained in runs 3, 8 and 20.

The rule presented in Fig. 19.8 exhibits a fractal like behavior, with gliders pe-
riodically emanating in both directions. The rule from run 8 presented in Fig. 19.9
shows frequent cell state change, where block-expanding overtakes in later phases.
However, as the first subplot shows, sometimes the rule does not converge at all.
Rule from run 20 shows a similar behavior, however it always converges to some
state. This explains its slightly higher performance.

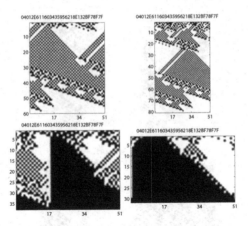

Fig. 19.4 Space-time plots of the 10th evolved rule
(04012E611603435956218E132BF78F7F).

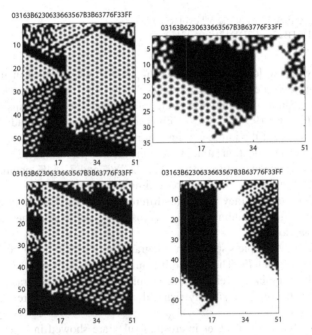

Fig. 19.5 Space-time plots of the 13th evolved rule
(03163B6230633663567B3B63776F33FF).

The results suggest that in the current formulation, our fitness function overemphasizes the cell change dynamics component, in the detriment of the rule performance. The function allows the development of poorly performing rules, provided that they have very strong dynamics (ex. 5th evolved rule). While the approach is successful in hindering the development of block-expanding strategies, the obtained

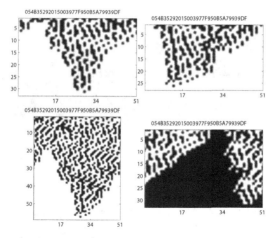

Fig. 19.6   Space-time plots of the 6th evolved rule
(054B35292015003977F950B5A79939DF).

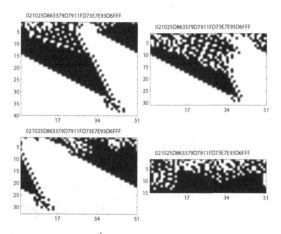

Fig. 19.7   Space-time plots of the 17th evolved rule
(021025D863379D7911FD73E7E93D6FFF).

dynamics do not always exhibit particle like structures, being somewhat chaotic.
High performance rules are characterized by both high fitness and high cell change
dynamics. A multiobjective formulation and optimization would ensure that the development of chaotic rules are not encouraged and both objectives are maximized
in tandem.

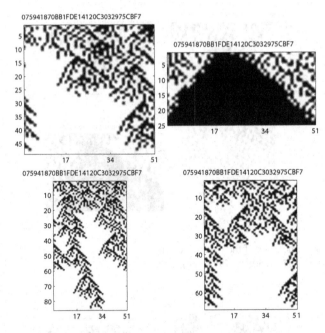

Fig. 19.8 Space-time plots of the 3rd evolved rule (075941870BB1FDE14120C3032975CBF7).

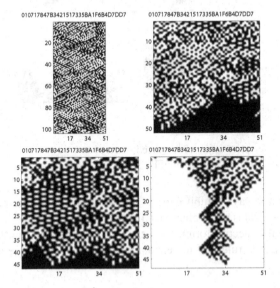

Fig. 19.9 Space-time plots of the 8th evolved rule (010717847B3421517335BA1F6B4D7DD7).

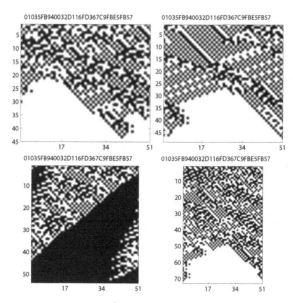

Fig. 19.10   Space-time plots of the 20th evolved rule (01035FB940032D116FD367C9FBE5FB57).

19.5   Conclusions and Future Work

The experiments have indicated that rewarding the non-compressibility under RLE of space-time diagrams has the potential to guide the evolutionary search away from basic block-expanding rules and help automatic CA programming. Without relying on any prefiltering, in all our experimental runs, interesting complex rules were detected. This is in contrast with the low discovery percentage of complex rule of a basic evolutionary search.

We believe a multiobjective optimization framework to be much more suitable for the simultaneous maximization of both classification fitness and cell state change dynamics. Another observation is, that in most cases, the newly emerged, more frequent transient regions are short lived, because they lack symmetry. Albeit performing better, it is still hard for these rules to propagate information on very large scales.

Therefore, future research will focus on developing computationally efficient pattern recognition methods and rewarding mechanisms, which coupled in a multiobjective optimization framework will hopefully facilitate the formation of transient regions exhibiting bilateral symmetry.

Acknowledgements. This research is supported by Grant PN II TE 320, Emergence, auto-organization and evolution: New computational models in the study of complex systems, funded by CNCS, Romania. David Iclănzan acknowledges the support of Sapientia Institute for Research Programs (KPI).

References

[1] Andre, D., Bennett III, F.H., Koza, J.R.: Discovery by genetic programming of a cellular automata rule that is better than any known rule for the majority classification problem. In: Proceedings of the First Annual Conference on Genetic Programming, GECCO 1996, pp. 3–11. MIT Press, Cambridge (1996)

[2] Chira, C., Gog, A., Lung, R., Iclanzan, D.: Complex Systems and Cellular Automata Models in the Study of Complexity. In: Studia Informatica series, vol. LV(4), pp. 33–49 (2010)

[3] Das, R., Mitchell, M., Crutchfield, J.P.: A genetic algorithm discovers particle-based computation in cellular automata. In: Davidor, Y., Männer, R., Schwefel, H.-P. (eds.) PPSN 1994. LNCS, vol. 866, pp. 344–353. Springer, Heidelberg (1994)

[4] de Oliveira, P.P.B., Bortot, J.C., Oliveira, G.: The best currently known class of dynamically equivalent cellular automata rules for density classification. Neurocomputing 70(1-3), 35–43 (2006)

[5] Ferreira, C.: Gene Expression Programming: A New Adaptive Algorithm for Solving Problems. Complex Systems 13(2), 87–129 (2001)

[6] Gacs, P., Kurdyumov, G.L., Levin, L.A.: One-dimensional uniform arrays that wash out finite islands. Probl. Perdachi. Inform. 14, 92–98 (1978)

[7] Gog, A., Chira, C.: Cellular Automata Rule Detection Using Circular Asynchronous Evolutionary Search. In: Corchado, E., Wu, X., Oja, E., Herrero, Á., Baruque, B. (eds.) HAIS 2009. LNCS, vol. 5572, pp. 261–268. Springer, Heidelberg (2009)

[8] Juille, H., Pollack, J.B.: Coevolving the ideal trainer: Application to the discovery of cellular automata rules. In: Koza, J.R., Banzhaf, W., Chellapilla, K., Deb, K., Dorigo, M., Fogel, D.B., Garzon, M.H., Goldberg, D.E., Iba, H., Riolo, R. (eds.) Genetic Programming 1998: Proceedings of the Third Annual Conference, University of Wisconsin, Madison, Wisconsin, USA, pp. 519–527. Morgan Kaufmann, San Francisco (1998)

[9] Juille, H., Pollack, J.B.: Coevolutionary learning and the design of complex systems. Advances in Complex Systems 2(4), 371–394 (2000)

[10] Koza, J.R.: Genetic Programming: On the Programming of Computers by Means of Natural Selection. MIT Press, Cambridge (1992)

[11] Land, M., Belew, R.K.: No perfect two-state cellular automata for density classification exists. Physical Review Letters 74(25), 5148–5150 (1995)

[12] Mariano, A.S., de Oliveira, G.M.B.: Evolving one-dimensional radius-2 cellular automata rules for the synchronization task. In: AUTOMATA 2008 Theory and Applications of Cellular Automata, pp. 514–526. Luniver Press (2008)

[13] Mitchell, M., Hraber, P.T., Crutchfield, J.P.: Revisiting the edge of chaos: Evolving cellular automata to perform computations. Complex Systems 7, 89–130 (1993)

[14] Mitchell, M., Crutchfield, J.P., Das, R.: Evolving cellular automata with genetic algorithms: A review of recent work. In: Proceedings of the First International Conference on Evolutionary Computation and Its Applications (EvCA 1996), Russian Academy of Sciences (1996)

[15] Mitchell, M., Thomure, M.D., Williams, N.L.: The role of space in the Success of Coevolutionary Learning. In: Proceedings of ALIFE X - The Tenth International Conference on the Simulation and Synthesis of Living Systems (2006)

[16] Jiménez Morales, F., Crutchfield, J.P., Mitchell, M.: Evolving two-dimensional cellular automata to perform density classification: a report on work in progress. Parallel Comput. 27, 571–585 (2001)

[17] Oliveira, G.M.B., Martins, L.G.A., de Carvalho, L.B., Fynn, E.: Some investigations about synchronization and density classification tasks in one-dimensional and two-dimensional cellular automata rule spaces. Electron. Notes Theor. Comput. Sci. 252, 121–142 (2009)

[18] Packard, N.H.: Adaptation toward the edge of chaos. In: Dynamic Patterns in Complex Systems, pp. 293–301. World Scientific, Singapore (1988)

[19] Pagie, L., Mitchell, M.: A comparison of evolutionary and coevolutionary search. Int. J. Comput. Intell. Appl. 2(1), 53–69 (2002)

[20] Tomassini, M., Venzi, M.: Evolution of Asynchronous Cellular Automata for the Density Task. In: Guervós, J.J.M., Adamidis, P.A., Beyer, H.-G., Fernández-Villacañas, J.-L., Schwefel, H.-P. (eds.) PPSN 2002. LNCS, vol. 2439, pp. 934–943. Springer, Heidelberg (2002)

[21] Wolz, D., de Oliveira, P.P.B.: Very effective evolutionary techniques for searching cellular automata rule spaces. Journal of Cellular Automata 3, 289–312 (2008)

[22] Wolfram, S. (ed.): Theory and Applications of Cellular Automata. Advanced series on complex systems, vol. 1. World Scientific Publishing, Singapore (1986)

[23] Zhao, Y., Billings, S.A.: Identification of the Belousov-Zhabotinsky Reaction using Cellular Automata Models. International Journal of Bifurcation and Chaos 17(5), 1687–1701 (2007)

[24] Cook, M.: Universality in elementary cellular automata. Complex Systems 15(1), 1–40 (2004)

[25] Wuensche, A.: Classifying cellular automata automatically: finding gliders, filtering, and relating space-time patterns, attractor basins, and the z parameter. Complex 4, 47–66 (1999)

[26] Wolfram, S.: Computation theory of cellular automata. Communications in Mathematical Physics 96(1), 15–57 (1984)

[27] Hanson, J., Crutchfield, J.: The attractorbasin portrait of a cellular automaton. Journal of Statistical Physics 66(5), 1415–1462 (1992)

[28] Hanson, J., Crutchfield, J.: Computational mechanics of cellular automata. PhD thesis, University of California, Berkeley, CA (1993)

[29] Steiglitz, K., Kamal, I., Watson, A.: Embedding computation in one-dimensional automata by phase coding solitons. IEEE Trans. Comput. 37, 138–145 (1988)

[30] Aizawa, Y., Nishikawa, I., Kaneko, K.: Soliton turbulence in one-dimensional cellular automata. Phys. D 45, 307–327 (1990)

[31] Crutchfield, J., Mitchell, M.: The evolution of emergent computation. Proceedings of the National Academy of Sciences 92(23), 10742 (1995)

Chapter 20
Using Genetic Algorithms and Model Checking for P Systems Automatic Design

Cristina Tudose, Raluca Lefticaru, and Florentin Ipate

Abstract. Membrane computing is a recently developed research field, whose models, P systems, are bio-inspired computational models, abstracted from the structure and the functioning of living cells. Their applications are various and have been reported in biology, bio-medicine, economics, approximate optimization and computer graphics. However, it is a difficult task to design a P system which solves a given problem, especially because of their characteristics, such as nondeterminism, parallelism, possible presence of active membranes. In this paper we propose an approach to P systems automatic design, which uses genetic algorithms and model checking. More precisely, we use a type of fitness function which performs several simulations of the P system, in order to assess its adequacy to realize the given task, complemented by formal verification of the model.

20.1 Introduction

Membrane computing is an emergent branch of natural computing, which proposes and investigates new computational models, namely *P systems*, initially inspired by the structure and functioning of the living cell [19]. Later on, other biological entities

Cristina Tudose
Department of Computer Science, University of Pitesti
Targu din Vale nr. 1, 110040, Pitesti, Romania
e-mail: cristina.tudose@upit.ro

Raluca Lefticaru
Department of Computer Science, University of Pitesti
Targu din Vale nr. 1, 110040, Pitesti, Romania
e-mail: raluca.lefticaru@upit.ro

Florentin Ipate
Department of Computer Science, University of Pitesti
Targu din Vale nr. 1, 110040, Pitesti, Romania
e-mail: florentin.ipate@ifsoft.ro

D.A. Pelta et al. (Eds.): NICSO 2011, SCI 387, pp. 285–302, 2011.
springerlink.com   © Springer-Verlag Berlin Heidelberg 2011

have served as inspiration and have given birth to other computing paradigms, such as tissue, colony or spiking neural P systems [21]. The key ingredients for basic cell-like P systems are: a *membrane structure* consisting of several membranes arranged in a hierarchical structure inside a main membrane (called the *skin*) and delimiting *regions* or compartments, *multisets of objects* contained in each region and *rules* which are applied in a maximal and parallel mode.

Similarly to other nature inspired computing models or evolutionary strategies, P systems have been employed to tackle NP-complete problems. The approach, different from the one of evolutionary strategies, is realized by P systems with active membranes [20]. The main idea behind is to trade time for space: using a division rule successively n steps, due to the maximal parallelism, 2^n copies of the same membrane are obtained. Furthermore, recent advances in distributed and parallel computing have motivated an intensive research effort in membrane computing, regarding for example their simulation using graphics processing units (GPUs) [3].

A large number of software tools for simulating P systems have been developed, many of them with the purpose of dealing with real world problems, such as those arisen from biology [2]. One of the most promising software projects in membrane computing, the P-Lingua framework [4, 18], proposes a new programming language, aiming to become a standard for the representation and simulation of P systems. This framework has been continuously developed, in order to accept more complex variants of P systems and to solve real world problems [2, 15].

In this context, of recent developments in P system simulators and their application in solving various problems, it is desirable to obtain P systems models able to realize given tasks. However, the process of designing a P system fit for solving a given problem or performing a certain task can be tedious and error-prone. Simulation tools can help observing whether the P system has some possible incorrect behaviour, model-checkers can help verifying if the desired properties are maintained on all possible computations [8, 14]. But the effort, based on trial and error, needed to design the P system, still remains an open problem.

In this paper we propose an approach to P system automatic design using genetic algorithms. The adequacy of the P system for solving a given problem or task is evaluated based on the results obtained in several simulations using P-Lingua (in case of nondeterministic P systems), complemented by a model verification realized with the model-checker Spin. When the P system is nondeterministic, an unknown number of different computations may exist. Since our algorithm will only perform the evaluation of a predetermined number of computations (which may be less than the total number of computations), this will be complemented by model checking, in order to ensure that the required property holds for all cases - not just for those for which simulation results are available.

This paper is structured as follows. We introduce the theoretical foundations of P systems in Section 20.2; the proposed approach is presented in Section 20.3. Some examples and the experimental obtained results are explained in Section 20.4, the related work is presented in Section 20.5 and finally the conclusions are drawn in Section 20.6.

20.2   Background

Membrane Computing

Before presenting our approach to P system verification, let us establish the notation used and define the class of cell-like P systems addressed in the paper. Basically, a P system is defined as a hierarchical arrangement of membranes, identifying corresponding regions of the system. Each region has an associated finite multiset of objects and a finite set of rules; both may be empty. Given a finite alphabet $V = \{a_1, ..., a_p\}$, a *multiset* is either denoted by a string $u \in V^*$, where the order is not considered, or by an associated vector of non-negative integers, $\Psi_V(u) = (|u|_{a_1}, ..., |u|_{a_p})$, where $|u|_{a_i}, 1 \leq i \leq p$ denotes the number of a_i occurrences in u.

The following definition refers to cell-like P systems with active membranes (see [20] for details).

Definition 20.1. A *P system* is a tuple $\Pi = (V, H, \mu, w_1, ..., w_n, R)$, where V is a finite set, called *alphabet*; H is a finite set of labels for membranes; μ is a membrane structure (a rooted tree), consisting of n membranes injectively labelled with elements of H; w_i, $1 \leq i \leq n$, are strings over V, describing the multisets of objects initially placed in the n regions of μ; R is a finite set of rules, where each rule is of one of the following forms:

(a) $[a \rightarrow v]_h^\alpha$, where $h \in H$, $\alpha \in \{+, -, 0\}$ (electrical charges), $a \in V$ and v is a string over V describing a multiset of objects associated with membranes and depending on the label and the charge of the membranes (evolution rules).

(b) $a[]_h^\alpha \rightarrow [b]_h^\beta$, where $h \in H$, $\alpha, \beta \in \{+, -, 0\}$, $a, b \in V$ (send-in communication rules). An object is introduced in the membrane, possibly modified, and the initial charge α is changed to β.

(c) $[a]_h^\alpha \rightarrow []_h^\beta b$, where $h \in H$, $\alpha, \beta \in \{+, -, 0\}$, $a, b \in V$ (send-out communication rules). An object is sent out of the membrane, possibly modified, and the initial charge α is changed to β.

(d) $[a]_h^\alpha \rightarrow b$, where $h \in H$, $\alpha \in \{+, -, 0\}$, $a, b \in V$ (dissolution rules). A membrane with a specific charge is dissolved in reaction with an object a (possibly modified).

(e) $[a]_h^\alpha \rightarrow [b]_h^\beta [c]_h^\gamma$, where $h \in H$, $\alpha, \beta, \gamma \in \{+, -, 0\}$, $a, b, c \in V$ (division rules). A membrane is divided into two membranes. The objects inside the membrane are replicated, except for a, that may be modified in each membrane.

The membrane structure, μ, is denoted by a string of left and right brackets ($[_l$, and $]_l^e$), each with the label l of the membrane; the electrical charge e of each membrane is also given.

The rules are applied in maximally parallel mode, which means that they are used in all the regions at the same time and in each region all the objects to which a rule *can* be applied *must* be the subject of a rule application [20]. However, a membrane can be subject of only one rule of types (b) – (e) in one computation step.

In type (e) (membrane division) rules, all the contents present before the division, except for object a, can be the subject of rules in parallel with the division. In this case we consider that in a single step two processes take place: first the contents are affected by the rules applied to them, and after that the results are replicated into the two new membranes. If a membrane is dissolved, its content (multiset and interior membranes) becomes part of the immediately external membrane which has not been dissolved at that computation step. The skin is never dissolved neither divided. When send-out communication rules are applied in the skin membrane, the transferred objects will become part of a region outside the skin, called *environment*.

A *configuration* of the P system Π is uniquely identified by the current membrane structure μ' and the contents of each region in μ'. We have a transition step from a configuration c_1 to a configuration c_2 if we can pass from c_1 to c_2 by using the maximal parallelism mode (as well as the rule restrictions stated above); this is denoted by $c_1 \Longrightarrow c_2$. In the set of all configurations, we will distinguish halting ones; c is a *halting configuration* if there is no region i such that its contents can be further developed.

Model Checking

Model checking is an automated technique that, given a model of a system and a property to verify, usually expressed as a temporal logic formula, exhaustively explores the state space and returns *yes* if the property is true and a counterexample otherwise. In the last years, a lot of research was done in the field of model checking based verification of P systems. The tools used so far are Maude [1], Prism [22], NuSMV [7, 12], Spin [8].

In this paper we use the Spin model checker, considered as the one of the most powerful model checkers available, to validate a P system obtained using genetic algorithms. We have previously proved that Spin works better than NuSMV on verifying P systems [8] and the automatic translation of a P system into Promela, the language used by the Spin model checker, was briefly described in [8, 14].

Genetic Algorithms

Genetic algorithms (GAs) [17] are a class of evolutionary algorithms, that use techniques inspired from biology, such as selection, recombination (crossover) and mutation, applied on a population of potential solutions, called chromosomes (or individuals).

Single-point crossover, the recombination operator used in this paper, randomly chooses a crossover point and exchanges the subsequences before and after that point between two individuals to create two new offspring. For example, the chromosomes 00000000 and 11111111 could be crossed over at the third locus to produce the two offspring 00011111 and 11100000. Recombination is applied with a probability (crossover rate) p_c. Depending on this rate, the next generation will contain the parents or the offspring.

The mutation operator randomly changes some genes in a chromosome. For example, the chromosome 00101101 could be mutated in its third position to yield 00001101. Mutation can occur at each gene in a chromosome with some probability p_m (mutation rate). Its main purpose is to introduce variation in the population.

The codification used in this paper for the P systems representation is binary, i.e. each gene 0, 1 corresponds to a single rule from a larger set, which could be selected or not in the current P system. All the P systems from the population share the same membrane structure, alphabet, initial multisets; they differ only by the set of rules, encoded in the binary chromosome. The chromosomes are created at random, each gene having an equal chance to be 0 or 1.

In this paper we complement the results obtained for P system generation using genetic algorithms with model checking. Previous approaches that use genetic algorithms and model checking have been reported in literature [9, 11]. In [11] is presented an automated finite state based model design strategy, which uses the model checker NuSMV to evaluate the fitness, taking into account the satisfied/unsatisfied specifications and the counterexamples provided in the last case.

Similarly, in this paper the fitness of one P system is measuring its adequacy to perform a given task (for example to compute something and return a certain result). Unfortunately, due to the nondeterminism of the P systems, some executions (computations) could produce the expected result, but others do not. Consequently, P-Lingua simulators are used, but complemented with a P systems verification. If a P system passes the simulations test, it is translated to Promela, the language accepted by the model checker Spin, and further verified. The fitness value is based on the results obtained from both the simulation and verification. The model checker copes with the GAs in the fitness evaluation phase, as it is further illustrated in Fig. 20.5.

20.3   P System Generation Strategy

20.3.1   Problem Formulation

In the followings we will describe the type of P system generation problems aimed to be solved using this approach.

Given: the alphabet V, the membrane structure μ, the initial multisets in each membrane $w_1, \ldots, w_n \in V^*$, a set of possible rules $\mathscr{R}$ and some desired properties of the P system.

Problem: find a P system $\Pi = (V, H, \mu, w_1, \ldots, w_n, R)$, $H = \{1, \ldots, n\}$, such that $R \subseteq \mathscr{R}$ and the computations of the P system satisfy the given properties.

Problem representation. For solving this problem, the chromosome representing the P system should only codify whether a given rule is selected or not in the current set of rules. Each P system $\Pi = (V, H, \mu, w_1, \ldots, w_n, R)$ is codified using a binary chromosome $(x_1, \ldots, x_m) \in \{0, 1\}^m$, where $m = |\mathscr{R}|$ and each gene $x_i \in \{0, 1\}$ is specifying whether or not $r_i \in R$.

The expected (desired) properties are expressed in a natural language. The fitness function encompasses them, more precisely it increases with a certain value corresponding to each unsatisfied requirement (or desired property), as it can be further observed from Fig. 20.1-20.5.

20.3.2　Fitness Evaluation

The fitness function is a key ingredient of the search algorithm, responsible for guiding the search, so we will consider several ways of computing the fitness and compare the results obtained in each case. However, some common aspects appear in all of them, as described below.

The fitness functions proposed in the following section are measuring how far are the P system computations from satisfying the expected properties. A value 0 corresponds to the case when all the P system computations satisfy the desired properties. The genetic algorithm will minimize the fitness function in order to find a solution for the problem.

Consider for example the problem presented in [5], i.e. to find, starting from an initial configuration ($w_1 = \lambda, w_2 = a^2bz_1$, where λ denotes the empty string, $V = \{a,b,c,z_1,z_2,z_3,z_4\}$, $\mu = [_1[_2]_2]_1$), a deterministic P system which provides in the halting configuration 16 objects of type c in the skin membrane. Naturally, the fitness function proposed by the authors of [5], later considered in [6], computes the value |number of c objects in the skin membrane $- 16$|. The authors evaluate this only for the terminal configuration (or the last configuration obtained after *maxSteps*) for a deterministic by construction P system.

Similarly to this evaluation, we can define for two configurations of a P system (the expected and the obtained one) a metric for the distance between them, which could be useful in other fitness evaluations.

For a 1-compartment P system (i.e. a P system with only one membrane, the skin) with $V = \{a_1,...,a_p\}$ and two configurations c_1,c_2, this can be defined by

$$| c_1 - c_2 | = \sum_{i=1}^{p}(| |c_1|_{a_1} - |c_2|_{a_1} | + \cdots + | |c_1|_{a_p} - |c_2|_{a_p} |,$$

where for a multiset u, $\Psi_V(u) = (|u|_{a_1},...,|u|_{a_p})$.

For a P system with n membranes, given two configurations $c_1 = (c_{11},...,c_{1n})$ and $c_1 = (c_{21},...,c_{2n})$, $| c_1 - c_2 | = \sum_{i=1}^{n} | c_{1i} - c_{2i} |$.)

This definition can be extended for two sequences $c = c_0,...,c_k, d = d_0,...,d_k$ of configurations, $k \geq 0$, by $| c - d | = \sum_{i=0}^{k} | c_i - d_i |$. (Alternatively, the distances between configurations can be weighted such that the weight for $| c_i - d_i |$ is larger than the weight for $| c_{i+1} - d_{i+1} |$.)

The type of fitness functions we present in Section 20.4 can incorporate the metrics proposed, in case an exact configuration or an exact computation of the P system are required. By exact configuration we understand that the number of each object in each region of the current configuration must be the expected one. On the contrary, in Problem 1 proposed in [5], the target is not an exact configuration, but a configuration which contains 16 objects of type c and can contain any number of other objects. An exact computation will be, similarly, a computation which presents all the exact configurations, which were expected.

20.4   Experimental Results

In the following we present the results obtained using a genetic algorithm implemented using the JGAP package [16], the P-Lingua simulator [18] and the Spin model checker. In these case studies we have considered one and two-membrane P systems, having evolution, communication and dissolution rules.

Problem 1: Generating a deterministic, halting P system which computes the square of a given number.

This problem has been used for experimentation in [5, 6] and consists in generating P systems of the following type: $\Pi = (V, H, \mu, w_1, w_2, R)$, where $V = \{a, b, c, z_1, z_2, z_3, z_4\}$, $H = \{1, 2\}$, the membrane structure is $\mu = [_1[_2 \ ]_2]_1$, the initial multisets are $w_1 = \lambda$, $w_2 = a^2 b z_1$. The desired properties of the P system are: (1) determinism; (2) termination, i.e. halting after at most $maxSteps$; (3) having in the terminal configuration a number of objects c in the skin membrane equal to 16.
   The set of rules R is a subset of $\mathcal{R}_1$, defined as:

$$\mathcal{R}_1 = \begin{cases} r_1 \equiv [a \to ab]_2 & r_7 \equiv [z_2 \to z_1]_2 & r_{13} \equiv [a \to \lambda]_1 \\ r_2 \equiv [b \to bc]_2 & r_8 \equiv [z_3 \to z_4]_2 & r_{14} \equiv [b \to \lambda]_1 \\ r_3 \equiv [c \to b^2]_2 & r_9 \equiv [z_1]_2 \to b & r_{15} \equiv [b \to c]_2 \\ r_4 \equiv [a \to bc]_2 & r_{10} \equiv [z_2]_2 \to a & r_{16} \equiv [c \to \lambda]_2 \\ r_5 \equiv [z_1 \to z_2]_2 & r_{11} \equiv [z_3]_2 \to c & r_{17} \equiv [z_4 \to z_1]_2 \\ r_6 \equiv [z_2 \to z_3]_2 & r_{12} \equiv [z_4]_2 \to a & r_{18} \equiv [z_4]_2 \to b \end{cases}$$

The search space of this problem has the size 2^{18} and one aspect which hinders the search is the nondeterminism of the P systems obtained by randomly selecting a subset of $\mathcal{R}_1$. In this approach we consider also the nondeterministic P systems, assign them fitness values and let them participate in the reproduction process, depending on their fitness. In the approach presented in [5], the nondeterministic P systems are replaced in the moment they are created by deterministic P systems: if two rules have the same left hand side, one of them is randomly dropped. This makes somehow the search easier.
   For a consistent comparison with previous approaches [5, 6] (even if there are some differences in other aspects), we consider the same parameters of the elitist genetic algorithm: population size = 30, maximum allowed evolutions = 30. Single-point crossover with the probability $p_c = 0.8$ and mutation with the rate $p_m \in \{0.1; 0.5\}$ were applied.
   In order to evaluate the adequacy of the P systems, we have proposed two fitness functions, f_1 and f_2, described in Fig. 1 and Fig. 2, respectively. The

$fitness \leftarrow 0$
Load the P system corresponding to the current chromosome
$NondetPairs \leftarrow$ number of rule pairs with the same left hand side
if $NondetPairs > 0$ **then**
    $fitness \leftarrow \delta * NondetPairs$
    return $fitness$
else
    {The P system is deterministic so we need to simulate only one computation}
    $step \leftarrow 0$
    while P system is not in a halting state $\wedge$ $step < maxSteps$ **do**
        evolve one step (move to the next configuration of the P system)
        $step \leftarrow step + 1$
    end while
    if P system is in a halting configuration **then**
        $fitness \leftarrow fitness + |$number of c obtained $- 16|$
    else
        $fitness \leftarrow fitness + |$number of c obtained $- 16| + \eta$
    end if
    return $fitness$
end if

Fig. 20.1 Algorithm describing the computation of the fitness function f_1 for a P system solving *Problem 1*. The values δ, η represent some penalty values added when the expected properties (determinism, termination) are not satisfied. In our experiments for f_1 we have considered $\delta = 25$, $\eta = 1$ and $maxSteps = 20$

$fitness \leftarrow 0$
Load the P system corresponding to the current chromosome
$NondetPairs \leftarrow$ number of rule pairs with the same left hand side
if $NondetPairs > 0$ **then**
    $fitness \leftarrow \delta * NondetPairs$
end if
$step \leftarrow 0$
while P system is not in a halting state $\wedge$ $step < maxSteps$ **do**
    evolve one step (move to the next configuration of the P system)
    $step \leftarrow step + 1$
end while
if P system is in a halting configuration **then**
    $fitness \leftarrow fitness + |$number of c obtained $- 16|$
else
    $fitness \leftarrow fitness + |$number of c obtained $- 16| + \eta$
end if
return $fitness$

Fig. 20.2 Algorithm describing the computation of the fitness function f_2 for a P system solving *Problem 1*. The values δ, η represent some penalty values added when the expected properties (determinism, termination) are not satisfied. In our experiments for f_2 we have considered $\delta = 1$, $\eta = 1$ and $maxSteps = 20$

main difference between them is that f_2 continues to simulate and evaluate the P system even when the nondeterminism has been detected. The numerical values for the penalties δ, η were experimentally obtained and they have the following intuitive explanation. For the fitness function f_1, the value of $\delta = 25$ is much higher in order to penalize the nondeterministic P systems, because the evaluation is interrupted immediately after $fitness \longleftarrow \delta * NondetPairs$. For the fitness function f_2, the value of $\delta = 1$ is much lower, because the evaluation continues and the P systems are penalized first with $fitness \longleftarrow \delta * NondetPairs$ and second by $fitness \longleftarrow fitness + |\text{number of } c \text{ obtained} - 16| + \eta$, where $\eta = 1$ is the penalty for non-halting computations.

Several P systems that satisfy the given conditions can be obtained, as presented in [5, 6], for example, having one of the following satisfactory set of rules:

$$R_{sat1} = \left\{ \begin{array}{ll} [a \rightarrow ab]_2 & [z_2 \rightarrow z_3]_2 \\ [b \rightarrow bc]_2 & [z_3 \rightarrow z_4]_2 \\ [z_1 \rightarrow z_2]_2 & [z_4]_2 \rightarrow a \end{array} \right\} \quad R_{sat2} = \left\{ \begin{array}{ll} [a \rightarrow ab]_2 & [z_2 \rightarrow z_3]_2 \\ [b \rightarrow bc]_2 & [z_3 \rightarrow z_4]_2 \\ [z_1 \rightarrow z_2]_2 & [z_4]_2 \rightarrow b \end{array} \right\}$$

$$R_{sat3} = \left\{ \begin{array}{ll} [a \rightarrow ab]_2 & [z_3 \rightarrow z_4]_2 \\ [b \rightarrow bc]_2 & [a \rightarrow \lambda]_1 \\ [z_1 \rightarrow z_2]_2 & [b \rightarrow \lambda]_1 \\ [z_2 \rightarrow z_3]_2 & [z_4]_2 \rightarrow b \end{array} \right\} \quad R_{sat4} = \left\{ \begin{array}{ll} [a \rightarrow ab]_2 & [z_3 \rightarrow z_4]_2 \\ [b \rightarrow bc]_2 & [a \rightarrow \lambda]_1 \\ [z_1 \rightarrow z_2]_2 & [b \rightarrow \lambda]_1 \\ [z_2 \rightarrow z_3]_2 & [z_4]_2 \rightarrow a \end{array} \right\}$$

$$R_{sat5} = \left\{ \begin{array}{ll} [a \rightarrow ab]_2 & [z_3 \rightarrow z_4]_2 \\ [b \rightarrow bc]_2 & [a \rightarrow \lambda]_1 \\ [z_1 \rightarrow z_2]_2 & [z_4]_2 \rightarrow a \\ [z_2 \rightarrow z_3]_2 & \end{array} \right\} \quad R_{sat6} = \left\{ \begin{array}{ll} [a \rightarrow ab]_2 & [z_3 \rightarrow z_4]_2 \\ [b \rightarrow bc]_2 & [a \rightarrow \lambda]_1 \\ [z_1 \rightarrow z_2]_2 & [z_4]_2 \rightarrow b \\ [z_2 \rightarrow z_3]_2 & \end{array} \right\}$$

$$R_{sat7} = \left\{ \begin{array}{ll} [a \rightarrow ab]_2 & [z_3 \rightarrow z_4]_2 \\ [b \rightarrow bc]_2 & [b \rightarrow \lambda]_1 \\ [z_1 \rightarrow z_2]_2 & [z_4]_2 \rightarrow a \\ [z_2 \rightarrow z_3]_2 & \end{array} \right\}$$

Let us illustrate the computation of a solution P system, having for example the satisfactory set of rules R_{sat1}. This corresponds to the 110011010001000000 chromosome, expressing that the set of rules contains only the subset $\{r_1, r_2, r_5, r_6, r_8, r_{12}\}$ of $\mathcal{R}_1$. For simplicity, we will denote a configuration of the P system by (μ, w_1, w_2), i.e. the current membrane structure μ and the multisets w_1, w_2 in each membrane.

The P system computation is the following: $([_1 [_2]_2]_1, \lambda, a^2 bz_1) \Longrightarrow ([_1 [_2]_2]_1, \lambda, a^2 b^3 cz_2) \Longrightarrow ([_1 [_2]_2]_1, \lambda, a^2 b^5 c^4 z_3) \Longrightarrow ([_1 [_2]_2]_1, \lambda, a^2 b^7 c^9 z_4) \Longrightarrow ([_1]_1, a^3 b^9 c^{16}, \lambda)$. It can be observed that in the last computation step the rule $[z_4]_2 \rightarrow a$ is applied.

This rule dissolves the inner membrane and all the objects are transferred in the skin membrane. The computation halts and the terminal multiset from the skin membrane contains 16 objects of type c.

Similar to this, the computation for the P system corresponding to the satisfactory set of rules R_{sat4} is: $([_1[_2]_2]_1, \lambda, a^2bz_1) \implies ([_1[_2]_2]_1, \lambda, a^2b^3cz_2) \implies ([_1[_2]_2]_1, \lambda, a^2b^5c^4z_3) \implies ([_1[_2]_2]_1, \lambda, a^2b^7c^9z_4) \implies ([_1]_1, a^3b^9c^{16}, \lambda) \implies ([_1]_1, c^{16}, \lambda)$. For all the solutions obtained the P systems are deterministic, they halt and the halting configurations contain c^{16}. The difference is given by other objects that could also appear in the skin membrane, the desired property in this case is not to obtain an exact configuration, but a configuration with 16 objects of type c.

For comparison, the experimental results obtained in [5, 6] are provided in Table 20.2. The success rates reported by the authors of the studies were $1/30 \approx 3\%$ and $12/30 = 40\%$. It can be observed that a higher rate of mutation increases the success rate of the genetic algorithm. Also, the fitness function f_2 has better results than f_1 and the functions used in [5, 6].

Problem 2: Generating a deterministic, never halting P system which computes all the square numbers.

The aim is to generate a P system which computes sequentially the squares of all natural numbers, i.e. each configuration from the P system contains a number of objects c equal to the square of a number. For example, there are 0 objects c in the initial configuration, but there are c^1, c^4, c^9 or c^{16} objects after $1, 2, 3$ or 4 computation steps, respectively. Considering known the alphabet $V = \{s, a, b, c, x\}$, the membrane structure $\mu = [_1]_1$, the initial multiset $w_1 = s$ and a set $\mathcal{R}_2$ of possible rules, find a P system such that:

- In each configuration, the number of c objects is the square of b, i.e. considering $conf$ the current configuration, $|conf|_c = |conf|_b \cdot |conf|_b$. In this case, it can be observed that b is a counter, representing the current step of the computation.
- The P system does not halt (the computation never ends).
- If $step \geq 1$ then in the current configuration the following property is true: $(|conf|_s = 0 \wedge |conf|_a = 1 \wedge |conf|_x = 1)$.

where

$$\mathcal{R}_2 = \left\{ \begin{array}{llll} r_1 \equiv [s \rightarrow a]_1 & r_7 \equiv [a \rightarrow bc]_1 & r_{13} \equiv [b \rightarrow bcc]_1 \\ r_2 \equiv [s \rightarrow ab]_1 & r_8 \equiv [a \rightarrow ab]_1 & r_{14} \equiv [x \rightarrow ab]_1 \\ r_3 \equiv [s \rightarrow c]_1 & r_9 \equiv [a \rightarrow cx]_1 & r_{15} \equiv [x \rightarrow ac]_1 \\ r_4 \equiv [s \rightarrow cx]_1 & r_{10} \equiv [b \rightarrow x]_1 & r_{16} \equiv [x \rightarrow a]_1 \\ r_5 \equiv [s \rightarrow abxc]_1 & r_{11} \equiv [b \rightarrow bc]_1 & r_{17} \equiv [x \rightarrow xc]_1 \\ r_6 \equiv [a \rightarrow b]_1 & r_{12} \equiv [b \rightarrow a]_1 & r_{18} \equiv [x \rightarrow bc]_1 \end{array} \right\}.$$

$fitness \leftarrow 0$
Load the P system corresponding to the current chromosome
$NondetPairs \leftarrow$ number of rule pairs with the same left hand side
if $NondetPairs > 0$ **then**
  $fitness \leftarrow \delta * NondetPairs$
  return $fitness$
else
  {The P system is deterministic so we need only one computation}
  $step \leftarrow 0$
  while P system is not in a halting state $\wedge$ $step < maxSteps$ **do**
    evolve one step (move to the next configuration of the P system)
    $step \leftarrow step + 1$
    $fitness \leftarrow fitness + |$number of $c - ($number of $b)^2| + |$number of $b - \text{step}| +$
    $+ |$number of $s| + |$number of $a - 1| + |$number of $x - 1|$
  end while
  if P system is in a halting configuration **then**
    $fitness \leftarrow fitness + \eta$
  end if
  return $fitness$
end if

Fig. 20.3 Algorithm describing the computation of the fitness function f_3 for a P system solving *Problem 2*. The values δ, η represent some penalty values added when the expected properties (determinism, termination) are not satisfied. In our experiments for f_3 we have considered $\delta = 25, \eta = 1$ and $maxSteps = 20$

Some examples of distinct solutions found are the P systems with the satisfactory sets of rules:

$$R_{sat1} = \left\{ \begin{array}{ll} [s \rightarrow abxc]_1 & [a \rightarrow ab]_1 \\ [b \rightarrow bc^2]_1 & [x \rightarrow xc]_1 \end{array} \right\}, R_{sat2} = \left\{ \begin{array}{ll} [s \rightarrow abxc]_1 & [a \rightarrow cx]_1 \\ [b \rightarrow bc^2]_1 & [x \rightarrow ab]_1 \end{array} \right\},$$

which generate the computation: $s \Longrightarrow abcx \Longrightarrow ab^2c^4x \Longrightarrow ab^3c^9x \Longrightarrow ab^4c^{16}x \Longrightarrow ab^5c^{25}x \Longrightarrow \ldots$ (this computation never halts). In the previous notation for the P system computation, because the P system has only one membrane, we have omitted the membrane structure $\mu = [_1]_1$ and have written only the current multiset. Similar to f_1, f_2 we have considered f_3, f_4, which are described in Fig. 3, 4 and were used with the genetic algorithm to find solutions for the problem.

$fitness \leftarrow 0$

Load the P system corresponding to the current chromosome

$NondetPairs \leftarrow$ number of rule pairs with the same left hand side

if $NondetPairs > 0$ **then**

    $fitness \leftarrow \delta * NondetPairs$

end if

$step \leftarrow 0$

while P system is not in a halting state $\wedge$ $step < maxSteps$ **do**

    evolve one step (move to the next configuration of the P system)

    $step \leftarrow step + 1$

    $fitness \leftarrow fitness + |$number of c $-$ (number of $b)^2| + |$number of b $-$ $step| + |$number of $s| + |$number of $a - 1| + |$number of $x - 1|$

end while

if P system is in a halting configuration **then**

    $fitness \leftarrow fitness + \eta$

end if

return $fitness$

Fig. 20.4 Algorithm describing the computation of the fitness function f_4 for a P system solving *Problem 2*. The values δ, η represent some penalty values added when the expected properties (determinism, termination) are not satisfied. In our experiments for f_4 we have considered $\delta = 1$, $\eta = 1$ and $maxSteps = 20$

Problem 3: Designing a nondeterministic P system, that generates the language $L = \{a^{2^n} b^{3^n} | n > 1\}$.

This P system is expected to generate various computations, such that in each halting configuration the multiset $a^{2^n} b^{3^n}, n > 1$ should be sent to the environment (the region outside the skin membrane). For the automatic generation we have considered a two-membrane structure $\mu = [_1 [_2]_2]_1$, the alphabet $V = \{a, b, z_1\}$, initial multisets $w_1 = \lambda, w_2 = ab$ and the set $\mathscr{R}_3$ of possible rules, defined as:

$$\mathscr{R}_3 = \begin{cases} r_1 \equiv [a]_1 \rightarrow []_1 a & r_7 \equiv [b \rightarrow ab]_1 & r_{13} \equiv [a]_2 \rightarrow []_2 b \\ r_2 \equiv [a]_1 \rightarrow []_1 b & r_8 \equiv [b \rightarrow a]_1 & r_{14} \equiv [a \rightarrow ab]_2 \\ r_3 \equiv [a \rightarrow ab]_1 & r_9 \equiv [a \rightarrow aa]_2 & r_{15} \equiv [b \rightarrow bbb]_2 \\ r_4 \equiv [a \rightarrow b]_1 & r_{10} \equiv [a \rightarrow aaz_1]_2 & r_{16} \equiv [b \rightarrow bbbz_1]_2 \\ r_5 \equiv [b]_1 \rightarrow []_1 b & r_{11} \equiv [z_1]_2 \rightarrow \lambda & r_{17} \equiv [b \rightarrow abb]_2 \\ r_6 \equiv [b]_1 \rightarrow []_1 a & r_{12} \equiv [a]_2 \rightarrow []_2 a & r_{18} \equiv [b]_2 \rightarrow []_2 b \end{cases}.$$

Similar to the previous defined fitness functions, f_5 described in Fig. 20.5 evaluates the adequacy of a P system for solving the given problem. The nondeterminism is in this case needed (the halting configuration of the P system can contain for example $a^4 b^9$ in the environment, as well as $a^8 b^{27}$). Consequently, several simulations are needed to evaluate the fitness of an individual. The fitness function checks if the properties are satisfied on each possible computation. This is achieved by:

$fitness \leftarrow 0$
for $i = 1 \rightarrow maxSimulations$ **do**
    Load the P system corresponding to the current chromosome
    $step \leftarrow 0$
    while P system is not in a halting state $\wedge$ $step < maxSteps$ **do**
        evolve one step (move to the next configuration of the P system)
        $step \leftarrow step + 1$
    end while
    if step = 0 **then**
        $fitness \leftarrow fitness + 2 * \delta$
    end if
    if step = 1 **then**
        $fitness \leftarrow fitness + 1 * \delta$
    end if
    if $step > 1 \wedge$ P system is in a halting configuration **then**
        $fitness \leftarrow fitness +$ |number of a obtained $- 2^{step-1}$| + |number of b obtained $-$ 3^{step-1}|
    else
        $fitness \leftarrow fitness + \eta$
    end if
end for
if $fitness = 0$ **then**
    Transform the P system for using it with Spin
    Run the Spin model checker against each property
    $fitness \leftarrow fitness +$ number of falsified properties
end if
return $fitness$

Fig. 20.5 Algorithm describing the computation of the fitness function f_5 for a P system solving *Problem 3*. The values δ, η represent some penalty values added when the expected properties are not satisfied. In our experiments for f_5 we have considered $\delta = 25$, $\eta = 10$, $maxSteps = 20$ and $maxSimulations = 5$

- Simulating a predefined number of computations (*maxSimulations*) with P-Lingua and increasing the fitness value when the desired properties are not satisfied.
- Verifying the P system with the Spin model checker, only if the current fitness value is 0 (i.e. the properties were satisfied on all the simulated computations).

The model checker is called only for well-suited P systems, due to the fact that the model verification (which involves an exhaustive search in the state space of the P system) is much more time consuming than a simple simulation of the P system. For verifying the P system model, each property is transformed in a temporal logic formula and the model checker is run against it. In case there exists a computation of the P system which invalidates the formula, it will be returned and the fitness function will increase accordingly. The properties verified were: (A) Globally, if the current configuration is halting, then the number of a objects in the environment

is 2^{step-1}; (B) Globally, if the current configuration is halting, then the number of
b objects in the environment is 3^{step-1}. For more details regarding the P system
transformation in Promela, specification of LTL properties and their verification
with Spin, [8, 14] can be consulted.

Some examples of satisfactory set of rules obtained for Problem 3 are:

$$R_{sat1} = \left\{ \begin{array}{ll} [a]_1 \rightarrow []_1 a & [a \rightarrow aa]_2 \\ [b]_1 \rightarrow []_1 b & [b \rightarrow bbb]_2 \\ [a \rightarrow aaz_1]_2 & [z_1]_2 \rightarrow \lambda \end{array} \right\}, R_{sat2} = \left\{ \begin{array}{ll} [a]_1 \rightarrow []_1 a & [a \rightarrow aa]_2 \\ [b]_1 \rightarrow []_1 b & [b \rightarrow bbb]_2 \\ [b \rightarrow bbbz_1]_2 & [z_1]_2 \rightarrow \lambda \end{array} \right\}.$$

Considering in this case a P system configuration described by (μ, env, w_1, w_2), i.e.
we add information about the multiset of objects in the environment, env, some
possible computations for the satisfactory set of rules R_{sat1} are:

- $([_1[_2]_2]_1, \lambda, \lambda, ab) \implies ([_1[_2]_2]_1, \lambda, \lambda, a^2 b^3 z_1) \implies ([_1]_1, \lambda, a^4 b^9, \lambda) \implies$
  $([_1]_1, a^4 b^9, \lambda, \lambda)$.
- $([_1[_2]_2]_1, \lambda, \lambda, ab) \implies ([_1[_2]_2]_1, \lambda, \lambda, a^2 b^3) \implies ([_1[_2]_2]_1, \lambda, \lambda, a^4 b^9) \implies$
  $([_1[_2]_2]_1, \lambda, \lambda, a^8 b^{27} z_1^2) \implies ([_1]_1, \lambda, a^{16} b^{81}, \lambda) \implies ([_1]_1, a^{16} b^{81}, \lambda, \lambda)$.

Discussion about the Experimental Results Obtained

For all three problems presented before and their corresponding fitness functions,
random search and genetic algorithms were applied in order to find a solution.

The results presented in Table 20.2 are averaged after 100 runs of an elitist genetic algorithm, having the parameter settings similar to the ones from [5], namely:
crossover rate $p_c = 0.8$, mutation probability $p_m \in \{0.1; 0.5\}$, population size 30,
maximum allowed generations 30. The GA implementation was realized using
JGAP, a Java Genetic Algorithms Package [16]. The first three rows are taken for
comparison from the papers [5, 6].

The results from Table 20.1 show the success rates obtained for the same problems using random search. They are obtained by generating random populations
of 30 individuals and letting the search to continue until the maximum number of
allowed generations was reached or a solution was found.

Table 20.1 Experimental results obtained using random search

	Problem 1	Problem 2	Problem 3
Success rates	3%	0%	2%
Search space	2^{18}	2^{18}	2^{18}

It can be observed that the success rates obtained with genetic algorithms are
higher than the results obtained in the same conditions by random search. Figure
20.6 presents how the evolution progresses across generations. The average best
fitness for each problem is presented versus the current generation.

Table 20.2 Experimental results obtained for different fitness functions, averaged over 100 runs

Problem	Fitness function	Mutation rate	Success rate	Avg. gen.	δ	η
Problem 1	from [5]	0.5	0%	-	-	-
Problem 1	from [5]	0.8	3%	-	-	-
Problem 1	from [6]	-	40%	20.97	-	-
Problem 1	f_1	0.1	7%	27.68	25	1
Problem 1	f_1	0.5	22%	25.77	25	1
Problem 1	f_2	0.1	41%	20.48	1	1
Problem 1	f_2	0.5	56%	18.37	1	1
Problem 2	f_3	0.1	8%	28.74	25	1
Problem 2	f_3	0.5	14%	27.69	25	1
Problem 2	f_4	0.1	4%	28.27	1	1
Problem 2	f_4	0.5	13%	27.16	1	1
Problem 3	f_5	0.1	8%	27.65	25	10
Problem 3	f_5	0.5	12%	27.07	25	10

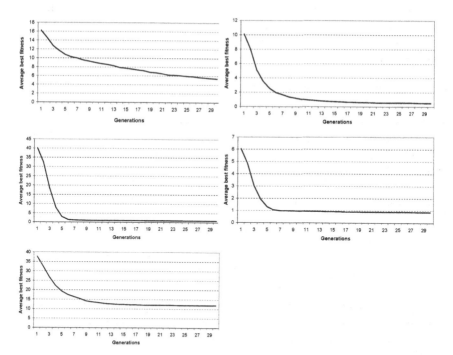

Fig. 20.6 The convergence characteristics of GAs: average best fitness value versus generation[a]

[a] The x-axis shows the generation, the y-axis shows the average best fitness value. The results were averaged over 100 runs. Note that the range of the y-axis may be different for different charts, which correspond to the fitness functions $f_1 - f_5$, in the order left to right. The results were considered for a mutation rate of $p_m = 0.5$, the values of δ, η are given in Table 20.2.

20.5   Related Work

Membrane computing and evolutionary algorithms are both nature inspired paradigms [10]. The literature concerning both of them could be classified in: (a) applications of evolutionary algorithms in membrane computing and (b) applications in
evolutionary algorithms inspired by the nature of P systems.

From the first category only a few papers have been reported, to the best of our
knowledge. One attempt is presented in [5] and consists of applying a genetic algorithm in order to obtain a P system which computes the square of a number. Unlike
the work of Escuela et al. [5], in which mutation is used to produce new rules, our
approach considers that the (potentially large) number of allowed rules is known
beforehand. This assumption is in many case realistic and produces much improved
results compared to [5].

A similar approach (regarding the fitness function evaluation) is given in [6], the
difference being the application of a quantum-inspired evolutionary algorithm. A
comparison between the results reported in these two papers [5, 6] suggests that
quantum-inspired evolutionary algorithms are considerably more effective than genetic algorithms in obtaining a P system solution to a given problem. However,
the results produced by our approach are comparable with the results produced by
quantum-inspired algorithms, as reported in [6]. An in-depth comparison between
the two approaches is, however, beyond the scope of this paper and will be the subject for further work.

Another application of genetic algorithms in membrane computing is realized in
[13], where the structure of the P system and the set of rules are fixed. The search
in this case is focused on determining some parameters which appear in the rules,
such as the given P system to solve a broadcasting problem.

The second type of papers propose evolutionary algorithms which have, apart
from the classical elements, some ingredients inspired by membrane computing.
In [23] a hybrid distributed evolutionary algorithm is employed for continuous optimization problems. The strategy presented in [23] is inspired from the way the
evolution rules are used in membrane systems. Other approaches, which combine
particle swarm optimization or ant colony optimization with membrane inspired
strategies, are given in [25] and [24], respectively.

20.6   Conclusions

This paper presents an approach on P system automatic design using genetic
algorithms and model checking. This approach is empirically evaluated by generating simple P systems which realise some given tasks or have some predefined
properties on their computations. Some fitness functions are proposed for each particular problem and the results obtained are compared with similar ones presented
in the literature.

The fitness evaluation is based on a predetermined number of simulations (which
may be less than the total number of possible computations) and, consequently, this

is complemented with model checking, in order to ensure that the required properties hold for all cases, not just for those for which simulation results are available.

The experimental results obtained are promising and future work concerns extending this approach, in order to generate more complex membrane systems.

Acknowledgements. This work was supported by CNCSIS - UEFISCSU, project number PNII - IDEI 643/2008. The authors would like to thank the anonymous reviewers for their comments and suggestions that allowed us to improve the paper.

References

[1] Andrei, O., Ciobanu, G., Lucanu, D.: Executable specifications of P systems. In: Mauri, G., Păun, G., Jesús Pérez-Jímenez, M., Rozenberg, G., Salomaa, A. (eds.) WMC 2004. LNCS, vol. 3365, pp. 126–145. Springer, Heidelberg (2005)

[2] Cardona, M., Colomer, M.A., Margalida, A., Pérez-Hurtado, I., Pérez-Jiménez, M.J., Sanuy, D.: A P system based model of an ecosystem of some scavenger birds. In: Păun, G., Pérez-Jiménez, M.J., Riscos-Núñez, A., Rozenberg, G., Salomaa, A. (eds.) WMC 2009. LNCS, vol. 5957, pp. 182–195. Springer, Heidelberg (2010)

[3] Cecilia, J.M., García, J.M., Guerrero, G.D., del Amor, M.A.M., Pérez-Hurtado, I., Pérez-Jiménez, M.J.: Simulating a P system based efficient solution to SAT by using GPUs. Journal of Logic and Algebraic Programming 79(6), 317–325 (2010)

[4] Díaz-Pernil, D., Pérez-Hurtado, I., Pérez-Jiménez, M.J., Riscos-Núñez, A.: A P-lingua programming environment for membrane computing. In: Corne, D.W., Frisco, P., Păun, G., Rozenberg, G., Salomaa, A. (eds.) WMC 2008. LNCS, vol. 5391, pp. 187–203. Springer, Heidelberg (2009)

[5] Escuela, G., Gutiérrez-Naranjo, M.A.: An application of genetic algorithms to membrane computing. In: Eighth Brainstorming Week on Membrane Computing, pp. 101–108 (2010), http://www.gcn.us.es/8BWMC/volume/08GEscuela.pdf

[6] Huang, X., Zhang, G., Ipate, F.: Evolutionary design of a simple membrane system. In: Proc. of Twelfth International Conference on Membrane Computing, CMC12 (to appear 2011)

[7] Ipate, F., Gheorghe, M., Lefticaru, R.: Test generation from P systems using model checking. Journal of Logic and Algebraic Programming 79(6), 350–362 (2010)

[8] Ipate, F., Lefticaru, R., Tudose, C.: Formal verification of P systems using Spin. International Journal of Foundations of Computer Science 22(1), 133–142 (2011)

[9] Johnson, C.: Genetic programming with fitness based on model checking. In: Ebner, M., O'Neill, M., Ekárt, A., Vanneschi, L., Esparcia-Alcázar, A.I. (eds.) EuroGP 2007. LNCS, vol. 4445, pp. 114–124. Springer, Heidelberg (2007)

[10] Kari, L., Rozenberg, G.: The many facets of natural computing. Communications of the ACM 51(10), 72–83 (2008)

[11] Lefticaru, R., Ipate, F., Tudose, C.: Automated model design using genetic algorithms and model checking. In: BCI 2009: Proceedings of the 2009 Fourth Balkan Conference in Informatics, pp. 79–84. IEEE Computer Society, Los Alamitos (2009)

[12] Lefticaru, R., Ipate, F., Gheorghe, M.: Model checking based test generation from P systems using P-Lingua. Romanian Journal of Information Science and Technology 13(2), 153–168 (2010)

[13] Lefticaru, R., Ipate, F., Gheorghe, M., Zhang, G.: Tuning P Systems for Solving the Broadcasting Problem. In: Păun, G., Pérez-Jiménez, M.J., Riscos-Núñez, A., Rozenberg, G., Salomaa, A. (eds.) WMC 2009. LNCS, vol. 5957, pp. 354–370. Springer, Heidelberg (2010)

[14] Lefticaru, R., Tudose, C., Ipate, F.: Towards automated verification of P systems using Spin. In: Ninth Brainstorming Week on Membrane Computing, pp. 237–250 (2011), http://www.gcn.us.es/9BWMC/volume/16Lefticaru_psys_spin.pdf

[15] Martínez-del-Amor, M.A., Pérez-Hurtado, I., Pérez-Jiménez, M.J., Riscos-Núñez, A.: A P-Lingua based simulator for tissue P systems. Journal of Logic and Algebraic Programming 79(6), 374–382 (2010)

[16] Meffert, K., Meseguer, J., Martí, E.D., Meskauskas, A., Vos, J., Rotstan, N.: JGAP-Java Genetic Algorithms and Genetic Programming Package, http://jgap.sf.net (last visited, July 2011)

[17] Mitchell, M.: An Introduction to Genetic Algorithms. MIT Press, Cambridge (1998)

[18] P-Lingua website, http://www.p-lingua.org (last visited, July 2011)

[19] Păun, G.: Computing with membranes. Journal of Computer and System Sciences 61(1), 108–143 (2000)

[20] Păun, G.: P systems with active membranes: Attacking NP-complete problems. Journal of Automata, Languages and Combinatorics 6(1), 75–90 (2001)

[21] Păun, G., Rozenberg, G., Salomaa, A. (eds.): The Oxford Handbook of Membrane Computing. Oxford University Press, Oxford (2010)

[22] Romero-Campero, F.J., Gheorghe, M., Bianco, L., Pescini, D., Jesús Pérez-Jímenez, M., Ceterchi, R.: Towards probabilistic model checking on P systems using PRISM. In: Hoogeboom, H.J., Păun, G., Rozenberg, G., Salomaa, A. (eds.) WMC 2006. LNCS, vol. 4361, pp. 477–495. Springer, Heidelberg (2006)

[23] Zaharie, D., Ciobanu, G.: Distributed evolutionary algorithms inspired by membranes in solving continuous optimization problems. In: Hoogeboom, H.J., Păun, G., Rozenberg, G., Salomaa, A. (eds.) WMC 2006. LNCS, vol. 4361, pp. 536–553. Springer, Heidelberg (2006)

[24] Zhang, G., Cheng, J., Gheorghe, M.: An approximate algorithm combining P systems and ant colony optimization for traveling salesman problems, Eighth Brainstorming Week on Membrane Computing, pp. 321–340 (2010), http://www.gcn.us.es/8BWMC/volume/25ACOPS_Latex.pdf

[25] Zhou, F., Zhang, G., Rong, H., Gheorghe, M., Cheng, J., Ipate, F., Lefticaru, R.: A particle swarm optimization based on P systems. In: Sixth International Conference on Natural Computation, pp. 3003–3007 (2010)

Chapter 21
Design of Evolvable Biologically Inspired Classifiers

Rinde R.S. van Lon, Pascal Wiggers, Lon J.M. Rothkrantz, and Tom Holvoet

Abstract. Complex systems are emergent, self-organizing and adaptive systems. They are pervasive in nature and usually hard to analyze or understand. Often they appear intelligent and show favorable properties such as resilience and anticipation. In this paper we describe a classifier model inspired by complex systems theory. Our model is a generalization of neural networks, boolean networks and genetic programming trees called computational networks. Designing computational networks by hand is infeasible when dealing with complex data. For designing our classifiers we developed an evolutionary design algorithm. Four extensions of this algorithm are presented. Each extension is inspired by natural evolution and theories from the evolutionary computing literature. The experiments show that our model can be evolutionary designed to act as a classifier. We show that our evolved classifiers are competitive compared to the classifiers in the Weka classifier collection. These experiments lead to the conclusion that using our evolutionary algorithm to design computational networks is a promising approach for the creation of classifiers. The benefits of the evolutionary extensions are inconclusive, for some datasets there is a significant performance increase while for other datasets the increase is very minimal.

Rinde R.S. van Lon
DistriNet Labs, Department of Computer Science, Katholieke Universiteit Leuven
e-mail: rinde.vanlon@cs.kuleuven.be

Pascal Wiggers
Man-Machine Interaction Group, Department of Mediamatics,
Delft University of Technology

Lon J.M. Rothkrantz
Man-Machine Interaction Group, Department of Mediamatics,
Delft University of Technology

Tom Holvoet
DistriNet Labs, Department of Computer Science, Katholieke Universiteit Leuven

D.A. Pelta et al. (Eds.): NICSO 2011, SCI 387, pp. 303–321, 2011.
springerlink.com   © Springer-Verlag Berlin Heidelberg 2011

21.1 Introduction

Many problems in the world can be formulated as classification tasks. In fact, humans are classifying things, sounds, objects, scents and feelings all the time. Most of these 'problems' are solved naturally by humans, while in fact some of them are quite hard. That is, computers have not been able to solve some of these problems.

Classifiers are typically implemented using parameterized models, e.g. Bayesian networks or neural networks, in which the model structure is specified by hand and the parameters are learned by collecting statistics over the data. Such models capture average behavior, but fail to model large variations and exceptions to the rule, they lack adaptivity. Looking at software systems in general it can be noted that they do not possess the adaptive properties which are so distinctive for natural systems. In the world around us, adaptation is everywhere. Every organism is constantly adapting to its environment. According to Salehie and Tahvildari [30] there is a need for a new approach for developing adaptive software.

In nature different adaptation strategies exist, the focus of this paper lies on so called *complex systems*. Examples of complex systems are as diverse as ant colonies [9], immune systems [14], the human mind [25] or a market economy. A common feature of these systems is that they consist of autonomous interacting entities, it is hard to pinpoint the location of the adaptivity in such a system. Yet their tremendous power is apparent when looking at them from an engineering point of view.

In this paper we use complex systems as an inspiration for the design of a classifier model. Our model is a variant of neural networks which we call computational networks (CN). A computational network is a graph with mathematical functions as nodes. These function nodes are similar to the functions in genetic programming trees [22]. The model is a generalization of random boolean networks, neural networks and genetic programming trees. A computational network allows more expressive functions, it poses no restrictions on the graph topology (recurrent loops are allowed) and it allows real values.

Designing computational networks by hand is nearly impossible due to some of their complex system properties (which we will discuss in section 21.2). As an alternative we turn to artificial evolution for the design task. An advantage of evolving classifiers based on complex systems compared to the traditional approach is that the classifier can be structurally adapted to a specific problem. If the problem requires a more adaptive model, evolution should be able to design the model to be more adaptive. Another motive for using evolution is that natural complex systems have emerged through natural evolution, as such it is plausible that complex systems can be designed using artificial evolution. An evolutionary algorithm (EA) is constructed that designs computational networks. Several extensions for this EA are proposed and investigated to verify their usefulness.

Our long term goal is to develop adaptive classifiers which are capable of handling dynamism such as is encountered in the real world. In this paper, however, we focus on creating static classifiers inspired by the theory of complex systems. We hope that in future work our model can be a starting point towards more adaptive classifiers.

First, in section 21.2 we discuss some properties of complex systems that serve as inspiration for our model and the design choices we made. In section 21.3 we present our model, so-called computational networks and in section 21.4 we present our evolutionary design algorithm. Section 21.5 describes the extension comparison experiment and the performance comparison experiment. Section 21.6 discusses the obtained experimental results and we conclude this paper in section 21.7.

21.2   Complex Systems

Research of complex systems is done in different disciplines. The field of Complex Adaptive Systems (CAS) [14] is concerned with an analysis of complex systems as there is currently not a complete mathematical model of it. Collective intelligence (CI), is more concerned with modeling the behavior of complex systems. For example, in [31] Schut identifies several characteristic properties of collective intelligence systems. Swarm intelligence (SI) is inspired by natural complex systems, and is a paradigm for implementing adaptive systems [2, 20]. Further, the field of artificial life (a-life) aims to construct artifacts that approach the complexity of biological systems [1]. Typical examples of a-life are the work of Sims [32] and the research field of cellular automata, where from simple rules complex patterns can emerge.

The remainder of this section presents a definition for complex systems and a set of properties based on an extensive study of the related literature. Additionally, several fundamental design choices are presented for the design of complex systems.

21.2.1   Definition

Our working definition of complex systems is:

Complex systems are adaptive systems composed of emergent and self-organizing structures.

This definition says that in order to obtain an adaptive system, emergence and self-organization are required. In the next two sections a total of six properties are identified for emergence and self-organization. In our view, these properties are required to obtain an emergent and self-organizing system. Consequently, these six properties need to be included in the model in order to obtain an adaptive system. Using these properties the design choices in section 21.2.2 are made.

21.2.1.1   Emergence

Emergence is a concept lacking a concise definition [16]. However, Corning [3] defines emergence as the arising of novel structures during the process of self-organization in complex systems. This shows the close relation between the topics of self-organization and emergence, just as in the proposed definition. A more simple definition of emergence is posed by Schut [31] as "the whole is more than the sum of its parts". This lack of a proper definition results in a lot of controversy how emergent properties are to be distinguished from non-emergent properties [4]. Nevertheless, three properties are presented which, in our opinion, belong to emergence.

The distinction between a *local and a global* perspective or level is often associated with emergence [6]. Local is the level of the individual which describes the behavior of the individual. Global is the system wide perspective, which describes characteristics on a system level. Emergence occurs when global behavior of a system is not evident from the local behavior of its elements [20].

Interaction between individuals is an important property for emergence. When parts interact in less simple ways, reduction does not work, the interaction as well as the parts have to be studied [16]. The stability of an economy can be seen as an emergent property caused by the interactions of the participants (e.g. buyers, sellers) in this economy [20].

This interaction between individuals gives rise to possible non-linear relations in a system. This *non-linearity* is often stated as: small causes can have large effects, and large causes can have small effects.

21.2.1.2 Self-organization

De Wolf and Holvoet [6] state that self-organizing systems appear to organize themselves without external direction, manipulation, or control. Di Marzo Serugendo [7] states that self-organizing systems give rise to emergent properties. We identify three properties which are in our opinion crucial to obtain self-organization.

Decentralization is an obvious property of self-organization, there is no central controller that supervises all the parts. All individuals are autonomous.

The interaction discussed in section 21.2.1.1 gives rise to *feedback*. Feedback is a circular cause-and-effect relation [13]. There is positive, self-reinforcing feedback which allows a deviation to enlarge in an explosive way. And there is negative, self-counteracting feedback which stabilizes the system.

Self-organizing systems sometimes appear to behave randomly. The occurrence of *randomness* in complex systems is arguably more a limitation of the observer than an intrinsic property of a system. If the observer would be able to perceive every part of the system and completely understand the relations between the parts, the randomness would disappear.

21.2.2 *Design of Complex Systems*

The design of a complex system is a delicate task, that is, the design of a useful complex system is complicated. As was stated in section 21.2.1.1, reduction does not work, it is very hard to assemble parts which on a global level yield desired behavior.

Several attempts have been made to create complex systems, for example, Mitchell and Hofstadter [26] have created an emergent architecture capable of making analogies. In subsequent research their idea was applied to music prediction, geometric puzzles [10], and auditory perception [8].

An alternative approach is automating the design of complex systems. Artificial evolution is widely used for this purpose. One of the most prominent examples is *genetic programming* (GP) pioneered by Koza [22]. GP evolves a population of computer programs (in the form of syntax trees) to perform a task. Another example is the earlier mentioned work of Sims [32], where robot behavior and morphology were evolutionary designed. Evolution has also successfully been applied for the design of hardware [35], the design of robots [29], the development of an acoustic communication system between robots [36] and for designing cellular automata [34]. Another automated approach is that of learning classifier systems pioneered by John Holland [15].

We made several design decisions for modeling complex systems. In the following sections these decisions are presented, the choices were listed in [37].

21.2.2.1   Centralization or Decentralization

The choice between a centralized or decentralized control structure of a complex system is clear. As discussed in section 21.2.1.2, self-organization and decentralization are important properties of complex systems. As such, a decentralized approach is recommended.

21.2.2.2   Differentiation

This choice concerns the homogeneity of the individuals in the complex systems. Natural complex systems of both types exist, ants in an ant colony are (largely) the same, while there are several types of antibodies in the immune system. Groß et al. [11] show that homogenous robots are capable of a division of labor, meaning that robots perform different kind of tasks. An advantage of the homogenous option over the heterogenous option is that it reduces the design task significantly. As compromise, we prefer a semi-homogenous option which reduces the design task but still allows a limited amount of variety among individuals.

21.2.2.3   Communication Structure

In order to enable signaling or communication between individuals, a communication structure has to be chosen. According to Uny Cao et al. [37] there are three types of communication: interaction via environment, interaction via sensing and interaction via communications. An example of environmental interaction is stigmergy as employed by ants. The difficulty of this type of communication lies in the definition of the environment. It is hard to define for more abstract problems. A similar problem arises for interaction via sensing. If the problem is difficult to translate to the physical domain, how can sensing be defined? The only communication structure which does not suffer from these potential physical domain translations is interaction via communications. This communication structure can be characterized as a network topology. The most famous example of this structure is of course the human brain, essentially a network of interconnected neurons. Because this latter option does not have the disadvantages of the other types of communication, interaction via communications is the preferred communication structure for a model of complex systems.

21.3   Model

Based on the choices made in the previous section a model for complex systems is constructed.

As was stated, the system needs to be decentralized, it must be comprised of semi-homogenous individuals connected in a network. Furthermore, the system needs to have a possibility to interact with the outside world, it needs inputs and outputs. A network of semi-homogenous individuals is decentralized as long as each individual is autonomous. Such a decentralized network can be represented by a graph. Several graph based representations already exist such as hidden Markov models (HMMs), Bayesian networks (BNs), neural networks (NNs), random Boolean networks (RBNs) and GP syntax trees. HMMs and BNs are both probabilistic graphical models. Since complex systems are not limited to mere statistics, these models are not adequate to model complex systems.

Neural networks and especially recurrent neural networks (RNNs) are interesting since they are a networked computational model, with semi-homogenous individuals. The choice for RNNs over NNs is because recurrent neural networks have the ability to memorize values through recurrent loops, which is of great importance for time sequence processing [28]. An interesting observation on recurrent neural networks is that knowledge in the system can be seen as an emergent property, it emerges from the structure of the network.

A random boolean network is a similar emergent computational model, it is basically a series of interconnected boolean gates such as AND, OR and XOR gates. An important difference with NNs is that RBNs only deal with boolean values, and that each gate (node) can implement a different boolean function.

As discussed earlier, in genetic programming syntax trees are evolved. In a syntax tree the nodes are programming functions and the branches connect the inputs and outputs of the functions. The evolved syntax trees can be converted to an executable program. This is in contrast to RNN and RBN which are graph based computational models, while GP trees are graph (tree) based genotypical representations.

We propose a novel model that combines properties of RNNs, RBNs and GP syntax trees, called *computational network* (CN). Computational networks can be seen as a generalization of NNs and RBNs, their properties are compared in Table 21.1. In the table it can be seen that CNs do not have to be structured in any kind of form, in contrast to the structures of RNNs which are usually designed in advance. The structure of a GP tree is designed by evolution, however, it is required that the structure is a tree (no cycles are allowed). Further, RNNs, RBNs and GP trees are all graphs of interconnected mathematical functions. In RNNs these functions are usually all similar while in RBNs and GP trees these can be different. Since *semi-homogenous* individuals are required, CNs allow the use of different mathematical

Table 21.1 Comparison of three different computational models: computational networks, recurrent neural networks and boolean networks and GP trees

	CN	RNN	RBN	GP tree
type	computational model		genotypic representation
structured	no	yes	no	yes
functions	diverse	similar	diverse	diverse
values	real	real	boolean	real

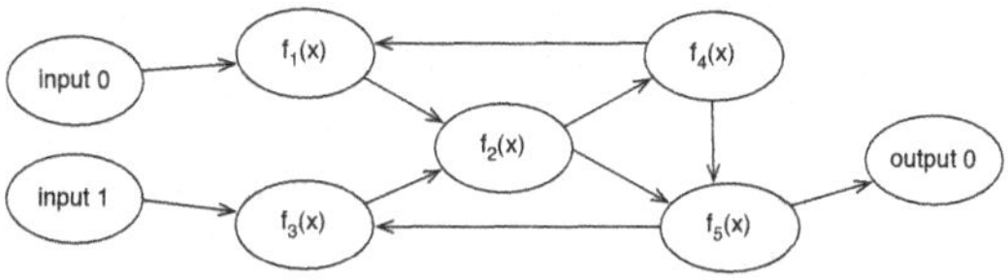

Fig. 21.1 Example of a computational network, with two inputs and one output, all weights and parameters are omitted.

functions using real values. RNNs and GP trees also use real values while RBNs are limited to booleans. An example of a computational network is shown in Figure 21.1. A node can be defined as a mathematical function $f(x_1, x_2, .., x_n)$ with a number of n inputs and one output, each node can implement a different function.

There are several reasons why CNs are chosen over RBNs, RNNs and GP trees. As said above, CNs are a generalization of RNNs and RBNs, while RNNs usually allow just one type of function in the network, a computational network does not pose restrictions on the functions. For example, consider a classification problem which looks for the minimum value of 10 real input values. In a computational network this can be represented using one single node containing the *min* function. When using neural networks at least an array of neurons with a sigmoid function is required to represent the same function, this is obviously much more complex compared to the CN solution. This means that computational networks are more expressive, more advanced computations can be done with a small number of nodes, thereby reducing the solution space. Furthermore, this particular problem cannot be solved by using boolean networks since it involves *continuous* values. Although it is possible to encode continuous values with boolean networks, the precision has to be defined on forehand which directly influences the network size. The more number of digits are needed for encoding the problem, the larger the network will be. Nodes in a GP tree share the same expressiveness as nodes in a CN graph. However, the interactions in a CN graph are unbound while the interactions in a GP tree are all in the same direction (there are no cycles).

An important aspect of computational networks is the order of computation. Circular dependencies are allowed, as it allows recurrent loops, which can be seen in Figure 21.1. Circular dependencies pose a problem because a node can only be executed when all input values are set. For this reason, when no executable node can be found, the first node that is found which is directly connected to an executed node (or input) is executed. For example in Figure 21.1 the first node to be executed would be node $f_1(x)$, even though not all its input values are set. When an input value is not set, the value of the previous time step is used instead.

Earlier it was said that complex systems are comprised of *autonomous* entities, as such, the chosen encoding should also have autonomous individuals. Nodes and their mathematical functions are in their simplicity not first to come to mind when thinking about autonomy. However, it is clear that their behavior is not dictated by some outside force. As such they can be considered autonomous.

Furthermore it has been proven that recurrent neural networks are equivalent to a universal Turing machine [18]. The same holds for a type of cellular automata (CA), which are a subclass of boolean networks [20]. Since computational networks are a generalization of RBNs and RNNs, CNs themselves are universal Turing machine equivalent. This means that a computational network is at least *capable* of representing a good classifier. Since RNNs have proven to be powerful classifiers, and CNs are a generalization of RNNs, it is expected that CNs are good classifiers themselves. A classifier can be described as a mathematical function which divides a problem space into several subspaces (classes), since a computational network is in essence a set of interconnected functions, it is appropriate to use it as a classifier.

21.4 Algorithm

This section discusses the evolutionary algorithm that we have devised to design computational networks. The purpose of this research is to design a complex system which acts as a classifier. A classifier predicts (classifies) the class of a *sample* based on its input features. A fitness score can be computed using the number of correct classified samples. The evolutionary algorithm evolves a population of classifiers (computational networks), each generation the classifiers are evaluated by testing them on a set of data samples.

Several possible extensions for evolutionary design are proposed. These extensions attempt to improve the evolutionary design process for computational networks. In Figure 21.2 an outline of the algorithm is shown, displaying the algorithmic extensions which usefulness is researched. The illustration shows that the increasing complexity extension modifies the input of the classifier and the fitness distance metric is a variant of the fitness function. EvoBoosting operates on the genotype and the embryogeny extension is a mapping between genotype and phenotype (the classifier). In the next sections each extension is discussed in more detail. A more detailed description is given in [24].

21.4.1 Representation

We designed two different representations, a direct representation and a generative representation. For the direct representation there is no distinction between genotype and phenotype. The variation operators (discussed in section 21.4.5) are directly

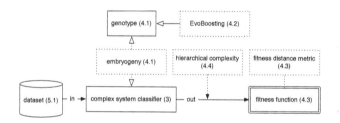

Fig. 21.2 Illustration of the evolutionary algorithm that designs a complex system classifier. The dotted blocks indicate algorithm extensions that are researched in this paper. For each block a section number is shown where a description is given about that block. The samples in the *dataset* are used as inputs of the *complex system classifier*. The classifier returns outputs which are evaluated by the *fitness function*. In the evolutionary algorithm the classifier is the *genotype*, or when the *embryogeny* extension is used, the genotype encodes the classifier (phenotype). The *EvoBoosting* extension operates on the *genotype*. The *hierarchical complexity* extension changes the outputs of the classifier. The *fitness distance metric* is an alternative fitness function.

applied to the graph representation of computational networks. Figure 21.3 shows an example of the direct representation. This representation can be encoded as $G = \langle V, E \rangle$, where V is the set of vertices (functions) and E is the set of edges in graph G.

The generative representation is called *embryogeny* which is a mapping between genotype and phenotype. The genotype is encoded as an tuple $\langle G_{seed}, R \rangle$. In this encoding G_{seed} is a graph called a seed and is encoded as above, and R is a set of rules. A rule is encoded as follows $r_i = \langle v_i, G_i \rangle$, v_i is a special identifier node and G_i is the substitution graph. Embryogeny is a function $f : \langle G_{seed}, R \rangle \rightarrow G$, that converts a genotype (a seed graph and a set of rules) into a phenotype (a graph). The *rules* in Figure 4(b) are used to substitute the nodes with the same name in the *seed* in Figure 4(a). Note that these substitutions are similar to grammatical rewriting systems such as created by Lindenmayer [23]. The seed and the rules produce the computational network presented in Figure 21.3. This technique also has similarities with automatically defined functions (ADFs) used in GP [22].

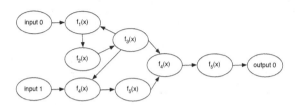

Fig. 21.3 Direct (phenotypic) representation of a computational network.

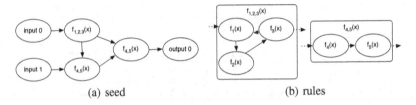

(a) seed   (b) rules

Fig. 21.4 Genotypic representation (embryogeny) of a computational network. The names in the boxes of (b) ($f_{1,2,3}$ and $f_{4,5}$) refer to the nodes of the seed in (a). By substituting nodes in (a) with their corresponding graph in (b), the graph in Figure 21.3 is obtained. In case a node in the seed has multiple outputs (for example $f_{1,2,3}$), the output node of the corresponding rule subgraph is connected to each output in the seed graph (see the outputs of f_3 in Figure 21.3).

As these two representations both encode computational networks, they can be compared to each other by conducting experiments. The advantage for using an embryogeny is the use of 'building blocks' [21]. These building blocks are implemented using the replacement rules visualized in Figure 4(b), which allow the reuse of substructures throughout the network.

21.4.2 EvoBoosting

EvoBoosting is a supervised learning algorithm inspired by the boosting concept of Kearns [19] who researched whether a group of weak classifiers can be combined to create a single strong classifier. EvoBoosting is similar to *incremental structure evolution*, but differs in its implementation [27]. Incremental structure evolution uses predefined structures, while EvoBoosting aims to do this on the fly. It is also similar to *complexification* used by Stanley and Miikkulainen [33].

EvoBoosting is activated after a certain number of generations has passed. When activated, the genotype with the highest fitness value, called G_{best} is selected from the last generation. In the next generation all genotypes in the population will be based on the G_{best} subgraph with some additional nodes randomly attached to it. The G_{best} subgraph is marked as 'frozen', meaning that it cannot and will not be changed during future generations. However, it is allowed to attach new nodes to the frozen part, stimulating the reuse of the previously evolved frozen subgraph.

In Figure 5(a) a computational network is shown which is the best of its generation (G_{best}) and which is used for EvoBoosting. Figure 5(b) shows one genotype in the next generation which contains the frozen G_{best} subgraph drawn in bold.

This method can recombine partial solutions into higher level (partial) solutions, which can theoretically result in increasingly more complex solutions. EvoBoosting can be applied on a conditional basis, for example after a number of generations or when a specific fitness level has been reached.

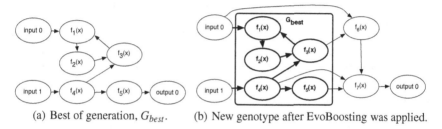

(a) Best of generation, G_{best}.   (b) New genotype after EvoBoosting was applied.

Fig. 21.5 Fictive EvoBoosting example, showing a best of generation genotype in (a) with nodes f_{1-5}. In (b) the same nodes are drawn in bold indicating that it is frozen, nodes f_6 and f_7 were randomly added as free nodes.

21.4.3 Fitness Distance Metric

Two fitness functions are proposed, the default fitness function simply counts the number of correct answers. It is denoted as

$$fitness_{default} = |\{o_i \in O | o_i = e_i\}| \qquad (21.1)$$

where O is the set of obtained outputs of a classifier, o_i is the output obtained for sample i and e_i is the expected (correct) output for sample i.

The second fitness function is called 'fitness distance metric'. This function rewards answers proportional to their distance to the correct answer:

$$fitness_{dist} = \frac{1}{|O|} \sum_{i=0}^{|O|} dist(o_i, e_i) \qquad (21.2)$$

where $dist(o_i, e_i)$ computes the distance (difference) between an obtained output and a correct output for a sample i. The fitness distance metric is expected to smooth the fitness landscape, making it easier for evolution to travel over irregular landscapes.

21.4.4 Hierarchical Complexity

This extension combines features of *increased complexity evolution* [35] and *exaptation*. Increased complexity evolution defines several stages in the evolutionary process, in each stage building blocks are used to reach higher levels of complexity. Exaptation also called *preadaptation* is the shifting of the function of a biological trait. It denotes that a trait originally 'intended' for a particular function can later be adapted to be used for another function. An example of exaptation given by Darwin [5] is that of the swimbladder in fishes, this organ is used for flotation, with it fish can stay at the current water depth without having to waste energy in swimming. Darwin [5] states that the swimbladder is 'ideally similar' in position and structure with the lungs of the higher vertebrate animals. He states that the swimbladder of fish has later evolved into lungs.

The hierarchical complexity extension divides the search space in several stages. This is done by initially limiting the number of outputs of the classifiers to two. During this initial stage, fitness is computed using only these two outputs. In later stages the number of outputs can be slowly increased until it matches the number of outputs which are required to solve the entire problem. This extension can be seen as slowly increasing the complexity of the problem. Figure 21.6 shows that this extension changes the number of outputs which are used for the fitness function, in a sense it is a hierarchic fitness function which gradually increases the fitness pressure.

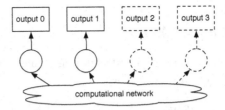

Fig. 21.6 Hierarchical complexity example. Showing a problem domain which requires four outputs. In the current view only two outputs are used for calculating the fitness, the other outputs (drawn with dotted lines) can be added in subsequent stages, resulting in an increasingly complex problem.

This extension can only be used with problems which require at least two independent outputs or at least three dependent outputs. Dependent outputs restrict each others values. For example, for a classifier which has to determine whether or not the light is on, this can be modeled with two outputs, one for the 'on' state, one for the 'off' state. It is clear that if one output has value 1 (activated), the other must have value 0 (deactivated). In this situation hierarchical complexity is not useful, solving this problem with one output already solves the problem for two outputs.

21.4.5   Variation and Selection

Mutation is used as a primary means for introducing variation in the population. Several mutation operators were designed and are used in conjunction. These operators can randomly reset the entire genotype, add or remove nodes, connections or rules (in the case of embryogeny). Also a mutation operator for changing the weights of the nodes and connections was used. When embryogeny is used a special crossover operator was used which exchanges rules between genotypes. For parent selection tournament selection is used.

21.5   Experiments

As stated earlier, the goal of this paper is to design classifiers. This section describes
the five datasets that are used and presents the two experiments that were performed.

21.5.1   Datasets

In Table 21.2 an overview of the properties of each dataset is shown. All datasets
except the Banana set are from the UC Irvine Machine Learning Repository[1].

Table 21.2 Overview of several properties of the five used datasets. The number of inputs
and outputs indicate how many inputs and outputs were used in the computational network
for that particular dataset.

Dataset	# instances	# inputs	# outputs
Banana set	500	2	1
Pima Indians Diabetes	768	8	1
Tic-tac-toe	958	9	1
Iris set	150	4	3
Connect-4	1349*	42	3

* The number of instances of the Connect-4 dataset is a subset of the original 67557
  instances.

Banana set.   The banana set is a two class problem which cannot be separated by a
simple linear classifier. Its name is derived from the banana shape of both classes
which is visible when plotted. It is an artificial dataset generated using PRTools[2],
a pattern recognition library for MatLab from Delft University of Technology.

Pima Indians Diabetes Database.   This dataset is a collection of measurements
and properties of female patients of at least 21 years old of Pima Indian heritage.
It has two classes, one is 'tested positive for diabetes' and the other is 'tested
negative for diabetes'. Based on the 8 attributes of the patients a classifier has to
determine to which class a patient belongs, i.e. if she has diabetes or not.

Tic-tac-toe endgame.   This dataset contains the complete set of possible board
configurations at the end of Tic-tac-toe games. The 9 inputs describe the play
field, each input can be in one of the following states: 'taken by player 1', 'taken
by player 2' or 'empty'. Based on these inputs a classifier has to decide whether
player 1 can win (two classes: 'positive' and 'negative').

Iris set.   The Iris dataset contains measurements extracted from a set of irises,
based on these values a classifier has to determine to which particular subclass
of the iris species it belongs. The set consists of 150 data points of four features
each: sepal length, sepal width, petal length and petal width. The data is evenly
divided among three classes, Iris Setosa, Iris Versicolour and Iris Virginica.

---

[1] http://archive.ics.uci.edu/ml/index.html
[2] http://prtools.org/prtools.html

Connect-4. This is a dataset containing all legal 8-ply positions in the game of connect-4 (also known as 'four in a row'). For each configuration neither player has won yet and the next move is not forced. Based on the information about the 42 positions, a classifier must determine whether player 1: wins, loses or if the game ends in a draw.

21.5.2 Extension Experiment

Four algorithmic extensions were discussed in section 21.4: embryogeny, EvoBoosting, fitness distance metric and hierarchical complexity. All extensions are compatible with each other and are tested on the five datasets discussed above, each extension can be on or off. For the Banana set, Pima Indians Diabetes set and Tic-tac-toe set only embryogeny and boosting are implemented, which means there are $2^2 = 4$ experiments. For the Iris and Connect-4 sets all four extensions are implemented resulting in a total of $2^4 = 16$ experiments. This experiment intends to find the best extension setting for each dataset. Because evolution is a stochastic process each experiment is repeated 10 times, each time with a different random seed. A population of 500 (genotype) individuals is used which is evolved for 500 generations.

21.5.3 Performance Experiment

Using the best extension settings determined in the extension experiment, a 'performance' experiment is conducted for each dataset. In this experiment computational networks are evolved for each dataset, the classification performance of these computational networks is then compared to classifiers from the Weka workbench [12].

Each dataset is divided in 10 equal parts using stratified sampling, ten fold cross validation is used resulting in 10 experiments. For each experiment the train set contains 9 folds and the tenth fold is used as a test set. For each of these 10 fold settings a computational network is constructed and all classifiers of the Weka workbench are executed on this setting. For each dataset this results in 10 scores (one for each fold setting) *per classifier*, these scores are then averaged and used as the final score of a classifier for that dataset. The best computational network is searched by using a population size of 5000 which is evolved for 500 generations

The computational networks are constructed using a population of 5000 (genotype) individuals which are evolved for 500 generations.

21.6 Results Analysis

21.6.1 Extension Experiment

Table 21.3 presents the results of the extension experiment. The first row shows the means and standard deviations of the baseline score for each dataset. The baseline

Table 21.3 Results of the extension experiment showing results for each extension on all five datasets. The μ and σ values indicate the mean and standard deviations of the 10 runs for each setting with different random seeds.

	Banana set		Pima indians		Tic-tac-toe		Iris		Connect-4	
	μ	σ	μ	σ	μ	σ	μ	σ	μ	σ
Baseline	0.962	0.018	0.761	0.027	0.775	0.033	0.816	0.056	0.852	0.009
Fit.dist.metric							0.993	0.006	0.821	0.004
EvoBoosting	0.955	0.025	0.765	0.024	0.762	0.022	0.718	0.255	0.836	0.011
Embryogeny	0.941	0.039	0.746	0.025	0.791	0.038	0.865	0.054	0.853	0.003
H.complexity							0.858	0.021	0.853	0.003

score is computed without using any extensions. The other rows show the scores obtained with that extension.

Banana set.   When examining the column of the Banana set in Table 21.3 it can be seen that the baseline already performs very good with a 0.962 score. In the same column it can be seen that both the EvoBoosting and the Embryogeny extensions individually degrade the performance of the classifier. However, this table does not show any possible interaction effects between the extensions. It turns out that there is a positive interaction effect, the score obtained using both the EvoBoosting and Embryogeny extensions is 0.969, which is slightly better compared to the baseline.

Pima Indians Diabetes Database.   For the Pima indians diabetes dataset the situation is different. The EvoBoosting extension improves slightly over the baseline while the Embryogeny extension degrades the solution. For this dataset no positive interaction effect exists.

Tic-tac-toe endgame.   Of the two extensions used for the Tic-tac-toe dataset, only the Embryogeny extension improves over the baseline. For this dataset also no positive interaction effect exists.

Iris set.   In the fourth column in Table 21.3 it can be seen that the Fit.dist.metric, Embryogeny and H.complexity extensions improve the solution compared to the baseline. Only for the EvoBoosting extension the performance is less than the baseline. The absolute best score is obtained using the Fit.dist.metric and H.complexity extensions together, yielding a score of 0.993.

Connect-4.   In the last column in the table, the scores for the Connect-4 dataset are shown. It can be seen that the Fit.dist.metric and the EvoBoosting extensions degrade the solution. Embryogeny and H.complexity slightly improve the score compared to the baseline. Remarkable is that using both the Embryogeny and the H.complexity extensions at the same time yields exactly the same score of 0.853. The interaction effect is exactly neutral.

21.6.2 Overview

Table 21.4 shows an overview of the previously described best performing extension settings. These settings are used in the performance experiment described in section 21.6.3. It is clear that there is not one specific setup that always performs best, the best extension setting is very data dependent. Each extension is part of the best setting for at least one dataset, but never for all datasets. Also, it is interesting that hierarchical complexity is present in the best settings for both datasets where it was tested.

Table 21.4 Overview of best performing extensions on the five tested datasets. The right column shows the difference in performance of that particular setting compared to the baseline.

Dataset	Best extension settings	Performance increase over baseline
Banana set	EvoBoosting and Embryogeny	0.007
Pima indians diabetes	EvoBoosting	0.004
Tic-tac-toe	Embryogeny	0.016
Iris	Fit.dist.metric and H.complexity	0.177
Connect-4	Embryogeny and H.complexity	0.001

In absolute terms, the improvement in performance for the extensions used for the Iris set are the most impressive. For the Tic-tac-toe dataset the increase is 1.6% which is reasonable. For the other three datasets the performance increase is minimal. Each of the extensions was designed to improve the generation of more complex structures. For some datasets it might be the case that very simple classifiers (with respect to structure) were required for good performance. This might explain the poor performance of the extensions for some of the datasets.

21.6.3 Performance Experiment

Table 21.5 shows the results obtained in the performance experiment. The results are compared with the best performing Weka classifiers[3] for each dataset. Table 5(a) shows the results for the Banana set. All scores are very high, indicating that this dataset is not challenging. The computational network classifier is outperformed by 9 of the Weka classifiers, although the performance difference is only 2%.

For the Pima Indians Diabetes dataset the results from Table 5(b) show that the evolved computational network significantly outperforms all Weka classifiers. This is an promising result and shows the potential of our method.

The performance of the computational network for the Tic-tac-toe dataset as shown in Table 5(c) is quite disappointing. The CN is outperformed by 32 Weka classifiers and the difference in performance is 14%. It is unclear why the performance is so miserable for this dataset as it is not regarded as particular challenging.

---

[3] A description for each of the Weka classifiers can be found at their website:
http://www.cs.waikato.ac.nz/ml/weka/index.html

Table 21.5 Performance experiment results for all datasets. In (a) and (c) the four best performing classifiers and the computational network are shown. In (b), (d) and (e) the top five performing classifiers are shown. Each μ and σ value indicates the mean and standard deviation, respectively, of the performance of that classifier measured over the ten different fold configurations.

(a) Banana set

#	Classifier	μ	σ
1	LibSVM	0.992	0.01
2	KStar	0.986	0.013
3	IB1	0.984	0.0080
4	IBk	0.984	0.0080
10	Computational network	0.976	0.013

(b) Pima Indians Diabetes

#	Classifier	μ	σ
1	Computational network	0.846	0.026
2	Logistic	0.777	0.026
3	MultiClassClassifier	0.777	0.026
4	RotationForest	0.772	0.027
5	DTNB	0.771	0.034

(c) Tic-tac-toe endgame

#	Classifier	μ	σ
1	IBk	0.987	0.0090
2	LibSVM	0.983	0.011
3	SMO	0.983	0.011
4	Ridor	0.983	0.011
33	Computational network	0.847	0.059

(d) Iris

#	Classifier	μ	σ
1	Computational network	1.0	0.0
2	Logistic	0.98	0.045
3	LibSVM	0.98	0.032
4	MultilayerPerceptron	0.973	0.047
5	SimpleLogistic	0.967	0.047

(e) Connect-4

#	Classifier	μ	σ
1	Computational network	0.919	0.011
2	MultilayerPerceptron	0.858	0.033
3	RotationForest	0.852	0.028
4	RandomCommittee	0.832	0.017
5	RandomForest	0.831	0.027

For the Iris dataset an exceptional result was obtained for the computational network: a 100% score for all folds, as shown in Table 5(d). The good performance of the Weka classifiers indicates that this dataset is unchallenging, as is widely known.

Table 5(e) shows that the CN for the Connect-4 dataset outperforms all Weka classifiers. Interestingly, the second best performing classifier is MultilayerPerceptron. This is a neural network which shares a very similar model with the CN. Nevertheless, the CN outperforms this neural network by more than 6%.

21.7 Conclusion

In this paper we presented a complex systems inspired classifier which is designed using an evolutionary algorithm. Four extensions for the evolutionary algorithm were presented and tested by evolving classifiers for five different datasets. This experiment shows that the performance of the extensions is data dependent. Also when an extension improved over the baseline, the performance gain varied greatly for the datasets. Another experiment was conducted assessing the quality of the produced classifiers. This experiment showed promising results for the evolutionary designed computational networks. For three datasets our solution performs better compared to all the Weka classifiers. However, for two datasets the Weka classifiers performed better than our evolved solution. The results show that evolving computational networks is a promising approach for creating competitive classifiers.

In future work the performance of the extensions on more datasets can be researched. Investigating a possible relation between complexity of the classifier and the effectiveness of the extensions is an intriguing research angle. It would be interesting to investigate whether our model which is inspired by complex systems is also capable of handling dynamism, just as natural complex systems. This can be researched by extending the work in [17, 36] by using CNs for the control of robots.

References

[1] Angeline, P.J., Pollack, J.B.: Coevolving High-Level Representations. Artificial Life III, 55–71 (1994)

[2] Bonabeau, E., Dorigo, M., Theraulaz, G.: Swarm intelligence: from natural to artificial systems. Oxford University Press, USA (1999)

[3] Corning, P.A.: The re-emergence of emergence: a venerable concept in search of a theory. Complexity 7(6), 18–30 (2002)

[4] Damper, R.: Emergence and levels of abstraction. International Journal of Systems Science 31(7), 811–818 (2000)

[5] Darwin, C.R.: On the Origin of Species. John Murray (1859)

[6] De Wolf, T., Holvoet, T.: Towards a Methodology for Engineering Self-Organising Emergent Systems. Self-Organization and Autonomic Informatics (I) 135, 18–34 (2005)

[7] Di Marzo Serugendo, G.: Engineering Emergent Behaviour: A Vision. In: Hales, D. (ed.) MABS 2003. LNCS, vol. 2927, pp. 1–7. Springer, Heidelberg (2003)

[8] Dor, R., Rothkrantz, L.J.M.: The Ears Mind An emergent self-organising model of auditory perception. JETAI, 1–23 (2008)

[9] Dorigo, M., Maniezzo, V., Colorni, A.: Positive feedback as a search strategy. Tech. Rep., Dipartimento di Elettronica, Politecnico di Milano, Milan, Italy (June 1991)

[10] Foundalis, H.E.: Phaeaco: A Cognitive Architecture Inspired by Bongard's Problems. PhD thesis (2006)

[11] Groß, R., Nouyan, S., Bonani, M., Mondada, F., Dorigo, M.: Division of Labour in Self-organised Groups. In: Asada, M., Hallam, J.C.T., Meyer, J.-A., Tani, J. (eds.) SAB 2008. LNCS (LNAI), vol. 5040, pp. 426–436. Springer, Heidelberg (2008)

[12] Hall, M., Frank, E., Holmes, G., Pfahringer, B., Reutemann, P., Witten, I.H.: The WEKA Data Mining Software: An Update. SIGKDD Explorations 11(1), 10–18 (2009)

[13] Heylighen, F.: The Science Of Self-Organization And Adaptivity. Knowledge Management, Organizational Intelligence and Learning, and Complexity. In: The Encyclopedia of Life Support Systems, EOLSS, pp. 253–280. Publishers Co. Ltd (1999)

[14] Holland, J.H.: Complex Adaptive Systems. Daedalus (1, A New Era in Computation) 121,17–30 (1992)

[15] Holland, J.H.: Hidden Order: How Adaptation Builds Complexity. Perseus Books, New York (1995)

[16] Holland, J.H.: Emergence: From Chaos To Order. Perseus Books, New York (1998)

[17] Hülse, M., Wischmann, S., Pasemann, F.: Structure and function of evolved neuro-controllers for autonomous robots. Connection Science 16(4), 249–266 (2004)
[18] Hyötyniemi, H.: Turing Machines are Recurrent Neural Networks. In: Alander, J., Honkela, T., Jakobsson, M. (eds.) STeP 1996 - Genes, Nets and Symbols, Finish Artificial Intelligence Society, Vaasa, Finland, pp. 13–24 (1996)
[19] Kearns, M.: Thoughts on hypothesis boosting (1988) (unpublished manuscript)
[20] Kennedy, J., Eberhart, R.C., Shi, Y.: Swarm Intelligence (The Morgan Kaufmann Series in Evolutionary Computation). Morgan Kaufmann, San Francisco (2001)
[21] Kicinger, R., Arciszewski, T., De Jong, K.A.: Evolutionary computation and structural design: A survey of the state-of-the-art. Computers and Structures 83(23-24), 1943–1978 (2005)
[22] Koza, J.R.: Genetic programming II. MIT Press, Cambridge (1994)
[23] Lindenmayer, A.: Mathematical models for cellular interactions in development, Part I: Filaments with One-sided Inputs. Journal of Theoretical Biology 18, 280–299 (1968)
[24] van Lon, R.R.S.: Evolving Biologically Inspired Classifiers. Msc thesis, Delft University of Technology, Delft (2010)
[25] Minsky, M.: The Society of Mind. Simon & Schuster, New York (1988)
[26] Mitchell, M., Hofstadter, D.R.: The Copycat Project: A Model of Mental Fluidity and Analogy-making. In: Holyoak, K., Barnden, J. (eds.) Advances in Connectionist and Neural Computation Theory. Ablex, Greenwich (1994)
[27] Pasemann, F., Steinmetz, U., Hu, M., Lara, B.: Robot Control and the Evolution of Modular Neurodynamics. Theory in Biosciences 120, 311–326 (2001)
[28] Figueira Pujol, J.C., Poli, R.: Efficient evolution of asymmetric recurrent neural networks using a PDGP-inspired two-dimensional representation. In: Banzhaf, W., Poli, R., Schoenauer, M., Fogarty, T.C. (eds.) EuroGP 1998. LNCS, vol. 1391, pp. 130–141. Springer, Heidelberg (1998)
[29] Römmerman, M., Kühn, D., Kirchner, F.: Robot design for space missions using evolutionary computation. In: 2009 IEEE Congress on Evolutionary Computation, pp. 2098–2105 (2009)
[30] Salehie, M., Tahvildari, L.: Self-adaptive software: Landcape and Research Challenges. ACM Transactions on Autonomous and Adaptive Systems 4(2), 1–42 (2009)
[31] Schut, M.C.: On model design for simulation of collective intelligence. Information Sciences 180(1), 132–155 (2010)
[32] Sims, K.: Evolving virtual creatures. In: Proceedings of the 21st Annual Conference on Computer Graphics and Interactive Techniques - SIGGRAPH 1994, July 1994, pp. 15–22 (1994)
[33] Stanley, K.O., Miikkulainen, R.: Competitive Coevolution through Evolutionary Complexification. Journal of Artificial Intelligence Research 21, 63–100 (2004)
[34] Terrazas, G., Siepmann, P., Kendall, G., Krasnogor, N.: An Evolutionary Methodology for the Automated Design of Cellular Automaton-based Complex Systems. Journal of Cellular Automata 2, 77–102 (2007)
[35] Torresen, J.: A Divide-and-Conquer Approach to Evolvable Hardware. In: Sipper, M., Mange, D., Pérez-Uribe, A. (eds.) ICES 1998. LNCS, vol. 1478, pp. 57–65. Springer, Heidelberg (1998)
[36] Tuci, E., Ampatzis, C.: Evolution of Acoustic Communication Between Two Cooperating Robots. In: Almeida e Costa, F., Rocha, L.M., Costa, E., Harvey, I., Coutinho, A. (eds.) ECAL 2007. LNCS (LNAI), vol. 4648, pp. 395–404. Springer, Heidelberg (2007)
[37] Uny Cao, Y., Fukunaga, A.S., Kahng, A.B.: Cooperative Mobile Robotics: Antecedents and Directions. Autonomous Robots 4, 7–27 (1997)

Chapter 22
Steel Sheet Incremental Cold Shaping Improvements Using Hybridized Genetic Algorithms with Support Vector Machines and Neural Networks

Laura Puigpinós, José R. Villar, Javier Sedano, Emilio Corchado, and Joaquim de Ciurana

Abstract. The complexity and difficulties in modelling the most of nowadays real world problems increase as the computational capacity does, specially in those processes where relatively new technology arises. One of such processes is the steel sheet incremental cold shaping. The steel sheet incremental cold shaping process is a new technique for shaping metal sheets when a reduced amount of pieces per lots should be manufactured. As it is a relatively new technique, there is a lack of knowledge in defining the operating conditions, so in order to fit them, before manufacturing a lot a trial and error stage is carried out. A decision support system to reduce the cost of processing and to assist in defining the operating conditions

Laura Puigpinós
Fundación Privada Ascamm, Avda. Universitat Autònoma,
23 08290 Cerdanyola del Vallés (SPAIN)
e-mail: lpuigpinos@ascamm.com

José R. Villar
University of Oviedo, Campus de Viesques s/n 33204 Gijón (SPAIN)
e-mail: villarjose@uniovi.es

Javier Sedano
Instituto Tecnológico de Castilla y León, Poligono Industrial de Villalonquejar,
Burgos (SPAIN)
e-mail: javier.sedano@itcl.es

Emilio Corchado
University of Salamanca, Salamanca (SPAIN)
e-mail: escorchado@usal.es

Joaquim de Ciurana
University of Girona, Escola Politècnica Superior Edifici PII, Avd. Lluís Santaló,
s/n 17071 Girona (SPAIN)
e-mail: quim.ciurana@udg.es

D.A. Pelta et al. (Eds.): NICSO 2011, SCI 387, pp. 323–332, 2011.
springerlink.com   © Springer-Verlag Berlin Heidelberg 2011

should be studied. This study focus on the analysis and design of the decision support system, and as it is going to be shown, the most suitable features have been found using a wrapper feature selection method, in which genetic algorithms support vector machines and neural networks are hybridized. Some facts concerning the enhanced experimentation needed and the improvements in the algorithm are drawn.

22.1 Introduction

Over recent years there have been a significant increase in the use of artificial intelligence and Soft Computing (SOCO) methods to solve real world problems. Many different SOCO applications have been reported: the use of Exploratory Projection Pursuit (EPS) and ARMAX for modelling the manufacture of steel components [2], EPS and Neural Networks (NN) for determining the operating conditions in face milling operations [10] and in pneumatic drilling process [11], genetic algorithms and programming for trading rule extraction [4] and low quality data in lighting control systems [15], feature selection and association rule discovery in high dimensional spaces [14] or NNs and principal component analysis and EPS in building energy efficiency [12, 13].

It is known that the complexity inherited in most of new real world problems, the steel cold shaping industrial process among them, increases as the computer capacity does. Higher performance requirements with a lower amount of data examples is needed due to the costs of generating new instances, specially in those processes where new technology arises.

In this sense, the steel cold shaping is a relatively new technology in the production of lots with a low quantity of pieces, which represents an effervescent area. NNs have been used to find relationships between the mechanical properties of the cold-rolled sheets of interstitial free and the chemical composition of the steel and the rolling and the batch annealing parameters [9]. NNs have been also applied for identification of the parameters for operating conditions [20, 21]. To the best of our knowledge, no specific study has been published in steel iterative cold shaping.

This study focuses on determining the main parameters in an steel sheet incremental cold shaping. The main objective is to find the most relevant feature subset; and the second objective is to obtain a decision support system in the operating conditions design, so the costs of producing such low amount of pieces in the lots is reduced. The next Section is concerned with the problem description. In Section 22.3 the algorithm used is detailed, while Sect. 22.4 deals with the experiments carried out and the results obtained. Finally, conclusions are drawn and future work goals are set.

22.2 Steel Sheet Incremental Cold Shaping

The metal incremental cold shaping is based on the concept of incremental deformation. This technology allows the manufacturing of pieces of metal sheet through the iteration of small sequential deformation stages until the desired shape is achieved and avoiding the axis-symmetric restrictions due to incremental rotatory deformation.

Comparing the incremental cold shaping with traditional deformation technologies it can be said that the former reduces the cost of specific machine tools and the manufacturing costs dramatically.

This type of technology has evolved from the well-known Rapid Manufacturing, allowing to generate pieces with complex geometries in a wide spread of materials without the need of frameworks or specific tools.

The main part of cold shaping has been controlled using numerical controlled tools in order to reduce as most as possible the fast, reliable, ans low-cost manufacturing of lots with an small amount of metal pieces and prototypes. The scheme of metal sheet incremental cold shaping process is shown in Fig. 22.1.

Fig. 22.1 The incremental cold shaping of a steel sheet. A sharpening tool is iteratively applied onto the metal sheet at a different depth. In the negative shaping only the sharp tool is moved, while in the positive shaping both the metal sheet and the sharp tool are moved.

The process of cold shaping starts with the design of a geometric shape in a 3D CAD file. This file should include as many layers as desired, each layer represents the bounds to be reached in each deforming step and are piled vertically. Consequently, the piece should be generated using the sequential and incremental layers, each one at a different depth and constraint within the defined bounds.

Plenty of parameters have to be fixed for the manufacture a metal piece, among them the force developed by the deforming head in each of the three dimensions, the speed change, the trajectory of the head, the surface roughness, the sheet pressure stress, the incremental step between layers, the number of steps or stages, the attack angle, the angle variation, the depth variation, etc.

22.2.1   The Problem Definition

The first aim of this study is to evaluate if it is possible to model the operating conditions so the suitability of the experiment could be established, in other words, to analyse whether the operating conditions would generate a faulty piece or not while the most relevant features involved are to be selected.

The second aim is to model the maximum suitable depth that can be reached with the given operating conditions. As in the former problem, the best feature subset is also required.

Therefore, there are two problems to solve, both including a feature selection process and a modelling process. The former is a two-class problem, while the second is a regression problem.

22.3 Feature Selection, Support Vector Machines and Neural Networks

In order to obtain a suitable feature subset some requirements are needed. As there are integer features, nominal features and real valued features, the algorithm should deal with any kind of data. Therefore, the same approach should be valid for the both subproblems, the two-class problem and the maximum depth. Besides, not only the best feature subset for each problem but also the best model are desired, a classifier in the former case and a regression model in he latter.

It is known that for this kind of problems the wrapper approach for feature selection performs better than filter solutions [3, 18]. These studies proposed wrapper feature selection methods using genetic algorithms (GA) for dealing with the feature subset selection, that is, each individual is a feature subset. To evaluate individuals a modeling technique has been applied: the former proposed a lazy learning model as the K-Nearest Neighbour, the latter made use of a NN model that iteratively fix the number of hidden neurons.

Different approaches as to how the NN is learnt have been studied. In [1] a GA approach to fingerprint feature selection is proposed and selected features are supplied as input to NN for fingerprint recognition, while in [17] a similar approach has been applied to automatic digital modulation recognition. Moreover, this type of approach has been reported to perform better than using statistical models [19]. Despite, Support Vector Machines (SVM) have been also used in conjunction with evolutionary feature selection to reduce the input space dimensionality [5, 6].

In this study, two different approaches are analysed. The first one is an specific GA wrapper feature selection method for each problem, the second approach is an evolution of the former, allowing different NN models to be found.

22.3.1 GA+SVM+NN Feature Selection

In this study we adopt two different solutions depending whether we are dealing with the two-class or the maximum depth problem. An hybridized method of GA evolving the feature subsets and a SVM classifier is chosen in the former case, while in the latter an hybridized method of GA evolving the feature subsets and a NN for modeling the desired output is used. In both modelling and feature selection problems the GA is an steady state approach with the percentage of elite individuals to be defined as a parameter. The algorithm has been implemented in MatLab [8], using both the NN and the SVM toolboxes.

The algorithm is outlined in Algorithms [1,2]. Algorithm [1] evaluates an individual (which is to say, a feature subset), while the latter, which is the main algorithm, shows the GA that evolves the feature subset and that calls Algorithm 1 whenever required.

For the sake of simplicity we have neither reproduce the algorithm for the SVM nor for the NN cases. Instead, we present the general case in the algorithms, and when it is said that a model is trained, the reader should consider which problem (the two-class or the regression problem) is related to the use of NN or SVM.

The typical steady state GA parameters, like the crossover and mutation probabilities, the number of generations, the population size and the elite population size, are all of them given for each experiment. The individual representation is the string of indexes of the chosen feature subset. The tournament selection is implemented and one point crossover is used. After each genetic operation the validity of the off-prints is analysis: repeated features indexes are erased and random indexes are introduced to fill the individual feature subset.

Third order polynomials are used as kernel functions for the SVM. The number of hidden nodes in the NN is set as a parameter. The NN models are generated randomly and trained using the corresponding Matlab library functions. In all cases, 10-fold cross validation is used, and the mean value of the mean squared error in each fold is the fitness of an individual.

Algorithm 22.1. IND_EVALUATION: Evaluates an individual

Require: I the input variables data set
Require: O the output variable data set
Require: ind the individual to evaluate, with its feature subset
  $model$ {the best model learned for ind}
  $mse = 0$ {the associated mean of Mean Square Error for ind}
  $indMSE = 0$ {best MSE found in the cross validation}
  for $k = 1$ to 10 **do**
    {run the k fold in the cross validation scheme}
    generate the train and test reduced feature data set
    initialize the model $indModel$
    train $indModel$ with the train data set
    $indKMSE \leftarrow$ calculate the MSE for $indModel$ with the test data set
    $mse+ = indKMSE$
    if $k == 1$ **or** $indMSE > indKMSE$ **then**
      $indMSE = indKMSE$
      $model = indModel$
    end if
  end for
  $mse = mse/10$
  return $[model, mse]$

Algorithm 22.2. GA$^+$ Feature Selection

Require: I the input variables data set
Require: O the output variable data set
Require: N the feature subset size
  $FS \leftarrow \{ \emptyset \}$ {the best feature subset}
  model {the model learned for FS}
  $mse = 0$ {the associated mean of Mean Square Error for FS}
  Generate the initial population, Pop
  for all individual *ind* in Pop **do**
    $[ind.model, ind.mse] = IND_EVALUATION(I,O,ind)$
  end for
  $g \leftarrow 0$
  while $g < G$ **do**
    while $size(Pop') < (popSize - |E|)$ **do**
      Generate new individuals through selection, crossover and mutation
      add valid individuals to Pop'
    end while
    extract the elite subpopulation $E \in Pop$
    for all individual *ind* in Pop' **do**
      $[ind.model, ind.mse] = IND_EVALUATION(I,O,ind)$
    end for
    $Pop = \{E \cup Pop'\}$
    sort Pop
    g++
  end while
  $FS \leftarrow Pop[0]$
  $[model, mse] \leftarrow$ corresponding model and MSE
  return $[FS, model, mse]$

22.3.2 Evaluating the Effect of Some Trainnig and Test Issues

Some variations were accomplished in order to evaluate the performance of the feature selection modelling. The main question is if the features where good enough to be chosen if we change the NN and its parameters.

For this purposes, a second collection of 10 fold cross validation data sets was generated independently to those prepared by Matlab. Moreover, a different NN was used for the second problem. In this case, a feedfoward NN with three hidden layers and one linear output unit, with 10 units in the first hidden layer, 15 units in the second hidden layer and 4 units in the third hidden layer is used; all hidden unit are hyperbolic tangent units. The NN was trained with the Lenverberg-Marquart method.

22.3.3 A Discussion on Cross Validation Schemes

The fitness of each individual is calculated, as outlined before, using a k-fold cross validation (cv) scheme. The main aim of this evaluation is to estimate the performance of a feature subset in operating conditions, when the sample given has not been presented for training. This has been found relevant in those problems for which the data set includes few samples [16].

Nevertheless, many different schemes can be used, the leave-one-out, the k-fold cv or the 5x2 cv among them. The selection of the cv relies on the data set dimensions.

The mean error among the whole set of models learned is proposed as the fitness function; for all cases in this study, the validation has been calculated with the test data set. However, it is worth noting that it is possible to use a validation data set for evaluating the mean error and so the fitness of the individuals.

Consequently, the cv should extract a three data sets from the original one: the train, the test and the validation data sets. Clearly, choosing the kind of cv will depend on the data set dimensionality; when there are not enough samples are available the solution will tend to do use leave one out cv without a specific validation data set. Automatic selection and developing of the cv scheme is also left for future work.

It is worth mentioning which model is kept for an individual. In this study we kept the best model found, that is, the one with lower mse, but if multi objective were used then a different criteria should be selected.

22.4   Experiments and Results

Generating data set samples is costly as the each one of the samples needs a real case to be carried out, that is, a sheet of steel has to be cold shaped; consequently, the smaller the number of experiments, the lower the cost and the smaller the data set size.

The data set comprises 19 samples, each one with the whole set of parameters values. Once the piece is processed as the corresponding sample establishes, then it is manually classified as {GOOD, BAD} according to the deformation or the quality faults that could appear in the piece. Besides, the maximum depth in each case is also measured. These two latter values are appended to each sample as the output variables.

As SVM and NN are to be used in the modelling part of the feature selection GA method, then the data set is normalized with means 0 and deviations 1.

In the first experimentation, the GA parameters has been fixed as follows: 50 individuals in the population, 100 of generations, the probability of crossover equals to 0.75, while the mutation probability is 0.25. An steady state GA evolutionary scheme is used, with a number of 5 elite individuals that will be kept in the next generation.

The size of the feature subset has been fixed to three. The SVM kernel function is fixed as third order polynomials and the feed forward back-propagation NNs includes 6 neurons in the hidden layer, with $\mu = 0.001$, $\mu_{dec} = 10^{-10}$, and $\mu_{inc} = 10^{-6}$. The parameters of the SVM and the NN have been kept constant during the feature selection and the model learning.

As stated in the previous section, the 10-fold cross validation schema is carried out. Only the validation results are used to compare individuals. For the two-class problem the mean of the classification errors among the folds is used to evaluate the models and the feature subsets. For the maximum depth estimation, each feature subset and its corresponding model are evaluated with the mean of the mean squared error on each fold.

In the case of the two-class problem, the best feature subset found includes the *step increment*, the *angle variation* and the *variation in depth*, with a mean classification error of 0.1%. For the second problem, the best feature subset found includes the variables *step increment*, the *number of stages* and *variation in depth*, with a mean error of 0.0096.

For the second approach experimentation a bayesian regularization was applied to the data sets [7]. All the parameters remain the same except for the NN learning parameters, which where fixed at $\mu = 1.0$, $\mu_{dec} = 0.8$, and $\mu_{inc} = 0.0001$. In this case, the obtained errors were meaningfully worse, and consequently, the feature subset chosen in this case was not the same as in the former experimentation. Only in the case of reducing the feature space to two variables both approaches proposed the same feature subset. Clearly, in order to obtain the best feature subset an automatic model parameters identification is needed so the obtained feature subset does not depend on the experiment decisions.

It is worth mentioning that the reduced number of data samples induces relatively high error values as the test data set includes only one or two samples. More accuracy should be obtained if a bigger number of samples is given. However, the cost of the data gathering increases considerably; this dilemma should be evaluated.

Moreover, the algorithms do not include local optimization of the models parameters. So it is possible that better performance of the models and the feature selection process should be achieved if such local optimization were implemented.

Finally, the maximum depth have been found regardless of the two-class problem, which was not the expected result in the expert opinion. It is though that the above mentioned local optimization of the models parameters should improve the performance and the experts confidence in the results.

22.5   Conclusions and Future Work

This study presents a feature selection method for choosing the best feature subset in a steel sheets cold shaping process divided in a two-class problem and a maximum depth estimation problem. Moreover, a genetic algorithm is hybridized, on the one hand, for the first case, with a support vector machine model to choose the best feature subset and on the other hand, for the second case, with a feed foward backpropagation neural network.

From the experimentation the best feature subset has been found for both problems, and some relevant facts have arisen. Firstly, the data set size should be increased in order to obtain better models fitness values. Secondly, local optimization for the models parameters should improve the obtained results. Besides, it could be desirable that the optimum number of features should be dynamically fixed, which represents an improvement in the individual representation. Finally, the analysis of the different cv schemes and the decision of which of them to use is left for future work.

Acknowledgements. This research has been funded by the Spanish Ministry of Science and Innovation, under project TIN2008-06681-C06-04, the Spanish Ministry of Science and Innovation [PID 560300-2009-11], the Junta de Castilla y León [CCTT/10/BU/0002] and by the ITCL project CONSOCO.

References

[1] Altun, A.A., Allahverdi, N.: Neural Network Based Recognition by Using Genetic Algorithm for Feature Selection of Enhanced Fingerprints. In: Beliczynski, B., Dzielinski, A., Iwanowski, M., Ribeiro, B. (eds.) ICANNGA 2007. LNCS, vol. 4432, pp. 467–476. Springer, Heidelberg (2007)

[2] Bustillo, A., Sedano, J., Curiel, L., Villar, J.R., Corchado, E.: A Soft Computing System for Modelling the Manufacture of Steel Components. Advances in Soft Computing 57(3), 601–609 (2008)

[3] Casillas, J., Cordón, O., del Jesus, M.J., Herrera, F.: Genetic Feature Selection in a Fuzzy Rule-Based Classification System Learning Process. Information Sciences 136(1-4), 135–157 (2001)

[4] de la Cal, E., Fernández, E.M., Quiroga, R., Villar, J., Sedano, J.: Scalability of a Methodology for Generating Technical Trading Rules with GAPs Based on Risk-Return Adjustment and Incremental Training. In: Corchado, E., Graña Romay, M., Manhaes Savio, A. (eds.) HAIS 2010. LNCS, vol. 6077, pp. 143–150. Springer, Heidelberg (2010)

[5] Fung, G.M., Mangasarian, O.L.: A Feature Selection Newton Method for Support Vector Machine Classification. Computational Optimization and Applications 28(2), 185–202 (2004)

[6] Huanga, C.-L., Wang, C.-J.: A GA-based feature selection and parameters optimization for support vector machines. Expert Systems with Applications 31(2), 231–240 (2006)

[7] Mackay, D.J.C.: Bayesian interpolation. Neural Computation 4(3), 415–447 (1992)

[8] The MathWorks, MATLAB - The Language Of Technical Computing (2011), http://www.mathworks.com/products/matlab/

[9] Mohanty, I., Datta, S., Bhattacharjeeb, D.: Composition-Processing-Property Correlation of Cold-Rolled IF Steel Sheets Using Neural Network. Materials and Manufacturing Processes 24(1), 100–105 (2009)

[10] Redondo, R., Santos, P., Bustillo, A., Sedano, J., Villar, J.R., Correa, M., Alique, J.R., Corchado, E.: A Soft Computing System to Perform Face milling Operations. In: Omatu, S., Rocha, M.P., Bravo, J., Fernández, F., Corchado, E., Bustillo, A., Corchado, J.M. (eds.) IWANN 2009. LNCS, vol. 5518, pp. 1282–1291. Springer, Heidelberg (2009)

[11] Sedano, J., Corchado, E., Curiel, L., Villar, J., Bravo, P.: The Application of a Two-Step AI Model to an Automated Pneumatic Drilling Process. Int. J. of Comp. Mat. 86(10-11), 1769–1777 (2008)

[12] Sedano, J., Curiel, L., Corchado, E., de la Cal, E., Villar, J.R.: A Soft Computing Based Method for Detecting Lifetime Building Thermal Insulation Failures. Int. Comp. Aided Eng. 17(2), 103–115 (2009)

[13] Sedano, J., Villar, J.R., Curiel, L., de la Cal, E., Corchado, E.: Improving Energy Efficiency in Buildings Using Machine Intelligence. In: Corchado, E., Yin, H. (eds.) IDEAL 2009. LNCS, vol. 5788, pp. 773–782. Springer, Heidelberg (2009)

[14] Villar, J.R., Suárez, M.R., Sedano, J., Mateos, F.: Unsupervised Feature Selection in High Dimensional Spaces and Uncertainty. In: Corchado, E., Wu, X., Oja, E., Herrero, Á., Baruque, B. (eds.) HAIS 2009. LNCS, vol. 5572, pp. 565–572. Springer, Heidelberg (2009)

[15] Villar, J.R., Berzosa, A., de la Cal, E., Sedano, J., Garcıa-Tamargo, M.: Multi-objective Simulated Annealing in Genetic Algorithm and Programming learning with low quality data. In publication for Neural Computing (2011)

[16] Villar, J.R., Sedano, J., Corchado, E., Puigpinós, L.: Soft Computing decision support for a steel sheet incremental cold shaping process. Accepted for the Proc. of the 12th Int. Conf. on Intel. Data Eng. and A. L, IDEAL 2011 (2011)

[17] Wong, M.L.D., Nandi, A.K.: Automatic digital modulation recognition using artificial neural network and genetic algorithm. Signal Proc. 84(2), 351–365 (2004)

[18] Yang, J., Honavar, V.: Feature Subset Selection Using a Genetic Algorithm. IEEE Intelligent Systems 13(2), 44–49 (1998)

[19] Zhang, P., Verma, B., Kumar, K.: Neural vs. statistical classifier in conjunction with genetic algorithm based feature selection. Pat. Recog. Letters 28(7), 909–919 (2005)

[20] Zhao, J., Cao, H.Q., Ma, L.X., Wang, F.Q., Li, S.B.: Study on intelligent control technology for the deep drawing of an axi-symmetric shell part. J. of Materials Processing Tech. 151(1-3), 98–104 (2005)

[21] Zhao, J., Wang, F.: Parameter identification by neural network for intelligent deep drawing of axisymmetric workpieces. J. of Materials Processing Tech. 166(3), 387–391 (2005)

Chapter 23
Real-Valued Genetic Algorithms with Disagreements

Andrei Lihu and Ştefan Holban

Abstract. This paper introduces a new mutation operator for real-valued genetic algorithms that refines the evolutionary process using disagreements. After a short introduction, we describe the new concept theoretically and then we exemplify it by defining a Gaussian distribution-based disagreements operator: the 6σ-GAD. We transform two common real-valued genetic algorithms into their disagreements-enabled counterparts and we conduct several tests proving that our newly obtained algorithms perform better because they gain strengthened neighborhood focus using partial disagreements and enhanced exploration capabilities through extreme disagreements.

23.1 Introduction

A *genetic algorithm* (GA) is a population-based optimization technique that mimics the process of Darwinian evolution ([4]). Due to their advantages, genetic algorithms, pioneered by Bremermann ([3]) and Fraser ([8]), and later developed by Holland ([11]), became a popular research area.

GAs belong to the family of *evolutionary algorithms* (EAs), but differentiate themselves from other evolutionary heuristics in terms of higher emphasis on recombination and treating mutation as a less important operator. Continuous research and development improved GAs in order to successfully solve most of the nonlinear and multi-objective problems from various fields of human activity.

Andrei Lihu
Department of Computer Science, Politehnica University of Timişoara, Bd. Vasile Pârvan 2, 300223 Timişoara, Romania
e-mail: andrei.lihu@gmail.com

Ştefan Holban
Department of Computer Science, Politehnica University of Timişoara, Bd. Vasile Pârvan 2, 300223 Timişoara
e-mail: stefan@cs.upt.ro

D.A. Pelta et al. (Eds.): NICSO 2011, SCI 387, pp. 333–346, 2011.
springerlink.com   © Springer-Verlag Berlin Heidelberg 2011

Briefly, a GA is better imagined as a population of individuals that try to find the optimal solution following the rules of evolution through iterations. At each iteration there is a *selection* process among the population that decides which are the best parents according to a fitness function f for the next generation. Then, parents form the offspring through a process called *recombination*. At the end of this process, there is a slight possibility that the offspring can be changed by *mutation*. This process is repeated until a termination criterion is met.

GAs' population members can be represented either in a binary form, or as real values. This paper deals with real-valued GAs; they are very convenient because we do not need an extra mapping function between genotype and phenotype when we want to work with real-valued problems.

Similar to other EAs' cases, when designing a genetic algorithm special attention should be given to accomplish two opposite goals: to achieve a good convergence rate and to maintain diversity across the population. Improving convergence leads to a faster discovery of the optimal solution and it is a general sign that the computational effort is low. Increasing the diversity of the population helps the algorithm to not be trapped into local minima. Usually, a compromise should be made to accommodate both aims. We approach this problem by introducing the metaphor of *disagreements* as a new mutation operator in real-valued genetic algorithms.

After a positive experience with the disagreements metaphor in *particle swarm optimization* (PSO)([13], [14]), we expanded the concept virtually to any population-based optimization technique and, particularly in this paper, to real-valued genetic algorithms. We think that any population has some members that disagree more or less with mainstream ideas or processes. In genetic algorithms, some individuals do not agree with current solutions and, since their genes represent the solutions, we gave them the ability to change themselves around their original characteristics using a new mutation type: *the disagreements operator*. If one may think that two clones of the same individual do not act or develop the same during lifetime because they have the possibility to choose, to agree or disagree — yet, for e.g. they resemble in appearance — then, the same way the disagreements operators create a "just around the corner" diversity. Or, as another example, let's think of teenagers that tend to contradict the generation of their parents and act differently. An important advantage of our technique is creating diversity within a low number of individuals.

The disagreements concept can have various implementations. In this paper, we introduce a Gaussian distribution-based mutation operator called 6σ-GAD. Testing on six popular optimization benchmarks has shown that the new method has good results. Utilizing partial disagreements, we enhance the exploitation and using the extreme disagreements, we increase the exploration.

In the field of genetic algorithms, related work include the normally distributed mutation in [5] and the continuous modal mutation ([10]). The first one is using a Gaussian distributed mutation, while the second work is a derivation of a Mühlenbein mutation ([15]) using a triangular probability distribution. Both of them lack a simplified neighborhood structure to boost exploitation and exploration, like 6σ-GAD does.

As an overview, in Section 23.2 we briefly present some general notions about real-valued genetic algorithms. Section 23.3 introduces the theoretical foundation of disagreements and the genetic algorithm with disagreements (GAD). In Section 23.4, we present the 6σ-GAD operator. In Section 23.5, we conduct the tests using the operators and analyze their results. Finally, in Section 23.6, we conclude that the new approach yields better results, especially for plateau functions and in multi-modal environments.

23.2   Real-Valued Genetic Algorithms

This section provides a short overview on real-valued GAs without going into details. More on this topic can be found in [9].

To describe GAs as part of the larger family of evolutionary algorithms, we use the notations and concepts from Bergh in [2]. At any iteration t, we consider a population of μ individuals that act as potential solutions in the hyperspace of solutions H^n:

$$P(t) = \left\{ x_1(t), x_2(t), ..., x_i(t), ...x_\mu(t) \right\}, \text{where } x_i(t) \in H^n \ . \tag{23.1}$$

Let f be the *fitness function* that measures the quality of a solution and $F(t) = \left\{ f(x_1(t)), f(x_2(t)), ..., f(x_i(t)), ...f(x_\mu(t)) \right\}$ that denotes the fitness of the whole population. Given the strategy parameters f, μ, λ, Θ_s, Θ_r and Θ_m, Bergh's adapted evolutionary framework from Bäck ([1]) can be described as follows:

```
General Evolutionary Framework
     t   ←  0
   P(t)  ←  intialise(μ)
   F(t)  ←  evaluate(P(t), μ)
   repeat
         P'(t)   ←  recombine(P(t), Θr)
         P''(t)  ←  mutate(P'(t), Θm)
         F(t)    ←  evaluate(P''(t), λ)
         P(t+1)  ←  select(P''(t), F(t), μ, Θs)
         t  ←  t+1
   until termination criteria is met
```

Θ_s, Θ_r and Θ_m are the probabilities to apply the operators of selection, recombination and mutation. λ represents the resulted offspring population, while μ refers to the parent population. Aside from some *elitist strategies*, when the best individuals are preserved in the next generation, in genetic algorithms the parent population is replaced entirely by children at each step. Selection, which is applied with a probability of Θ_s, is generally a variant of *tournament selection*: an iterative selection process of best individuals from a randomly selected subset of the population of size $q \leq |P(t)|$, until μ parents have been selected.

Genetic algorithms were initially represented in a binary form. To convert them into elements of the search space, one should have defined a mapping function to convert the binary genotype into its equivalent phenotype. To overcome this disadvantage, *real-valued GAs* were introduced [6].

Recombination, controlled by the parameter Θ_r, mixes the genotype traits normally from two parents and gives birth to the offspring. Because GAs primarily rely on recombination, its probability rate is high (around 70%). For real-valued GAs, *arithmetic crossover* operators are employed. An example of a simple arithmetic crossover operator can be found in [2]: if $x_a(t)$ and $x_b(t)$ are two parents, then the two children can be obtained as follows:

$$x_a(t+1) = r_1 x_a(t) + (1.0 - r_1)x_b(t)$$
$$x_b(t+1) = r_1 x_b(t) + (1.0 - r_1)x_a(t) \ , \tag{23.2}$$

where $r_1 \sim \mathcal{U}(0,1)$ (an uniformly distributed random variable).

The mutation perturbs the genotype of one given individual according to the *mutation rate* Θ_m. Mutation is a "second-class citizen" in GAs, therefore it has a small probability rate (usually 10%).

In the above-described evolutionary framework for genetic algorithms, in the next two sections, we introduce the disagreements-based mutation operator.

23.3 Disagreements

In a social context, disagreements are a common phenomenon that leads to diversity within a population.

After we obtained good results with disagreements for PSO in [13] and [14], we expanded the idea to any population-based heuristic. In this case, we enriched GAs with "a social note": we imagined that at each iteration a degree of disagreements related to current solutions can occur. Therefore, we applied them to GAs (which are population-based) to increase the exploration and to strengthen the exploitation.

In PSO, disagreements reside only in the social component of the updating principle, but in GA, not having such a separation, we applied the disagreements as a mutation.

We will introduce the disagreements concept as we did in [13], but we adapt it for mutation in GAs. A disagreement is defined as a function that takes values in the search hyperspace H^n. One can define the disagreements as the set of functions:

$$F_D = \{D : H^n \to H^n | \forall z \in H^n. \ D(z) \neq z\} \ . \tag{23.3}$$

We write the power set of all possible disagreement functions as follows:

$$\Delta_{all} = \mathscr{P}(\{id\} \cup F_D) \ , \tag{23.4}$$

where *id* is the identity function (no disagreement), also written $\emptyset_D$.

Let ρ be *a decision function*, called a "disagreement selector", that takes a set of disagreements $(\Delta_v \subseteq \Delta_{\text{all}})$ as argument at iteration t (from all iterations t_{all}) for an individual i (the number of all individuals is s) and decides which disagreement is invoked:

$$\rho : \{\Delta_{\text{all}} \times t_{\text{all}} \times s\} \rightarrow \Delta_{\text{all}}, \quad \rho(\Delta_v, t, i) = D_i, \quad D_i \in \Delta_v \ . \tag{23.5}$$

Let G_i be a genetic algorithm from G_{all}, which is the set of all GAs. We define the "disagreement injector" that gives the G_i's extension with disagreements as follows:

$$\Psi_{\text{GA}} : \{G_{\text{all}} \times \Delta_{\text{all}} \times \rho_{\text{all}}\} \rightarrow G_{\text{all}}, \quad \Psi_{\text{GA}}(G_i, \Delta_v, \rho) = G_{i\text{D}} \ , \tag{23.6}$$

where $G_{i\text{D}}$ is the resulting *genetic algorithm with disagreements*.

After we apply the injection operator Ψ_{GA} on a GA, we obtain a new genetic algorithm, called a genetic algorithm with disagreements (GAD).

In this way, we introduced the concept of disagreements as a special mutation operator that can be applied to any GA.

One may generalize (23.6) from GA to any population-based EA, as follows:

$$\Psi_{\text{EA}} : \{E_{\text{all}} \times \Delta_{\text{all}} \times \rho_{\text{all}}\} \rightarrow E_{\text{all}}, \quad \Psi_{\text{EA}}(E_i, \Delta_v, \rho) = E_{i\text{D}} \ , \tag{23.7}$$

where E_{all} is the set of all population-based evolutionary algorithms, E_i is one of such EAs and $E_{i\text{D}}$ is the resulting *evolutionary algorithm with disagreements* (EAD).

Although such generalizations are tempting because the idea can be applied to any population-based technique, it is always better to adapt the concept to each algorithm's particularities to have both good results in solving optimization problems and ideological consistency.

23.4 6σ-GAD Operator

23.4.1   Definition

In this section, we take a practical approach to disagreements for real-valued GAs: the "6σ-GAD operator" we developed to solve minimization problems. In terms of (23.6), it is defined as:

$$\Psi_{6\sigma-\text{GAD}}(rG_i) = \Psi_{\text{GA}}(rG_i, \Delta_{r6\sigma}, \rho_{r6\sigma}) \ , \tag{23.8}$$

where r is not a coefficient, it means "real-valued".

The employed subset of disagreements resemble those from [13]:

$$\Delta_{r6\sigma} = \{\emptyset_\text{D}, D_{6\sigma1}, D_{6\sigma2}\} \ . \tag{23.9}$$

Let $\lambda_1, \lambda_u \in \mathbb{R}_+^*, \lambda_1 < \lambda_u$. Let a be a vector containing the lower bounds of the search space H^n, $b - a$ vector containing the upper bounds in H^n, and $b - a$ is a vector containing the search space's ranges. We define the first type of disagreement, $D_{6\sigma1}$ — the partial disagreement:

$$D_{6\sigma1}(z) = z + q_i \otimes \frac{b - a}{2}, \quad q_i \sim \mathcal{U}(-\lambda_l, +\lambda_l), \quad z \in H^n , \qquad (23.10)$$

where q_i is the i–th component of q.

The second type of disagreement is the extreme disagreement $D_{6\sigma2}$:

$$D_{6\sigma2}(z) = z + w_i \otimes \frac{b-a}{2}, \quad w_i = w_{i_1} + sgn(w_{i_1}) \cdot \lambda_l , \qquad (23.11)$$
$$w_{i_1} \sim \mathcal{U}(-(\lambda_u - \lambda_l), +(\lambda_u - \lambda_l)) ,$$

where w_i is the i–th component of w and w_{i_1} is an uniformly distributed random number generated each time for w_i.

23.4.2 Design Philosophy

The design philosophy for disagreements provides better neighborhood exploitation through $D_{6\sigma1}$ and enhanced exploration using amplified values generated by $D_{6\sigma2}$. A simple visualization for the concept is in Fig. 23.1.

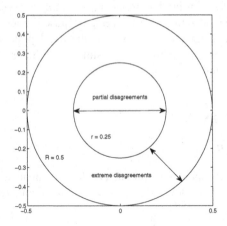

Fig. 23.1 Partial and extreme disagreements in 2 dimensions. Areas inside central circle and between circles show where q_i and w_i are generated for partial disagreements and for extreme disagreements, respectively ($\lambda_l = 0.25, \lambda_u = 0.50$).

The concept of partial and extreme disagreements is related to neighborhoods. Partial disagreements are values in the vicinity of the original value, while extreme disagreements are either more distant values or opposite ones.

As a further explanation, if one may think of the alphabet, some partial disagreements for letter M can be the letters N, O, P or L (a radius of partial disagreement should be defined), while some extreme disagreements for letter A can be Z (the dichotomy "first-last letter", although it can also be partial disagreement if we consider a circular alphabet) or even a non-letter, like "!"; also, for letter A, an extreme disagreement can be defined as a more distant character, like H. This example is given to show that genetic algorithms with disagreements can be used also on discrete domains.

23.4.3 Selector Function

At each iteration t, for each individual i, we generate $\theta(t,i) \sim \mathcal{N}(\mu_{6\sigma}, \sigma_{6\sigma}^2)$ a Gaussian distributed random variable with a chosen mean $\mu_{6\sigma}$ and a standard deviation $\sigma_{6\sigma}$. We define the following Gaussian regions:

1. The first region accounts for approx. 68.2 % of the bell curve (first 2 σs):

$$R_{1,2\sigma} = (\mu_{6\sigma} - \sigma_{6\sigma}) \cup (\mu_{6\sigma} + \sigma_{6\sigma}) \ . \tag{23.12}$$

2. The second region accounts for approx. 27.2 % of the bell curve (next 2 σs):

$$R_{3,4\sigma} = (\mu_{6\sigma} - 2\sigma_{6\sigma}, \mu_{6\sigma} - \sigma_{6\sigma}] \cup [\mu_{6\sigma} + \sigma_{6\sigma}, \mu_{6\sigma} + 2\sigma_{6\sigma}) \ . \tag{23.13}$$

3. The third region accounts for approx. 4.6 % of the bell curve (next 2 σs and the rest of what remains under the graphic of the Gaussian function):

$$R_{5,6\sigma} = (-\infty, \mu_{6\sigma} - 2\sigma_{6\sigma}] \cup [\mu_{6\sigma} + 2\sigma_{6\sigma}), +\infty) \ . \tag{23.14}$$

Based on (23.9), (23.10), (23.11) and the above-defined Gaussian regions, the selector function is defined as:

$$\rho_{r6\sigma}(\Delta_{r6\sigma}, t, i) = \begin{cases} \emptyset_D & \text{if } \theta(t,i) \in R_{1,2\sigma} \\ D_{6\sigma 1} & \text{if } \theta(t,i) \in R_{3,4\sigma} \\ D_{6\sigma 2} & \text{if } \theta(t,i) \in R_{5,6\sigma} \end{cases} \ . \tag{23.15}$$

23.5 Experimental Results

23.5.1 Setup

In order to test the effectiveness of the 6σ-GAD operator, we have chosen two common real-valued GAs: without (which we simply called GA) and with elitism (abbreviated eGA), respectively. We compared the results obtained by running the original algorithms and their disagreements-enabled counterparts on six popular optimization benchmarks.

For both two algorithms we applied the above-defined injection functions in order to obtain the disagreements enabled algorithms: $\Psi_{6\sigma-GAD}(GA) = GAD_{6\sigma}$ and $\Psi_{6\sigma-GAD}(eGA) = eGAD_{6\sigma}$, with $\lambda_l = 0.25, \lambda_u = 0.50$.

The original algorithms used a Gaussian-based mutation:

$$z'(t+1) = z(t+1) + \frac{b-a}{2} \cdot \mathcal{N}(0, h(t+1)) , \qquad (23.16)$$

where $z(t+1)$ is the chosen individual at iteration $t+1$, $z'(t+1)$ is the resulted individual after mutation and h' is the mutation step size (initially set to 0.1) that was updated at each iteration as follows:

$$h(t+1) = h(t) \cdot e^{\tau_1 \cdot \mathcal{N}(0,1)} , \qquad (23.17)$$

where τ is a coefficient that we set to 0.15.

Both algorithms used an arithmetic crossover as described in (23.2), had a crossover rate of 0.7, a mutation rate of 0.1, a population of *only 50 individuals* and used tournament selection. After transformation, both algorithms contain the disagreements mutation operator. In each test, we measured the mean fitness value and its standard deviation across 50 runs, in a hyperspace of 30 and 50 dimensions, respectively. Algorithms terminated after 20000 fitness evaluations.

The benchmark problems we used were Generalized Rosenbrock (F_2), Shifted Rastrigin (F_{14}, from CEC 2005), Shifted Schwefel (F_{21}, from CEC 2005), Griewank (F_5), Ackley(F_8) and Levy's function (F_{15}).

1. Generalized Rosenbrock can show GADs act on plateau functions:

$$F_2(X) = \sum_{i=1}^{n-1} (100(x_{i+1} - x_i^2)^2 + (1-x_i)^2), \ X \in \mathbb{R}^n . \qquad (23.18)$$

2. Shifted Rastrigin checks for behavior of convergence, because it is "single funnel" and provides a basin of attraction in the presence of many local optima:

$$F_{14}(X) = \sum_{i=1}^{n} (z_i^2 - 10\cos(2\pi z_i) + 10), \ X \in [-5, +5]^n, \ Z = X - o . \qquad (23.19)$$

3. Shifted Schwefel can give a hint on algorithm's robustness, as here we do not have a global basin of attraction:

$$F_{21}(X) = \sum_{i=1}^{n} \left(\sum_{j=1}^{i} z_j \right)^2 , \ X \in [-100, +100]^n, \ Z = X - o . \qquad (23.20)$$

4. Griewank determines how the new behavior helps escaping local minima:

$$F_5(X) = \sum_{i=1}^{n} \frac{x_i^2}{4000} - \prod_{i=1}^{s} \cos\left(\frac{x_i}{\sqrt{i}}\right) + 1, \ X \in [-600, 600]^n . \qquad (23.21)$$

5. Finally, with Ackley we checked if the new behavior helps search advancing in the big funnel that has so many local minima to stop in:

$$F_8(X) = -20 \cdot \exp\left(-0.2\sqrt{\tfrac{1}{n}\sum_{i=1}^n x_i^2}\right) - \exp\left(\tfrac{1}{n}\sum_{i=1}^n \cos(2\pi \cdot x_i)\right) +$$
$$+ 20 + e, \quad X \in [-32768,\ 32768]^n \ . \tag{23.22}$$

6. Finally, we checked our algorithms in test with Levy's ([12]) function:

$$F_{15}(X) = 0.1\left(\sin^2(3\pi x_1) + \sum_{i=1}^{n-1}(x_i - 1)^2[1 + \sin^2(2\pi x_{i+1})] +\right.$$
$$\left. +(x_n - 1)^2[1 + \sin^2(2\pi x_n)]\right), \quad X \in [-5, +5]^n \ . \tag{23.23}$$

We used the evolutionary framework Java EvA2 ([7]) both for implementing and for testing the above described GAs. The 6σ-GAD operator corresponds to the 'GAD' algorithm we have developed in our modified version of Java EvA2 library. Sources are currently available at https://github.com/andrei-lihu/Eva2-AL.

23.5.2   Results

Tables 23.1 - 23.3 show the results of our experiments and the figures Fig. 23.2 - 23.4 show the convergence graphs for half of the benchmark functions.

Table 23.1 Benchmark results for F_2 and F_{14}.

Algorithm	n	F_2 Mean	F_2 Std.dev.	F_{14} Mean	F_{14} Std. dev.
GA	30	13916.1397	12735.2698	273.0893	43.9950
	50	148501.1414	131132.9555	609.5590	71.1674
$GAD_{6\sigma}$	30	4234.7216	4677.4851	258.2947	41.4271
	50	56389.6761	58667.0893	571.7937	64.6468
eGA	30	21696.4295	20809.2549	269.6063	37.9522
	50	70914.2907	78052.2751	631.6585	58.3298
$eGAD_{6\sigma}$	30	2696.8574	1956.3036	261.8499	41.5853
	50	43632.4251	37715.4879	577.9603	59.7983

From Table 23.1 we can notice that on plateau functions like F_2 our concept works well both for the elitist and non-elitist variants, in all considered dimensions (as it can be seen in the convergence graph from Fig. 23.2 and the results from Table 1). On F_{14}, the newly derived algorithms preserve the behavior of convergence, even if they were disturbed by multiple disagreements, and they converge slightly faster and towards better solutions. In Table 23.2, for F_{21} we marked in bold the one of the weaker results (which is only marginally weaker) we obtained when we run the benchmarks. With Shifted Schwefel, in the absence of a main attraction basin, we can notice that the modified algorithm retains the original GA's robustness. Again, all tests, except the one pointed in bold, were completed with better results. The results for F_5 illustrate how GADs not only help escape local minima, but also converge faster (Fig. 23.3).

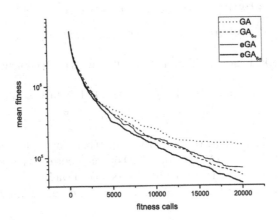

Fig. 23.2 Multi-run convergence graph for F_2 (50 dimensions). Comparing in pairs, it can be noticed that GADs behave better on plateau functions.

Table 23.2 Benchmark results for F_{21} and F_5.

Algorithm	n	F_{21} Mean	Std.dev.	F_5 Mean	Std. dev.
GA	30	**22046.9245**	**7618.3465**	29.1324	14.3562
	50	82454.2908	23888.7937	86.36725	58.3114
GAD$_{6\sigma}$	30	**24127.1678**	**6669.4884**	8.8745	4.7531
	50	80064.4803	25315.9253	39.2899	18.9912
eGA	30	26480.9251	10881.8832	31.0741	20.8923
	50	89631.3400	16099.5169	87.1203	54.2829
eGAD$_{6\sigma}$	30	23493.9514	7824.3470	8.5371	4.1517
	50	80977.3497	19898.3889	41.4706	23.8938

When we tested with Ackley function (Table 23.3), a long single-funnel with many local minima, both GADs got slightly better results than the original algorithms. Again, we have marked with bold the situation in which old algorithms were slightly better.

Finally, the results for Levy's function complete our tests and reconfirm the fact that applying disagreements operator can yield better results (Fig. 23.4).

All results show that the disagreements-enabled GAs are overall better in those six considered benchmarks. This is because partial disagreements, the exploitation is increased and through extreme disagreements, exploration is enhanced.

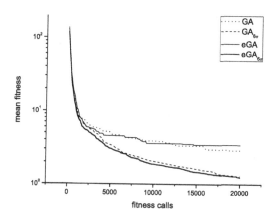

Fig. 23.3 Multi-run convergence graph for F_8 (30 dimensions). GADs are good at escaping local minima.

Table 23.3 Benchmark results for F_8 and F_{15}.

Algorithm	n	F_8 Mean	Std.dev.	F_{15} Mean	Std. dev.
GA	30	11.5019	2.5832	913.2887	291.0097
	50	**15.1179**	**3.6192**	2819.8395	667.8208
$GAD_{6\sigma}$	30	11.2211	2.5770	876.2631	270.2672
	50	**16.8094**	**2.4821**	2423.7095	597.8937
eGA	30	11.3187	2.1705	969.4400	344.0771
	50	16.7000	3.1279	2808.6859	622.8523
$eGAD_{6\sigma}$	30	11.1611	2.1993	791.8242	259.8484
	50	15.2551	3.3083	2318.3230	471.5483

23.5.3 Disagreements vs. Mutation

In order to prove the superiority of disagreements versus the Gaussian mutation, we compared the performance in benchmarks between them. We took our previously used common GA with elitism (eGA) and its disagreements counterpart, $eGAD_{6\sigma}$. For the common real-valued GA, we increased the mutation rate to 32%, which is the rounded amount of population affected by the disagreements in $eGAD_{6\sigma}$, and we named it "mGA". If the disagreements metaphor is the same as regular mutation, then the results for mGA and $eGAD_{6\sigma}$ must be very close. We run each test 50 times in 50 dimensions on a plateau function — Generalized Rosenbrock — and Griewank — a multi-modal function. During 20000 function evaluations, we measure the mean fitness, its standard deviation, the median and the success rate. The results are given in Table 23.4:

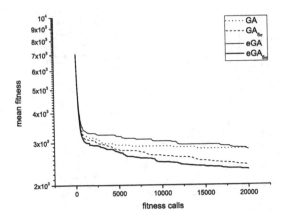

Fig. 23.4 Multi-run convergence graph for F_{15} (50 dimensions). GADs behave well in multi-modal environments.

Table 23.4 Benchmark results for disagreements and mutation.

Benchmark	Algorithm	Mean	Std.dev.	Median	SR
Rastrigin	mGA	29234.8152	44855.0816	12789.1867	0.00
	eGAD$_{6\sigma}$	51.6928	12.6612	48.8616	0.00
Griewank	mGA	23.4703	25.1055	10.0402	0.00
	eGAD$_{6\sigma}$	0.0211	0.0295	0.0010	0.50

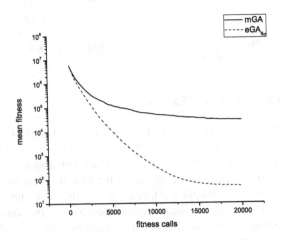

Fig. 23.5 Disagreements vs. a high-rate of mutation for Generalized Rosenbrock.

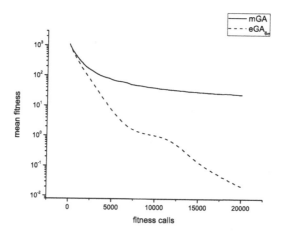

Fig. 23.6 Disagreements vs. a high-rate of mutation for Griewank.

Without any doubt, there is a huge gap in performance between disagreements and the real-valued Gaussian mutation. Both Table 23.4 and the convergence graphs from Fig. 23.5 and Fig. 23.6 prove that disagreements are a better mutation. With disagreements we obtained far superior results on the considered benchmarks.

23.6 Conclusion

Based on our previous work in PSO, we introduced the metaphor of disagreements in genetic algorithms as a mutation operator. We implemented and tested a disagreement mutation operator based on the normal distribution and a 3-tier neighborhood structure: the 6σ-GAD operator for real-valued GAs.

All tests showed that the new approach is good in optimization benchmarks. It has been empirically proven that any real-valued genetic algorithm that is transformed into a genetic algorithm with disagreements converges faster to a better solution. This is due to the fact that we enhance both exploration and exploitation with partial and extreme disagreements, respectively. Our technique performs well for GAs with a small population size, dramatically reducing the computational cost.

This paper applied the new idea of disagreements to real-valued genetic algorithms with good results. In the future work we will improve the current operator and we will develop, analyze and test new GAD operators.

Acknowledgements. This work was developed in the frame of PNII-IDEI-PCE-ID923-2009 CNCSIS–UEFISCSU grant and was partially supported by the strategic grant POSDRU 6/1.5/S/13-2008 of the Ministry of Labor, Family and Social Protection, Romania, co-financed by the European Social Fund – Investing in People.

References

[1] Bäck, T., Fogel, D.B., Michalewicz, T.: Basic Algorithms and Operators. In: Evolutionary Computation, vol. 1. Institute of Physics Publishing, Bristol (1999)

[2] Bergh, F.: An Analysis of Particle Swarm Optimizers (PhD thesis). University of Pretoria, Pretoria (2001)

[3] Bremermann, H.J.: The evolution of intelligence. The nervous system as a model of its environment, Technical report, no.1, contract no. 477(17), Dept. Mathematics, Univ. Washington, Seattle (1958)

[4] Darwin, C.H.: On the Origin of Species by Means of Natural Selection or the Preservation of Favoured Races in the Struggle for Life. Murray, London (1859)

[5] Deb, K.: Multi-Objective Optimization using Evolutionary Algorithms. Wiley and Sons Ed., Chichester (2001) ISBN: 978-0-471-87339-6

[6] Deb, K., Agrawal, R.: Simulated Binary Crossover for Continuous Search Space. Complex Systems 9, 115–148 (1995)

[7] EvA2 Project Homepage, http://www.ra.cs.uni-tuebingen.de/software/EvA2 (accessed in January 2011)

[8] Fraser, A.: Simulation of genetic systems by automatic digital computers. I. Introduction. Australian Journal of Biological Sciences 10, 484–491 (1957)

[9] Goldberg, D.E.: Genetic Algorithms in Search, Optimization and Machine Learning. Addison-Wesley, Reading (1989)

[10] Herrera, F., Lozano, M., Verdegay, J.L.: Tackling real-coded genetic algorithms: operators and tools for behavioural analysis. Artificial Intelligence Review 12(4), 265–319 (1998)

[11] Holland, J.: Adaptation in Natural and Artificial Systems. University of Michigan Press, Ann Arbor (1975)

[12] Levy, A., Montalvo, A.: The tunneling algorithm for the global minimization of functions. SIAM Journal on Scientific and Statistical Computing 6, 15 (1985)

[13] Lihu, A., Holban, Ş.: Particle Swarm Optimization with Disagreements. In: Tan, Y., Shi, Y., Chai, Y., Wang, G. (eds.) ICSI 2011, Part I. LNCS, vol. 6728, pp. 46–55. Springer, Heidelberg (2011)

[14] Lihu, A., Holban, Ş.: Particle Swarm Optimization with Disagreements on Stagnation. In: Semantic Methods for Knowledge Management and Communication - 3rd International Conference on Computational Collective Intelligence - Technologies and Applications, ICCCI 2011, Gdynia, Poland, September 21-23. SCI, vol. 7092. Springer, Heidelberg (2011)

[15] Mühlenbein, H., Schlierkamp-Voosen, D.: Predictive Models for the Breeder Genetic Algorithm, I. Continuous Parameter Optimization. Evolutionary Computation 1(1), 25–49 (1993)

Author Index